# Sustainable Supercapacitors

**Scrivener Publishing**
100 Cummings Center, Suite 541J
Beverly, MA 01915-6106

*Publishers at Scrivener*
Martin Scrivener (martin@scrivenerpublishing.com)
Phillip Carmical (pcarmical@scrivenerpublishing.com)

# Sustainable Supercapacitors

## Next-Generation of Green Energy Storage Devices

Edited by

**Basheer Ahamed**
*Dept. of Physics, BSA Cresent Institute of Science and Technology, Vandalur, Chennai, India*

and

**Chaudhery Mustansar Hussain**
*Dept. of Chemistry & Environmental Sciences at New Jersey Institute of Technology, Newark, New Jersey, USA*

Scrivener Publishing

WILEY

This edition first published 2025 by John Wiley & Sons, Inc., 111 River Street, Hoboken, NJ 07030, USA and Scrivener Publishing LLC, 100 Cummings Center, Suite 541J, Beverly, MA 01915, USA

For more information about Scrivener publications please visit www.scrivenerpublishing.com.



**Wiley Global Headquarters**
111 River Street, Hoboken, NJ 07030, USA

For details of our global editorial offices, customer services, and more information about Wiley products visit us at www.wiley.com.



***Library of Congress Cataloging-in-Publication Data***

ISBN 978-1-394-23787-6

Front cover images courtesy of Wikimedia Commons
Cover design by Russell Richardson

Set in size of 11pt and Minion Pro by Manila Typesetting Company, Makati, Philippines

Printed in the USA

10 9 8 7 6 5 4 3 2 1

# Contents

# Preface

The evolution and expansion of electronic products and devices have led to rapid development in modern science and technology. As this expansion increases briskly, so does the requirement for energy-storing devices that can handle the necessary power for their functioning. The miniaturization of electronic devices has sparked the evolution of energy-storing devices in their most portable forms. Integration of min- or micro-powering devices with tiny electronic devices has led to the simultaneous evolution of nanomaterials and, correspondingly, nanotechnology. Thus, procuring nanomaterials has emerged as an essential branch in synthetic chemistry. In recent years, various techniques have emerged that improve the composition control, shape, and size of nanomaterials in order to reach their greatest potential for dedicated applications.

The nanotechnology evolution has provided the control and ability to restructure matter at the atomic and molecular levels on a scale of l-100 nm. Nanotechnology also involves exploiting different properties and phenomena at that scale, in contrast to the properties exploited by single atoms or molecules of bulk behavior. Nanotechnology primarily aims to create materials, devices, and systems that exhibit fundamentally new properties and functions. As such, nanotechnology and functionalized nanomaterials have proven to be the ultimate frontier in the production of novel materials that have manufacturing longevity and cost-efficiency.

The integration of nanotechnology to produce functionalized nanomaterials and energy storage from electrochemical principles has established a new platform for science and technology. The integration of two technologies does not compromise their fundamentals and principles, but instead results in novel and high-performance supercapacitors.

This book consists of 11 chapters that review state-of-the-art technologies.

Chapter 1 deals with the developments in flexible fabric-type energy storage devices, as well as hybrid fabrics for energy storage and harvesting in flexible wearable electronics.

Chapter 2 elaborates on the role of electrolytes in the development of sustainable supercapacitors and the performance optimizations associated with them. The interaction between electrodes and electrolytes is fundamental to supercapacitor performance, as outlined in this chapter. It also discusses factors such as mobility, diffusion coefficient, ionic conductivity, ion solvation, thermal stability, viscosity, ionic size, electrochemical stability, dielectric constant, and dispersion interaction that affect the overall device performance.

A vast amount of contemporary research is dedicated to developing green supercapacitors as sustainable energy storage devices. Energy storage system that are based on green supercapacitors allow the storage of electricity during lean hours at a relatively cheaper value and later delivery. Chapter 3 highlights the recent developments in this area.

With escalating international demands for cost-effective and clean energy storage systems, supercapacitors have attracted more attention than ever. The strategies, synthesis methods, applications, and electrochemical studies supply the basis for the development of sustainable supercapacitors. The materials used in sustainable supercapacitors, such as novel transition metal oxides, metal-organic frameworks, conductive polymers, and biomass-based, as well as their composites (binary and ternary), are explored in Chapter 4, which also highlights the different materials used to develop and fabricate sustainable supercapacitors.

Chapter 5 focuses on the role of material selection and fabrication approach in achieving optimal performance and sustainability in supercapacitors. This section includes a detailed discussion on the significance of material selection, emphasizing the properties and characteristics required for sustainable electrode materials.

Battery technologies are widely used technology, but they offer several disadvantages like weight, volume, large internal resistance, poor power density, and poor transient response. On the other hand, advancements in material and technology have made increased the promise of supercapacitors, ultracapacitors, and electrostatic double-layer capacitors (EDLC). Chapter 6 explores how these solutions offer a more significant transient response, power density, low weight, low volume, and low internal resistance, making them suitable for several applications.

Chapter 7 explains how sustainable supercapacitors have steadily gained traction due to their potential for non-invasive health monitoring, while chapters 8, 9, and 10 outline the current state of sustainable supercapacitors. These chapters provide a thorough introduction to the various electrolytes and electrode materials used in the most recent supercapacitor technologies. The current challenges of the fabrication and utilization of

SCs are critically examined and potential solutions are analyzed in this section. Plus, the performance inconsistencies that arise from the improper use and extreme diversity of performance evaluation and reporting methods are highlighted.

Chapter 11 discusses the development opportunities and the challenges faced by supercapacitors, such as technical problems, establishing electrical parameter models, consistency testing, and establishing industry standards.

Overall, this is a timely handbook for researchers and scientists in pursuit of new advancements in supercapacitor technology. The handbook is ideal for a broad audience working in the fields of electrochemical sensors, analytical chemistry, chemistry and chemical engineering, materials science, nanotechnology, energy, environment, green chemistry, sustainability, electrical and electronic engineering, solid-state physics, surface science, device engineering and technology, etc.

Furthermore, this book will be an invaluable reference source for libraries in universities and industrial institutions, government and independent institutes, individual research groups, and scientists working in supercapacitors. We are tremendously grateful to all the authors for their outstanding enthusiastic efforts in making this book possible. Finally, our special thanks to Martin Scrivener and Laura Mohr of Scrivener Publishing for their continued support and help during this project.

**The Editors**
August 2024

1

# Flexible Sustainable Supercapacitors

**S. Siva Shalini, R. Balamurugan, I. Ajin and A. Chandra Bose***

*Nanomaterials Laboratory, Department of Physics, National Institute of Technology, Tiruchirappalli, Tamil Nadu, India*

***Abstract***
The intense growth in renewable energy sources and the global drive towards meeting the net-zero target have encouraged the development of long-cycle life-sustainable electrochemical energy storage devices. Despite many ongoing research and advancements, challenges still remain in commercial applications such as boosting power and energy density, improving scalable fabrication techniques, developing safety, increasing cycling lifetime, and securing comfortable wearing experiences. For wearable and flexible applications that demand high capacitance, high power ratings, long life cycles, stretchability, portability, and durability, flexible supercapacitors are a very promising development. Applications for smart textiles include communications devices, thermoregulated clothing, heat storage, thermoelectric energy harvesting, electroluminescence applications, and energy storage. For the above-mentioned applications, sustainable production of the power source is crucial. There are many ways that energy storage can be accomplished. Supercapacitors offer better performance in addition to a wide range of applications, including energy harvesting devices, hybrid cars, and smartphone components. In this chapter, we systematically summarize the developments in flexible fabric-type energy storage devices, as well as their hybrid fabrics for energy storage and harvesting in flexible wearable electronics.

***Keywords***: Flexible electrodes, sustainability, energy storage, biomass derived, portable hybrid supercapacitors, superior electrochemical performance

**Corresponding author*: acbose@nitt.edu

Basheer Ahamed and Chaudhery Mustansar Hussain (eds.) Sustainable Supercapacitors: Next-Generation of Green Energy Storage Devices, (1–32) © 2025 Scrivener Publishing LLC

## 1.1 Introduction

The development of energy storage and conversion devices using fossil fuels (coal, oil, gas, etc.) holds certain drawbacks such as skin diseases, excess pollution, etc. Recently, with the continuous increase in the world's population and the endless modernization of technologies, an intelligence era has arrived [1]. The history and advancements of supercapacitor are shown in Figure 1.1. Right now, wearable/portable electronics have obtained a fast growth in various fields such as sports, environmental, biomedical, and electronics applications [2]. One of the upcoming technologies with a significant potential influence in textile electronics, which offers wearable, flexible, and comfortable electrical systems [3, 4]. Their primary requirements are flexible energy conversion and storage systems that exhibit superior mechanical durability, high safety, low weight, and outstanding electrochemical performance [5]. Flexible supercapacitors are a desirable option because of their strong flexibility, increased safety, and nearly constant performance under a range of mechanical deformations [6]. However, they are required to possess a high energy density, long cycle life, and an excellent electrical conductivity [7].

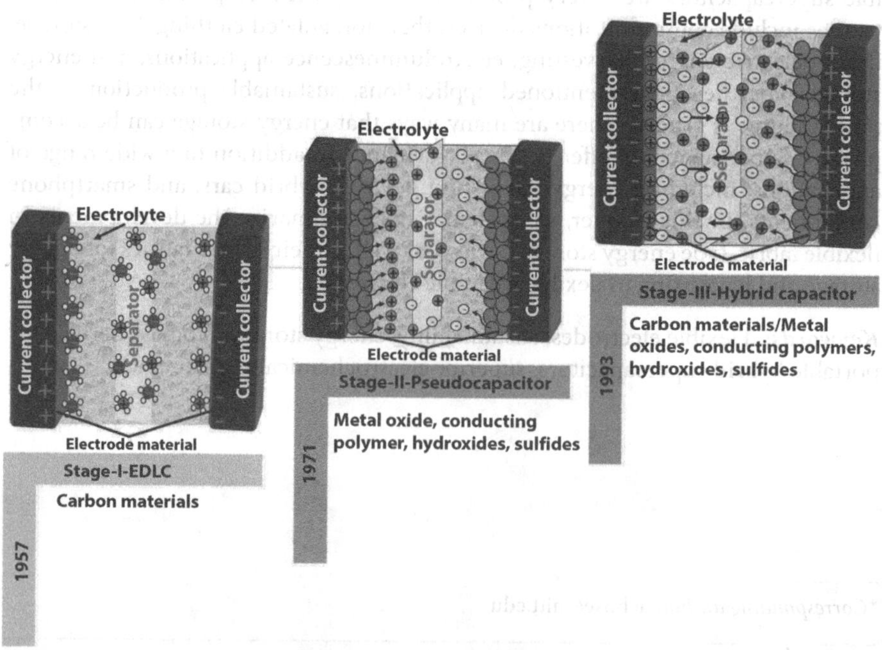


**Figure 1.1** History and advancement of supercapacitors.

The components of a flexible supercapacitor (FSC) usually include a flexible electrode with superior electrochemical qualities, a separator in a flexible assembly, and an electrolyte that is compatible [8]. In this chapter, we systematically summarized starting from the electrode materials to the assembly of flexible supercapacitors, with a focus on various types of electrode materials for flexible supercapacitors, their improving electrochemical performance, fabrication of the flexible electrode, and frequently used electrolytes [9].

## 1.2 Flexible Electrodes

The performance of flexible supercapacitors is significantly determined by the electrode. In particular, the flexible supercapacitors are better in overall performance, including their lifetime, power density, and specific capacitance [10]. The materials used for the electrodes affect the energy density and flexibility. Numerous nanomaterials, including conductive polymers, metal compounds, and carbon materials have played a key role in the development of flexible, smart, and self-sustaining FSCs [11]. The electrode materials for FSCs are categorized based on the energy storage mechanisms. The pseudocapacitor includes transition metal compounds, and conductive polymers have unique benefits and drawbacks to the use of flexible supercapacitors, and the electric double layer capacitor (EDLC) uses carbonaceous materials [12].

## 1.3 Electrode Materials

EDLCs uses carbonaceous electrode materials that consist of graphene, carbon nanotubes, activated carbon, and carbon nanofibers [13]. They serve as a viable candidate for constructing flexible supercapacitor electrodes due to their excellent stability, electrical conductivity, huge surface area, and strong mechanical performance [14]. However, their usage is constrained by the very low specific capacitance of these carbonaceous materials reduces the energy density. Surface modification techniques like heteroatom doping, surface activation, and exfoliation were utilized to significantly increase the capacitance [15]. Transition metal compounds are used as a pseudocapacitive electrode material because they are affordable, have a high theoretical specific capacitance, easy to handle, and chemically stable. These compounds include transition metal hydroxides, carbides, oxides, sulfides, and nitrides [16]. However, most transition metal oxides have short cycle lives and limited rate capabilities due to their poor reversibility and poor electrical conductivity. The $M_{n+1}X_nT_x$ MXenes have hydrophilic surfaces, a high surface-to-volume ratio, and strong electrical conductivity. X can be carbon or nitrogen, and $T_x$ is the surface termination. The electrochemical capacities can be improved by adding quantum dots to nanostructured transition metal oxide [17].

Conductive polymers such as polypyrrole (PPy), polythiophene, polyaniline (PANI), and their derivatives are an excellent choice for pseudocapacitors. Certain benefits of pseudocapacitive materials include their remarkable environmental stability, ease of synthesis, and electrical conductivity [18]. In this regard, the specific capacities of PPy, poly (3,4-ethylenedioxythiophene), and PANI are 480 F $g^{-1}$, 210 F $g^{-1}$, and 1284 F $g^{-1}$, respectively. The polymerization procedure, dopant concentration, structural morphology, and ionic diffusion length all have an impact on conductive polymers specific capacitance [19]. However, the doping and dedoping processes that occur during the charge and discharge course lead the conductive polymers to expand and contract, which causes mechanical deterioration and poor cycling stability [20]. In addition, recently developed innovative electrode materials include POMs (polyoxometalates), MOF (metal-organic framework), and BP (black phosphorus) [21]. Owing to its porous structure, significant surface area, and abundance of active redox sites that contain metal, MOF is one of them that is attracting the most attention as a supercapacitor electrode material [22].

## 1.4 Modifying Techniques to Enhance Electrochemical Performance

To ensure rate performance and high capacity, the ideal electrode materials for flexible supercapacitors need to have a large ion accessible surface area and rapid electron transfer kinetics [23]. Therefore, flexible supercapacitor electrode materials can achieve superior electrochemical performance by increasing their surface area, mass loading, reducing the path length for the electrolyte ion diffusion, and improving the ion transport [24]. Many strategies, like building diverse electrode material microstructures and creating composites with a well-balanced composition, are used to improve the electrochemical performance of the flexible electrode materials [25].

## 1.5 Flexible Supercapacitors

Xiaojing Lv *et al.* fabricated a flexible laterally configured electrochromic supercapacitor through a fully solution process. Figure 1.2a schematically illustrates the structure of flexible electrodes, and Figure 1.2b shows the detailed step-by-step fabrication process of the flexible laterally configured electrochromic supercapacitor device. For transparency and flexibility, polyethylene plastic film (PET) was used as a substrate. The supercapacitor's current collectors need high electrical conductivity. However, PET is a resistive material. For inducing high electrical conductivity, the conducting layer of the AgNWs/Aa-PDA complex and PEDOT:PSS was produced using spraying in the sequence method. On the above conducting layer, poly (3,4-propylenedioxythiophene) (magenta electrochromic polymer) and the gel electrolyte PMMA/PC/$LiBF_4$/BMIMOTF/ACN were cast sequentially to fabricate the asymmetric device PProDOT/PEDOT:PSS. It demonstrated strong cyclic stability, retaining 85.65% of its initial areal capacitance even after 1000 cycles of mechanical deformation, and a maximum areal capacitance of 2.15 mF/cm$^2$ at a current density of 0.05 mA/cm$^2$. Therefore, the electrochromic supercapacitor device with a flexible laterally configuration shows the potential for use in wearable and portable displays like smart watches and electronic paper [26].

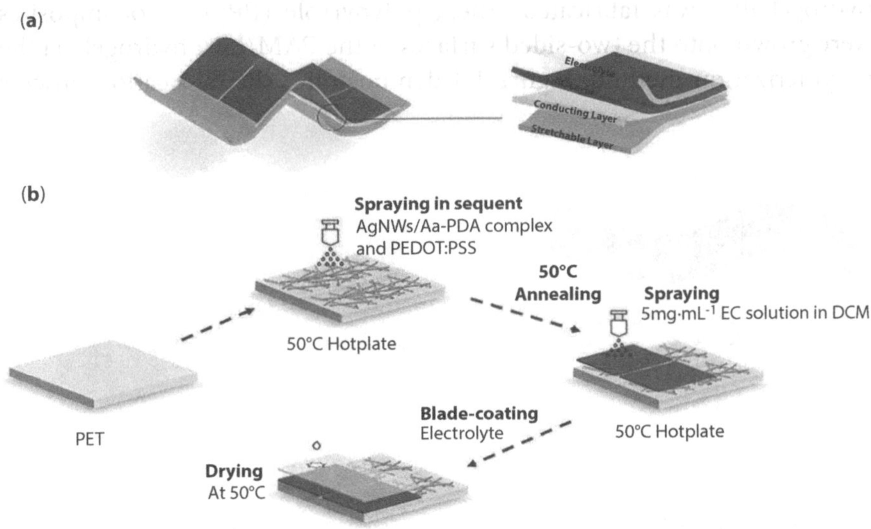


**Figure 1.2** (a) Schematic illustration of the structure and (b) the detailed fabrication process of the flexible laterally configured electrochromic supercapacitor device.

Qing Xin *et al.* optimized the ionic conductivity of flexible supercapacitor's electrolytes and electrode material with glycerol-Mo (Gly-Mo). Figure 1.3 shows a schematic illustration of the hydrogel with partially substituting water molecules and Gly-Mo. The electrolytic solution resistance of the hydrogel was optimized by increasing the concentration of Gly-Mo and investigating the solution resistance of the hydrogel at various temperatures (–40, 50 °C, and room temperature). The result indicates that an 8-mL substitution of Gly-Mo in 10 mL of hydrogel (PVA/(Gly-Mo)$_8$) provided low solution resistance at a wide range of temperatures from –40 to 50 °C. The ionic conductivity of the hydrogels was examined at a wide range of temperatures for various concentrations of Gly-Mo. PVA/(Gly-Mo)$_8$ provides an excellent ionic conductivity in extremely low temperature and high temperature. The flexibility of the hydrogel was investigated at room temperature and –20 °C. By increasing the concentration of Gly-Mo, predominantly hydrogel's flexibility increases drastically. The supercapacitor's specific capacitance retention rate was 64.2 % after 50 charge–discharge cycles at –10 °C, indicating that it is still effective. The above results show that the organohydrogel PVA/(Gly-Mo)$_8$ is suitable for flexible supercapacitors due to its good flexibility and conductivity at a wide temperature range of – 40 to 50 °C [27].

Shuzhen Cui *et al.* fabricated a flexible and hydrophilic hydrogel film-based all-in-one supercapacitor through a simple polymerization method. At first, through the polymerization process, PAM (polyacrylamide)/LiCl hydrogel film was fabricated. Then, polypyrrole (PPy) nanocomposites were grown onto the two-sided surfaces of the PAM/LiCl hydrogel via the polymerization method. Figure 1.4 demonstrates the fabrication process

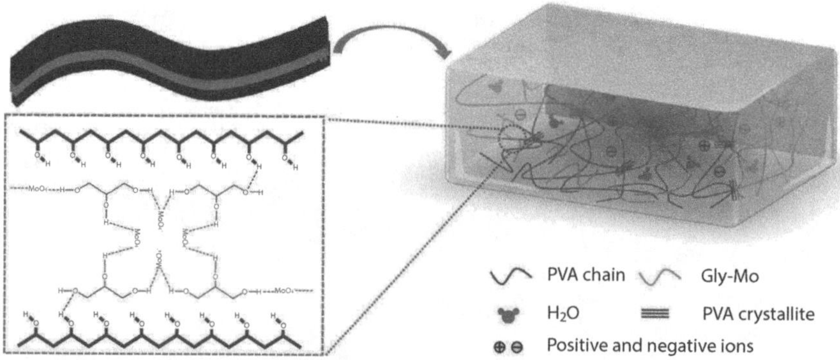


**Figure 1.3** Schematic illustration of the hydrogels.

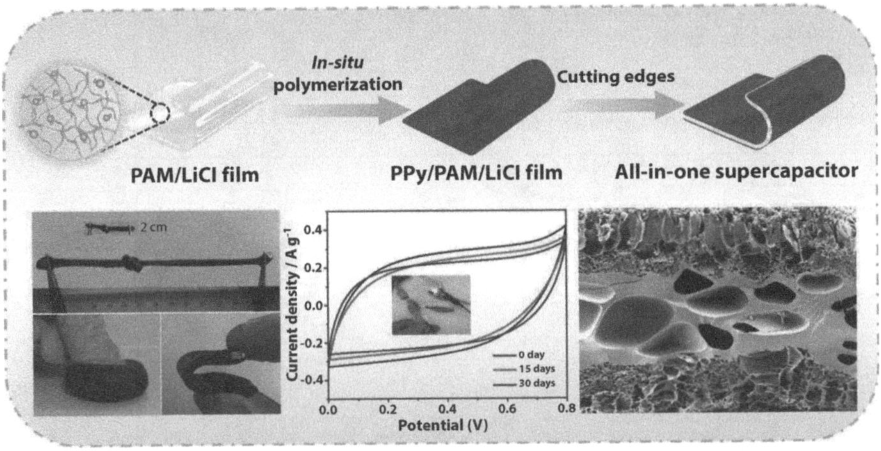


**Figure 1.4** Schematic illustrates the fabrication of the all-in-one flexible supercapacitor, flexibility of the device, its electrochemical performance, and its cross-sectional SEM image.

of an all-in-one flexible supercapacitor. Mechanical properties of the PPy/PAM/LiCl hydrogel were examined by stretching, twisting, knotting, folding, compressing, and bending the device, which is shown in Figure 1.4. Additionally, the all-in-one flexible supercapacitor demonstrated exceptional cycling stability, an appealing capacitive performance, and the ability to prevent electrolyte and electrode delamination under deformations. The all-in-one supercapacitor is found to have high ionic conductivity (29 mS cm$^{-1}$), high specific capacitance of 229.8 mF cm$^{-2}$ (0.8 mA cm$^{-2}$), and excellent flexible stretching up to 650 % of the original length. The successful coating of electrode materials onto hydrophilic hydrogel electrolytes offers remarkable opportunities for the creation of a flexible all-in-one supercapacitor, thereby promoting the advancement of wearable electronics [28].

Soomin Suh *et al.* demonstrated an ultrafast flexible supercapacitor. Figure 1.5a represents the scheme of an as-fabricated flexible PEDOT:PSS supercapacitor consisting of the electropolymerized PEDOT:PSS (EP-PEDOT:PSS) thin-film electrodes on Au/Ti/Ti-foil current collectors, along with a PTFE separator permeated with a sulfuric acid-polyvinyl alcohol ($H_2SO_4$-PVA) polymer gel electrolyte. Figure 1.5b demonstrates the electropolymerization experimental setup's scheme. The current response and mass loading on the current collector with respect to time are represented in Figure 1.5c,d, respectively. The EP-PEDOT:PSS-based flexible SC demonstrated high-performance EDLC behaviors, such as triangle-shaped galvanostatic charge/discharge (GCD) curves and rectangular cyclic

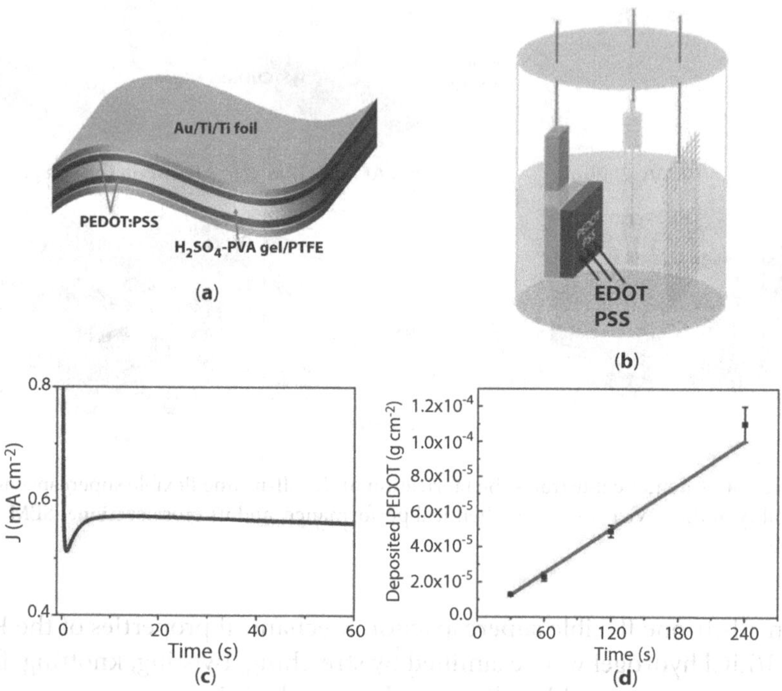


**Figure 1.5** Schematic diagrams of (a) all-solid-state flexible supercapacitor, and (b) three-electrode cell configuration for electropolymerization; (c) electropolymerization chronoamperometric current response at a constant applied voltage; (d) mass loading on current collector during electropolymerization.

voltammetry (CV) curves up to 300 V $s^{-1}$. It also demonstrated an excellent cyclic stability (~95 % capacitance retention over 50,000 cycles) and a high rate capability (greater than 90 % capacitance retention from 1 to 10 mA $cm^{-2}$), and offers remarkable mechanical adaptability, with bending radii between 4.0 and 0.6 mm, and resilience, maintaining approximately 94 % of its capacitance after 1,000 bending cycles [29].

Bhagya Dharmasiri *et al.* fabricated a flexible carbon fiber-based structural supercapacitor. In this, the electrical conductivity and pseudocapacitive activity of the carbon fiber (CF) were improved by the surface functionalization of CF with a covalently grafted layer of redox-active poly(o-phenylenediamine). The electrical conductivity of the solid electrolyte was optimized by incorporating SIL (solvated ionic liquid) with epoxy resin (RIM935 +RIMH936). Increasing the SIL concentration in solid electrolytes increases the electrical conductivity and flexibility but reduces the tensile strength. Seventy percent of SIL mix produced the optimized electrical conductivity with the required tensile strength. Using this electrical conductivity-optimized solid electrolyte and surface-functionalized CF with glass fibers as a separator, they fabricated the flexible supercapacitor. Its structural prototype is illustrated in Figure 1.6. The device's flexibility was examined at different bending angles, and it exhibited an increased specific capacitance at the bending angles of 90° and 135°. It also exhibited a maximum specific capacitance of 1439 mF $g^{-1}$ at 0.5 mA $g^{-1}$, a maximum energy density of 181.79 mW h $kg^{-1}$, and a maximum power density of 6.18 W $kg^{-1}$ [30].

Hengyu Zheng *et al.* used a wide temperature-tolerant and self-healing flexible supercapacitor with a ternary network organo-hydrogel electrolyte. A self-healing ternary network electrolyte was fabricated by freeze-drying method using polyvinyl alcohol (PVA), graphene, dimethyl sulfoxide, boric acid (BA), and sulfuric acid. Its self-healing capability was tested at room temperature, as shown in Figure 1.7a. Ten seconds after cutting, it joined together, and its mechanical stability was retained after 5 min. Its microscopical images (Figure 1.7b) showed that after the self-healing process, some cracks are found in graphene/BA/PVA organo-hydrogel. After self-healing, its temperature sustainability was examined, as shown in Figure 1.7c. Its electrical conductivity was investigated before and after the self-healing process by lighting LED, as shown in Figure 1.7d. This self-healing organo-hydrogel was used as an electrolyte and polyaniline

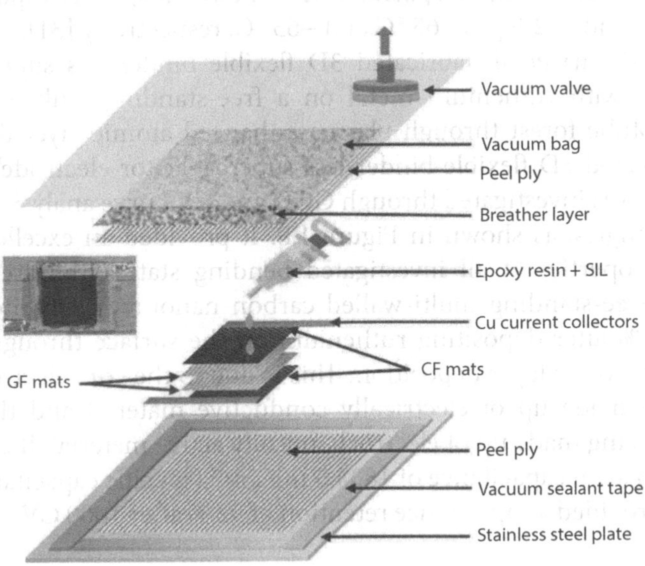


**Figure 1.6** Schematic illustration of each layer of the flexible supercapacitor.

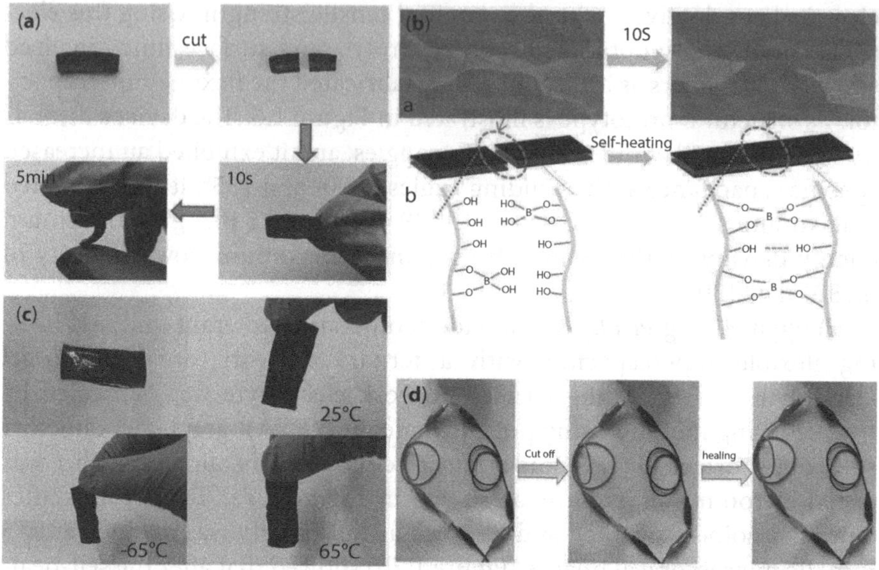


**Figure 1.7** (a) Optical images of self-healing graphene/BA/PVA organo-hydrogel, (b) optical microscopic images, (c) self-healed organo-hydrogel at various temperatures, and (d) optical images of blue LED lit after cutting and self-healing of organo-hydrogel.

fibers coated on carbon cloth were used as electrodes to fabricate the flexible symmetric supercapacitors. It exhibited a specific capacitance of 237.8 F $g^{-1}$ and 152 F $g^{-1}$ at 65 °C and –65 °C, respectively [31].

Hanie Kazari *et al.* fabricated 3D flexible binder-less supercapacitor electrodes with ruthenium metal on a free-standing multi-walled carbon nanotube forest through plasma-enhanced atomic layer deposition. The fabricated 3D flexible binder-less supercapacitor electrode's bending capability was investigated through GCD and CV curve analysis at various bending angles, as shown in Figure 1.8. It provided an excellent charge storage properties at all investigated bending states. The areal capacitance of free-standing multi-walled carbon nanotubes exhibited greater than 400 % after depositing ruthenium on the surface through plasma-enhanced atomic layer deposition. This is due to the core electrode material being made up of electrically conductive material and the surface material being made up of electrochemically active material. It exhibited a maximum areal capacitance of 1834.5 mF $cm^{-2}$ (specific capacitance of 393 F $g^{-1}$). It provided a capacitance retention of 92 % after 1000 CV cycles [32].

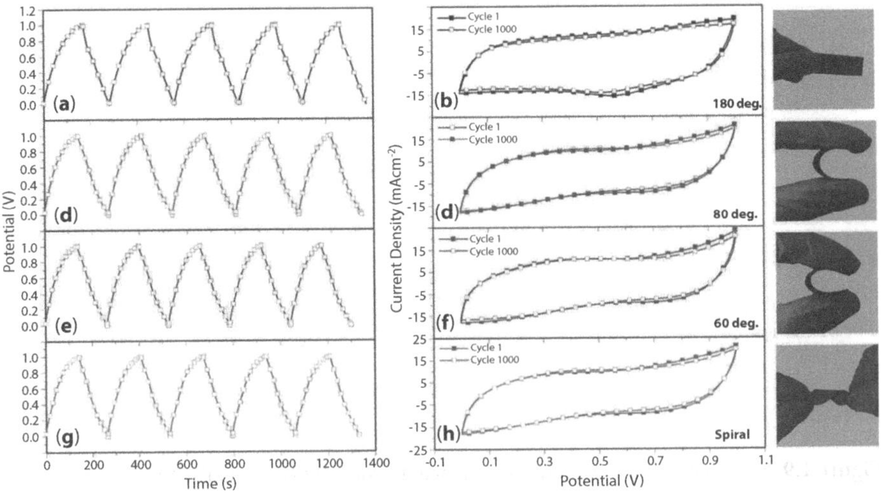


**Figure 1.8** GCD and CV curves of the 3D flexible electrode at various bending angles (a, b) 180°, (c, d) 80°, (e, f) 60°, and (g, h) spiral.

Jiawei Guo *et al.* fabricated a hierarchical porous 3D $Ni_3N$-CoN/NC heterojunction nanosheet with nitrogen vacancies to create a high-performance flexible supercapacitor. For high-performance flexible electrode fabrication, carbon cloth was used as a flexible and electrically conductive current collector. On the carbon cloth, porous nickel cobalt metal-organic framework (MOF) nanosheets were grown using the solvothermal method. Then, to improve the electrochemical activity of MOF, it is annealed in the ammonia environment and produced hierarchical porous 3D $Ni_3N$-CoN/NC heterojunction nanosheets with nitrogen vacancies. Due to the base material, MOF's porous nature and nanosheet-like structure, resultant nitrides provided a rapid ion diffusion path with the support of nitrogen vacancies, as shown in Figure 1.9. It provided a maximum specific capacity of 468.3 mA h $g^{-1}$ at 3 A $g^{-1}$ current density. With this porous $Ni_3N$-CoN/NC acting as an anode, and AC/CC acting as a cathode with PVA/KOH electrolyte, they fabricated a flexible high-performance supercapacitor ($Ni_3N$-CoN/NC/CC//AC/CC). It exhibits the highest energy density of 0.2144 mWh $cm^{-2}$ and a maximum power density of 80 mW $cm^{-2}$. It also exhibited excellent cycling stability over 1,500 charge–discharge cycles [33].

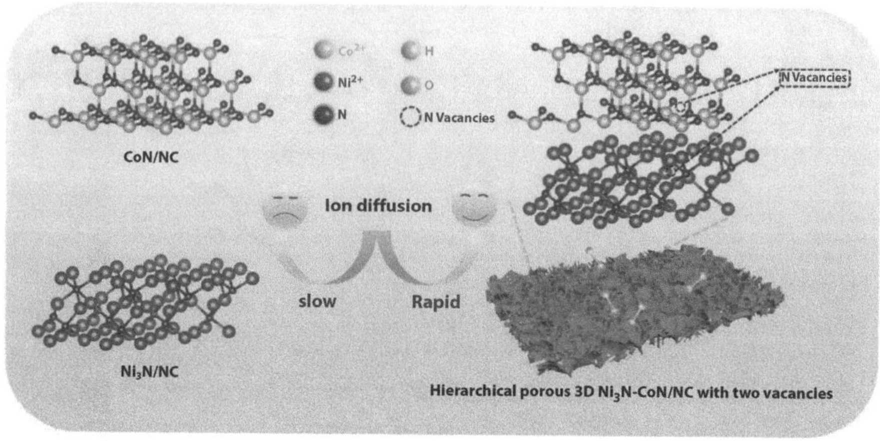


**Figure 1.9** Schematic illustration of hierarchical porous 3D Ni3N-CoN/NC heterojunction nanosheets with nitrogen vacancies.

Serkan Demirel *et al.* fabricated a flexible supercapacitor using stretchable and self-healable PVA–Nickel–Borax (PNB) matrix electrodes. In this, PNB electrodes were fabricated through the solution self-curing method. Its self-healing ability was tested, as shown in Figure 1.10a. It confirmed that after 8 min of disjoining, PNB was completely rejoined together. Homogeneity of the as-fabricated electrode was attained through proper

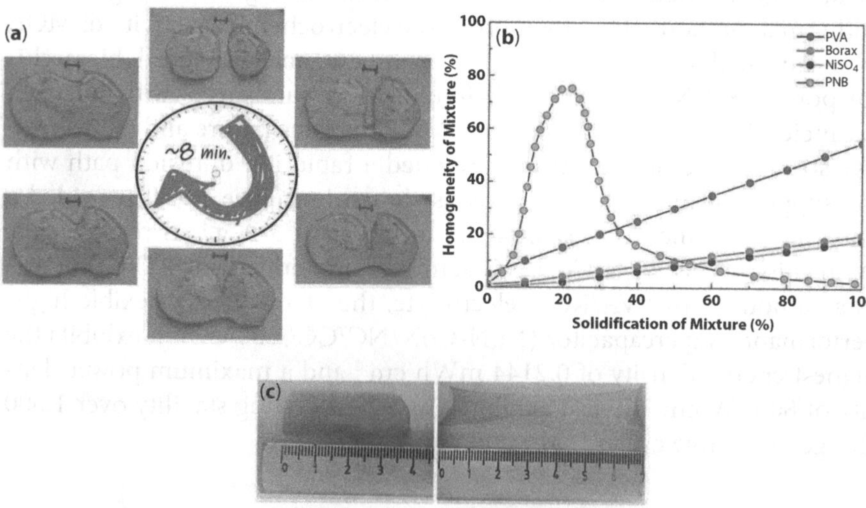


**Figure 1.10** (a) Self-healing of PNB, (b) homogeneity mixture identification, and (c) stretchablility of PNB.

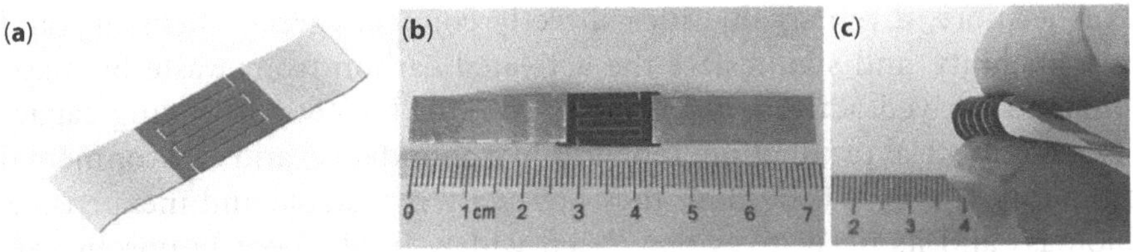


**Figure 1.11** (a) Structure diagram of the interdigitated micro-supercapacitors, optical images of (b) appearance size and (c) flexure flexibility test.

solidification of the mixture. It is confirmed by finding the homogeneity of the mixture with different solidification percentages of the mixture, as shown in Figure 1.10b. The stretching ability of the PNB matrix was demonstrated, as shown in Figure 1.10c. In this, dI/dV analysis was carried out for a more precise determination of redox peak values of PNB samples. Using these PNB electrodes, the fabricated devices provided a maximum specific capacitance of 88.95 F $g^{-1}$ with an energy density of 49.42 W h $kg^{-1}$ at a power density of 18.75 kW $kg^{-1}$ [34].

Shaoyi Lyu *et al.* fabricated a flexible supercapacitor using polymer–inorganic hybrid nanocomposite aerogel film electrodes. The flexible electrode was manufactured using cellulose nanofibers (CNFs), MXene, and carboxylic single-walled carbon nanotubes (SWCNT). First, the homogeneous mixture of all three materials was dispersed well in aqueous to prepare hydrogels. Then, the hydrogels were dried using the supercritical point drying method to produce aerogels. Due to the advantages of all three precursor materials, it provided an excellent electrical conductivity, flexibility, and porosity. Using this aerogel, the interdigitated micro-supercapacitors are fabricated, as shown in Figure 1.11a. Figure 1.11b displays the dimensions of the fabricated micro-supercapacitor. Figure 1.11c demonstrates the flexibility of the supercapacitor. It exhibited a maximum areal capacitance of 746.68 mF $cm^{-2}$ with an energy density of 9.04 W h $cm^{-2}$. It provided an outstanding cycle stability over 10,000 cycles with a capacity retention rate of 91.23 % [35].

## 1.6 Sustainable Supercapacitors

Activated carbon is the perfect negative electrode material for supercapacitor applications because of its special physical and electrochemical properties, which include high electrical conductivity, large specific surface area (SSA), flexible porous architectures, and high chemical stability.

Furthermore, a remarkable idea directly connects green chemistry and sustainability, and synthesizes the activated carbon from waste biomass. Biomass-derived activated carbon has various benefits, including capacity for industrial production, renewability, abundance, and environmental friendliness [36–39]. Because they are easily repeatable and inexpensive, biomass and its derivatives have been widely used to create porous carbon materials [40–42]. Using mixed egg yolk/white and rice waste as pre-carbonization catalysts, a two-step technique of pre-carbonization and pyrolytic activation was used to create an *in situ* nitrogen–oxygen co-doped porous carbon for supercapacitor applications [43]. Activated carbons from red pepper industrial waste (RPW) were used to create green and sustainable supercapacitor electrodes utilizing traditional chemical activation with $ZnCl_2$ at varied carbonization and activation temperatures [44]. Using coconut husk as a green and natural resource was used to create reduced graphene oxide (rGO) for supercapacitor (SC) applications [45]. Cotton-based activated carbon fiber was prepared by a urea-enhanced low-temperature hydrothermal carbonization and activation method used for supercapacitor applications [46]. Spent tea wastes were used to create porous carbon for supercapacitor applications [47]. Resin-based carbon materials were used for creating honeycomb-like carbon nanosheets [48]. Cashew nut shells [49], dry almond leaves [50], pineapple peel [51], purple leaf plum kernel shell [52], pine needles [53], Abutilon theophrasti steam [54], Sterculia foetida fruit shell [55], and subbituminous coal feedstock [56] were used for making sustainable electrode materials.

Zhu *et al.* use the renewable lignin as a precursor for the coproduction of porous carbon and calcium chloride. Homogeneously mixing the CaO with the lignin precursor to create the porous carbons, CaO acts as a hard template for pore creation. The combined sustainable production of lignin-derived porous carbon (LPC) and $CaCl_2$ was obtained through a simple preparation method. The method of preparation is shown in Figure 1.12. The lignin precursor was mixed with CaO, and carbonization was carried out; after carbonization, it was sonicated in an HCl solution, followed by filtration. The filter residue was dried to obtain the LPC; the filtrate was collected in a Petri dish and dried to get $CaCl_2$ powder. This was a cheap technique for producing LPC, which was used in supercapacitors; the templating agent CaO was recycled as $CaCl_2$. The porous carbon obtained in this process has rich pores and a high SSA. These pores create the most abundant sites for electrochemical reaction, reduce the ion mass transfer distance, and improve the rate performance capacity of the hybrid supercapacitors. The obtained LPC carbon was assembled with a zinc ion

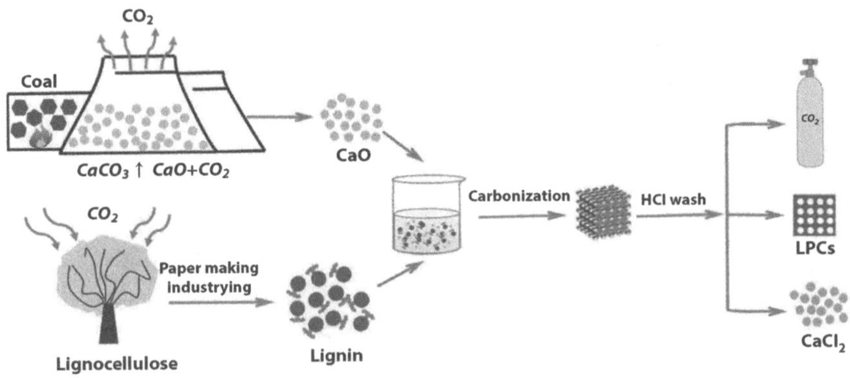


**Figure 1.12** Schematic representation of the sustainable production technology for the synthesis of LPC and $CaCl_2$.

hybrid capacitor, which has high cycle stability, better rate performance, and high specific capacitance [57].

Kitamoto *et al.* use a kraft lignin carbon precursor obtained from the kraft pulp for supercapacitor applications. The cross-linked kraft lignin was dissolved in double-distilled water with a NaOH activating agent to form a homogeneous suspension. Then, the above mixture was spray-dried to form a lignin/NaOH composite particle. Finally, it was carbonized, followed by acid wash to get the porous carbon particles. The synthesis route is shown in Figure 1.13. The amount of NaOH plays a significant role in the morphology of the kraft lignin. In the absence of NaOH and a low mass ratio of NaOH, the surface of kraft lignin was in perfect spherical shape. Gradually increasing the activating agent, the morphology changes from spherical to non-spherical with the development of many holes on the surface. At a 1:1 mass ratio of lignin/NaOH composite, the high amount of NaOH decomposes the carbon walls and completely changes the regular porous structure into hollow, non-spherical morphology. Thus, the high activating agent (NaOH) containing composite has a higher surface area than the low NaOH-containing composites. Even though it has a high SSA, the value of gravimetric specific capacitance, volumetric specific capacitance, and electrode density was high for the low SSA lignin/NaOH composite. The low-mass ratio NaOH/lignin composite has more spherical particles and was tightly bound to each other. As the NaOH/lignin mass ratio increases, voids develop between the particles. Hence, high specific capacity was due to the composite's high packing density [58].

Wang *et al.* fabricated the symmetric supercapacitor device using discarded aramid fibers (AF). AF was dipped into the $H_3PO_4$ solution,

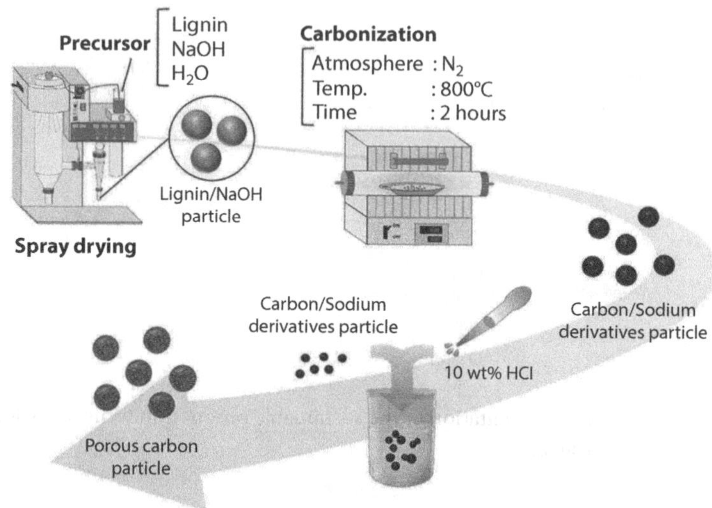


**Figure 1.13** Porous carbon particle synthesizing procedure.

followed by activation in a $N_2$ environment, and the sample was noted as AFP. Above, AFP was further submerged into a KOH solution with carbon mass/alkali ratios varying between 2 and 5, followed by activation in a $N_2$ environment. The sample was noted as AFPK-n, where n represents the alkali mass ratio. The sample's morphology changes during the different KOH activations, as shown in Figure 1.14a–d. AFP has a smooth surface.

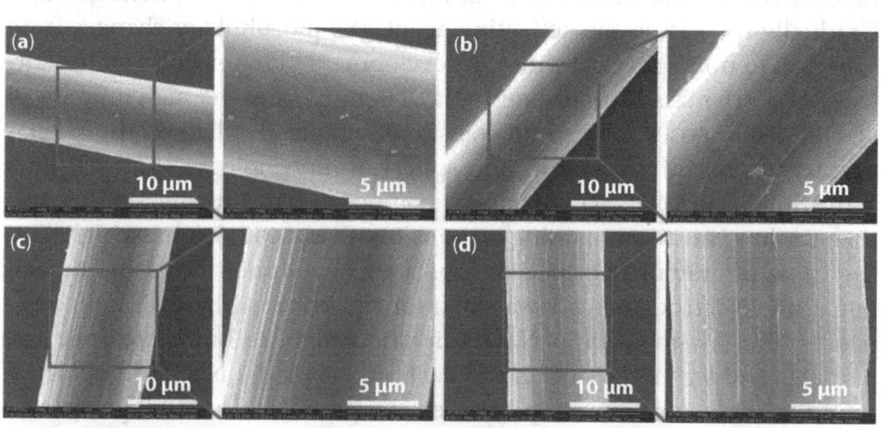


**Figure 1.14** SEM images of (a) AFP, (b) AFK-2, (c) AFPK-2, and (d) AFPK-4.

While increasing the alkali mass ratio, the streak grooves along the axial direction on the surface, and there are many mesoporous structures on the surface of the AFPK-4 sample. Figure 1.14d AFPK-4 has a lot of streaks on the surface. The SSA of all the samples was measured using BET characterization; AFPK-4 exhibits the highest SSA. It has been demonstrated that efficient charge storage and quick ion diffusion benefit from large SSA and hierarchical pore structures. The as-assembled aqueous symmetric supercapacitor based on AFPK-4 had an excellent energy density of 25.26 W h $kg^{-1}$ at a power density of 450 W $kg^{-1}$ [59].

Mehdi *et al.* produce activated carbon (AC) materials using biochar generated by pyrolyzing date seeds (DS) and activating them with $H_2SO_4$. For getting activated carbon, the initial step was pyrolysis of DS at 600 °C (DSBC 600 °C), followed by the carbonization process at 700, 800, and 900 °C temperatures; corresponding samples were represented as DSAC 700 °C, DSAC 800 °C, and DSAC 900 °C, respectively. The carbon content in DSBC 600 °C, DSAC 700 °C, DSAC 800 °C, and DSAC 900 °C was 62.86%, 68.83%, 74.56%, and 71.3%, respectively. From Raman spectra, the sample with the lowest ($I_D/I_G$) value, the DSAC-700, likely underwent more graphitization and developed ordered graphene layers. It suggests that the sample has more conductivity and better electrolyte ion mobility, resulting in an increased electrochemical performance. From the SEM image, the sample DSBC 600 °C has bulk particles with less porosity before activation. All the samples showed a three-dimensional structure; however, only DSAC-700 had macro- and mesopores. The performance of supercapacitors was enhanced by the existence of macro- and mesopores, which offer plenty of space for charge storage, electrolyte access, and pathways for quick charge transfer [60].

Narayanan *et al.* proposed a novel technique for producing agricultural waste-based electrode materials for supercapacitor applications. This work obtained activated carbon from the areca nut (AR) husk using KOH as the activating agent. The synthesis procedure was described in Figure 1.15a. Figure 1.15b illustrates the carbonization followed by the KOH activation of the samples; KOH activation created the pores on the surface of the activated carbon material. The activation was carried out in three different temperatures, 600, 700, and 800 °C, and the samples are denoted as AR-600, AR-700, and AR-800, respectively. Figure 1.15c displays the XRD curve of three samples; two peaks at 23.7° and 43.5° correspond to the (002) and (100) planes respectively of the amorphous natured carbon materials. Figure 1.15d–f indicates the FESEM images of AR-600, AR-700, and AR-800. As the temperature increases from 600 °C to 800 °C, the pores in

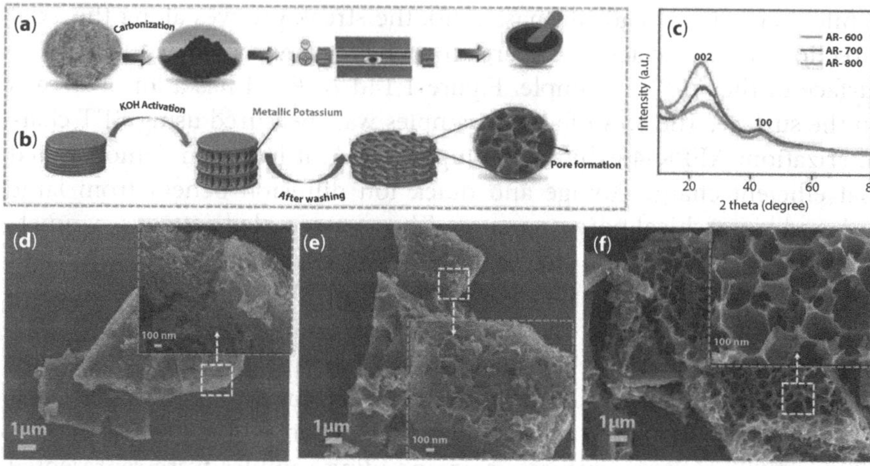


**Figure 1.15** (a) AR husk-derived porous carbon synthesis procedure, (b) KOH activation process, (c) XRD patterns of AR carbon, (d–f) FESEM images of activated carbon at different temperatures 600 °C, 700 °C, and 800 °C, respectively.

the material surface increased gradually, and the sample activated at 800 °C had more pores than the other samples. Further, the samples surface area and pore volume are measured using BET analysis, and AR-800 has more pore volume and surface area than the other samples. Porous materials with large surface area produce more active sites for establishing electrical double layers, increasing the pore size and capacitance [61].

Shaku *et al.* used marula nutshell as the sustainable precursor for supercapacitor electrode material. Figure 1.16 represents the synthesis procedure of nitrogen-doped KOH-activated carbon. Before activation, the sample had a smooth agglomerated surface; after activation, the surface had a lot of holes; these holes provide the path for the electrolyte ions during charge storage. The sample activated using a 1:1 ratio of carbon material and activating agent gives a larger surface area than the sample activated with a high proportion of activating agents. Excess KOH destroys the carbon structure rather than increasing the surface area. The sample activated using a 1:1 ratio of carbon material and activating agent was further doped by nitrogen, creating some new micropores that facilitate the charge storage, improving the electron transfer.

The nitrogen doping increased the sample's surface area from 1266 to 1427 $m^2\ g^{-1}$. This increase in surface area shows that the melamine decomposition, which covers the surface with C and N atoms but does not fill the holes created by the KOH reaction with carbon, caused the surface to become more porous. Before nitrogen doping, the device fabricated using

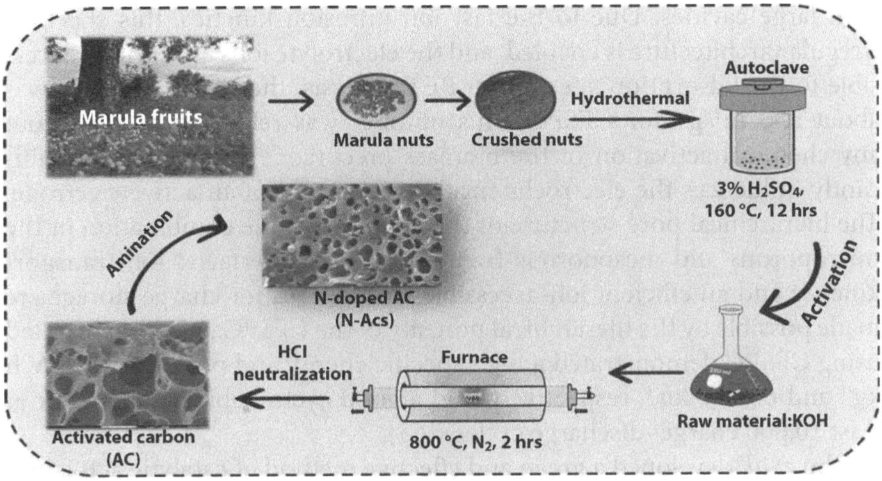


**Figure 1.16** Synthesis route to prepare the nitrogen-doped activated carbon material.

the as-prepared sample exhibits the energy density of 12.5 W h $kg^{-1}$ in 2.5 M $KNO_3$ electrolyte with a 1.8 V potential window. After nitrogen doping, the device can provide an energy density of 17.2 W h $kg^{-1}$ [62].

Arjunan and Kim used a single-step direct-carbonization procedure to create porous carbon material from the renewable source of Cassia fistula flower petals. The synthesis procedure is shown in Figure 1.17. The activation method used here was the physical activation method; the resultant sample was denoted as Cassia fistula flower-petal-derived sheetlike porous carbon (CFSPC). FE-SEM image of CFSPC has a sheetlike morphology

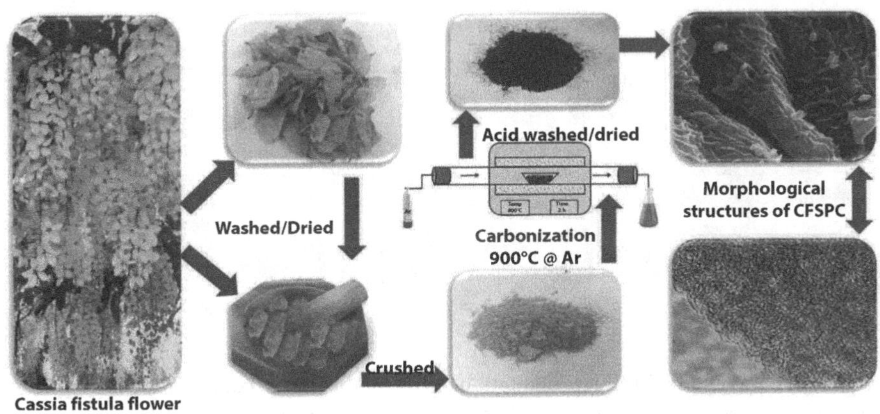


**Figure 1.17** Schematic illustrations of the synthesis procedure of CFSPC.

with large cavities. Due to the fast ion diffusion kinetics, this sheetlike irregular architecture is enabled, and the electrolyte ions are entirely accessible to the adsorption sites. From BET analysis, the specific surface was about 556 $m^2$ $g^{-1}$ for CFSPC. This number was relatively high without any chemical activation of the biomass precursor. The pore size significantly influences the electrochemical performance of an active electrode. The hierarchical pore structure of the CFSPC was the combination of the microporous and mesoporous framework. Fast interfacial ion-transport kinetics and an efficient ion-accessible surface area for charge storage are made possible by the hierarchical porosity of the CFSPC. Device fabricated using CFSPC demonstrated a high specific energy and power of 48.2 W h $kg^{-1}$ and 687 W $kg^{-1}$, respectively, and a good cycle stability (95.3% for at least 10,000 charge–discharge cycles) [63].

Zhu *et al.* developed a green and effective method of catalytic activation caused by trace KOH to produce activated carbons (AC) from oaks for electrochemical capacitor applications to utilize oak resources more effectively. The synthesis process of AC from oaks is shown in Figure 1.18a–c. Figure 1.18d depicts the tracheid-coated cytoderm of oak cells. The xylem conducting tissue in plants, which includes the tracheid, was crucial for transporting water and nutrients to plants. In addition to this, hemicellulose, cellulose, and lignin were the constituents of the biomass cytoderm. These cytoderm constituents were the main carbon sources during the carbonization process.

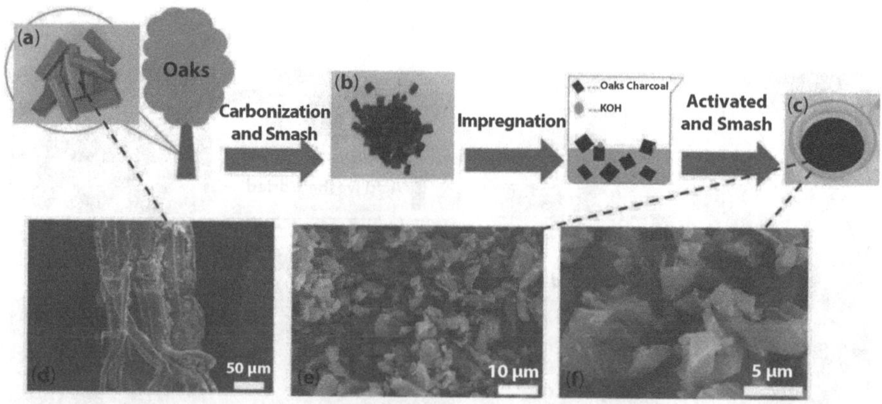


**Figure 1.18** (a–c) Graphical representation of the synthesis procedure for the AC from oaks, SEM images of (d) raw oaks, (e, f) activated carbon using KOH.

According to Figure 1.18e, f, the generated carbon had an uneven shape and ranges in size from 5 to 10 μm. The entire smashing process was responsible for the homogeneous particle size, which was advantageous for making carbon-based electrodes. To create a logical hierarchical porous structure, trace KOH can restrict mesopore growth and promote micropore development. The resulting sample had a large pore volume of 0.448 $cm^3$ $g^{-1}$ and a high SSA of 985 $m^2$ $g^{-1}$. As a result, the electrochemical performances of the sample were enhanced. The resulting sample exhibits outstanding cycling performance with 97% retention over 10,000 cycles at 2 A $g^{-1}$ and a high specific capacitance of 157.5 F $g^{-1}$ at a current density of 1 A $g^{-1}$ [64].

Wei *et al.* had been utilizing mantis shrimp (MS) shell waste as a precursor, and a simple self-template coupled dual-hydroxide (NaOH/KOH) activation approach successfully produced the honeycomb-like hierarchical porous carbons (HPCs). The synthesis process is shown in Figure 1.19; in the same procedure, the carbonization was carried out at three different temperatures, 700, 800, and 900 °C, and the samples are denoted as MSHPC700, MSHPC800, and MSHPC900, respectively. MS shells are mostly made of chitin fibers, minerals, colors, and proteins, which are put together in a particular combination. The activation temperature strongly

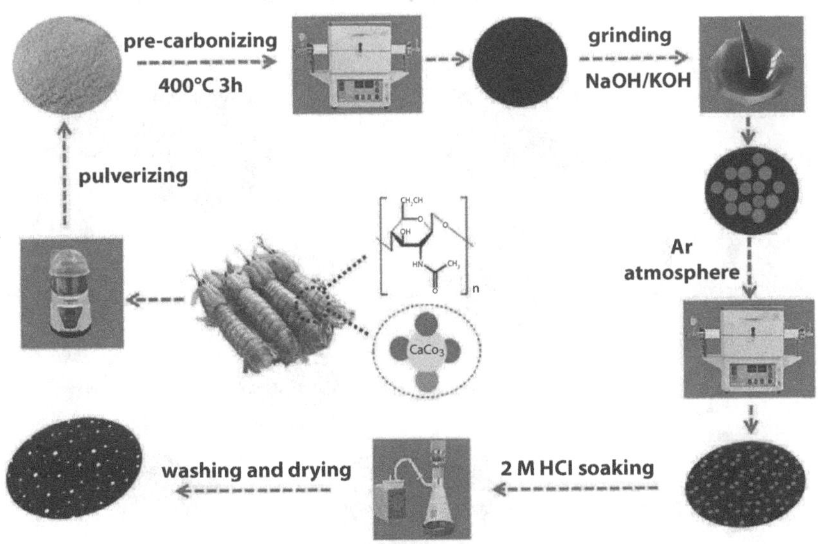


**Figure 1.19** Schematic diagram of the preparation process of porous carbons from MS shells.

influences surface functional groups, SSA, pore size distribution, and microstructure of as-obtained carbon products, which was a crucial component of preparing porous carbon materials. These factors significantly impact the electrochemical performance of carbon-based energy storage devices. Among the three samples, the MSHPC800 electrodes had the longest discharge times, attesting to their highest specific capacitance values. The hierarchical pore structure provides numerous paths for the transfer of electrolyte ions and can contribute to electrolyte penetration [65].

Qiu *et al.* developed a simple and effective approach to convert discarded bamboo shavings into petal-like hierarchical porous carbon electrode materials with an extremely high SSA. The combined catalyzed hydrothermal pretreatment method and co-activation by KOH/melamine was performed to obtain N, O co-doped biomass. Figure 1.20 displays a process flow diagram for synthesizing N, O co-doped hierarchical porous carbon. In this synthesis, melamine was employed as an N-element dopant and a pore modifier for porous carbon materials, while KOH primarily functions as a porogenic agent. The intercross-linked porous network structure of bamboo shavings-based porous carbon results in a multidimensional transport path for electrolyte ions and dramatically lowers the ion diffusion resistance. The hydrophilicity and accessible SSA of carbon-based materials can be greatly increased by the presence of N- and O-containing functional groups. Additionally, by participating in redox processes with

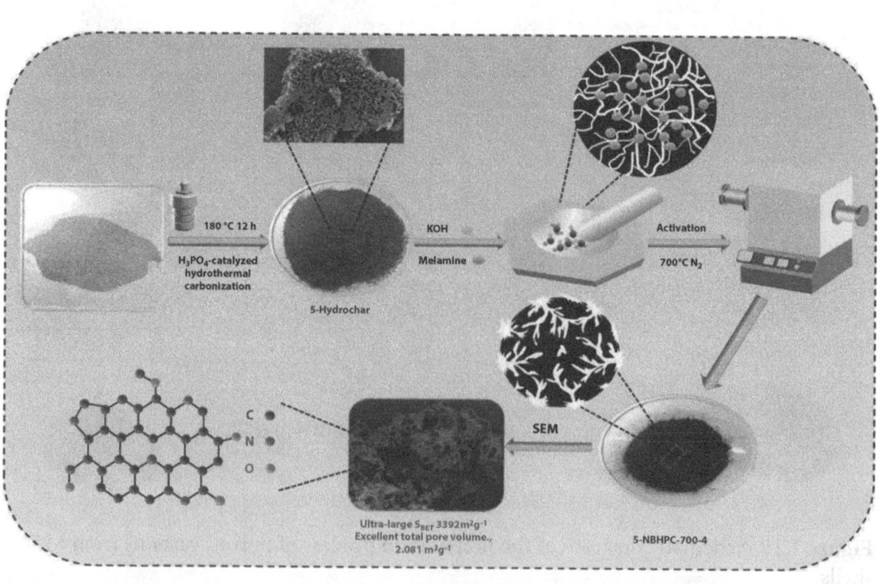


**Figure 1.20** Synthesis road map of N, O co-doped hierarchical porous carbon.

the electrolyte on the materials surfaces, these functional groups can contribute to a portion of the pseudocapacitance. The resultant porous carbon materials exhibit great qualities such as smooth ion diffusion channels, a lot of charge storage space, and significant pseudocapacitance because of the excellent hierarchical porous structure, exceptional SSA, and abundant N and O doping [66].

Cevik *et al.* reported fabricating highly ion-conducting, partly cross-linked hydrogels with sodium carboxymethyl cellulose (C) intercalated at varied fractions by Hibiscus sabdariffa (H). Both C and H served as the ion source to create stable blends, allowing $H_3O^+$ and $Na^+$ ions from the blend to migrate where no external salt was added. The construction of activated carbon-based supercapacitors with considerably better energy storage capacity than EDLC-based devices was made possible using highly safe biopolymer hydrogels. Carbon composite electrodes with conductive additives and activated carbon were employed to create symmetric supercapacitors. The bio-polymer hydrogel electrolyte (PHE) plays a dual role such as an electrolyte and a separator in fabricating highly flexible supercapacitors, which leads to a highly effective device. The PHE provides a relatively peaceful atmosphere that allows for an excellent ion transport. In the supercapacitor charge–discharge phenomenon, the polarization of the electrolyte molecules and the development of a double layer capacitor are depicted in Figure 1.21. The supercapacitor electrodes were supplied with sodium ions ($Na^+$) from the C and hydronium ($H_3O^+$) ions probably derived from H organic acids. H intercalates in the polymer structure through hydrogen bonding to create a negatively charged region throughout the electrolyte. During the charge–discharge process, unbound organic acids contribute to the environment as negatively charged anions [67].

Zhang *et al.* made the hybrid supercapacitor (HSC) device using the amorphous carbon/iron silicate (C/FeSi) anode and cobalt-nickel silicate (CoNiSi) cathode materials that are naturally obtained from bamboo leaves (BLs). CoNiSi cathode C/FeSi anode and materials made from natural BLs are prepared and it was shown in Figure 1.22. The morphological change and the sample's structural evolution happened during calcination. After two hours of calcination, the bamboo leaves' 3D-organized porous constructions are created. C/FeSi was an exceptionally well-known electrode material because of its significant surface area and remarkable porous properties. The C/FeSi and CoNiSi electrodes developed here have better electrochemical characteristics than current silicates-based electrodes. The assembled C/FeSi//CoNiSi HSC device achieves an excellent areal specific capacitance of 492 mF $cm^{-2}$ (96 F $g^{-1}$) at 4 mA $cm^{-2}$, an outstanding energy density of 3.35 W h $m^{-2}$ (6.55 W h $kg^{-1}$) at a power density of

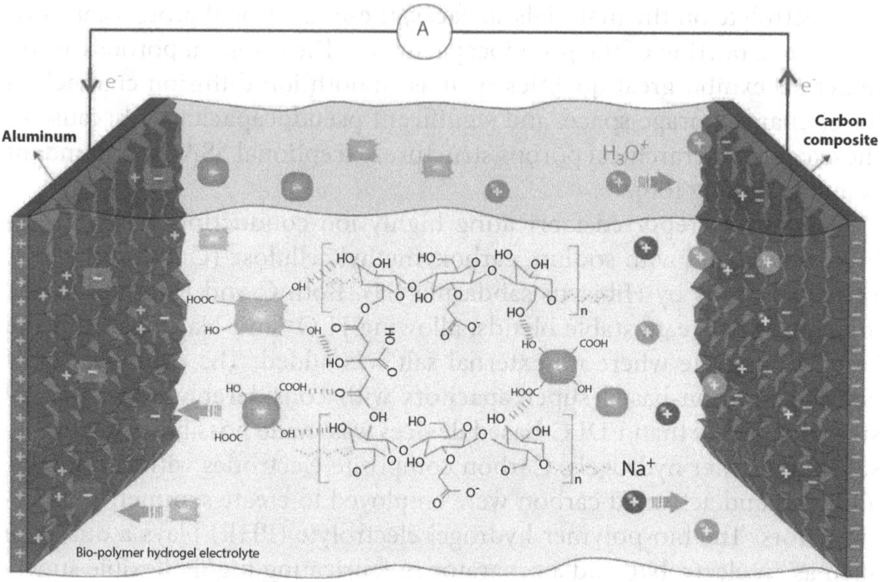


**Figure 1.21** Schematic working mechanism of supercapacitor.

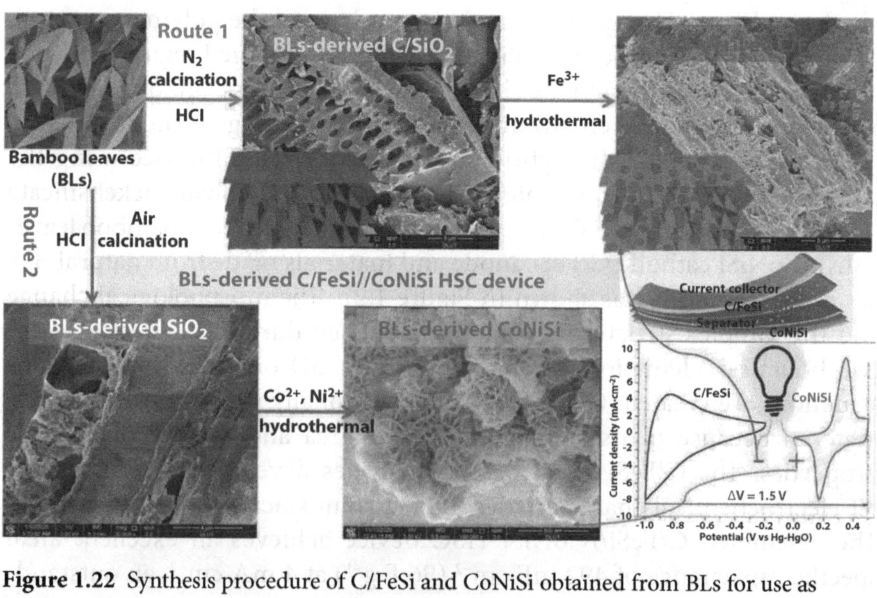


**Figure 1.22** Synthesis procedure of C/FeSi and CoNiSi obtained from BLs for use as anode and cathode materials in a high-performance HSC device.

3.5 W $m^{-2}$ (68.36 W $kg^{-1}$), and a remarkable cycle stability of 76% retention after 5000 cycles. These materials perform well because of the 3D hierarchical architectures of C/FeSi and CoNiSi, which facilitate electron transport and electrolyte movability [68].

## 1.7 Conclusions

Flexibility and sophisticated features are prerequisites for modern wearable and portable electronics. Undoubtedly, the energy storage device is the crucial and core unit in portable/wearable electronics, and supercapacitors are the most suitable ones in this field. To meet these requirements, flexible, smart, and self-sustainable supercapacitors, with suitable electrode material, have received extensive attention and made remarkable progress in the last decade. In this chapter, we provide an in-depth analysis of the design of flexible, intelligent, and self-sustaining SCs for wearable and portable electronics. Developing electrodes in a variety of forms, such as fiber, yarn, fabric, paper, foam, hydrogel, aerogel, etc., could result in high flexibility. These fundamental research studies are crucial for the development of smart materials, appropriate device configurations, and innovative working mechanisms, all of which are key components of smart supercapacitors. As different green energy-harvesting devices have been developed over the past five years, numerous researchers have combined them with supercapacitors into a single, compact configuration to create all-in-one, self-sustaining supercapacitors. They have the capacity to continuously power wearable and portable electronics by simultaneously converting and storing ambient energy as electricity. Naturally, demand for wearable and portable electronics will rise as artificial intelligence and the "internet of things" advance. As a result, flexible, smart, and self-sustaining SCs gained a lot of attention and have produced some interesting outcomes.

## References

1. Raabe, D., The materials science behind sustainable metals and alloys. *Chem. Rev.*, 123, 5, 2436–2608, 2023, https://doi.org/10.1021/acs.chemrev.2c00799.
2. Jha, S., Mehta, S., Chen, Y., Ma, L., Renner, P., Parkinson, D.Y., Liang, H., Design and synthesis of lignin-based flexible supercapacitors. *ACS Sustain. Chem. Eng.*, 8, 1, 498–511, 2019, https://doi.org/10.1021/acssuschemeng.9b05880.
3. Xuan, R., Li, J., Huang, D., Wang, T., Zhang, B., Niu, F., Zhang, G., Sun, R., Wong, C.P., Stretchable and self-healable polyelectrolytes for flexible and

sustainable supercapacitor. *J. Power Sources*, 487, 229394–229403, 2021, https://doi.org/10.1016/j.jpowsour.2020.229394.

4. Huang, H., Han, L., Wang, Y., Yang, Z., Pan, L., Xu, M., A powder self-healable hydrogel electrolyte for flexible hybrid supercapacitors with high energy density and sustainability. *Small*, 17, 10, 2006807–2006817, 2021, https://doi.org/10.1002/smll.202006807.
5. Yang, G., Huang, J., Wan, X., Zhu, Y., Liu, B., Wang, J., Hiralal, P., Fontaine, O., Guo, Y., Zhou, H., A low cost, wide temperature range, and high energy density flexible quasi-solid-state zinc-ion hybrid supercapacitors enabled by sustainable cathode and electrolyte design. *Nano Energy*, 90, 106500–106510, 2021, https://doi.org/10.1016/j.nanoen.2021.106500.
6. Madhu, R., Periasamy, A.P., Schlee, P., Hérou, S., Titirici, M.M., Lignin: a sustainable precursor for nanostructured carbon materials for supercapacitors. *Carbon*, 207, 172–197, 2023, https://doi.org/10.1016/j.carbon.2023.03.001.
7. Abdulwahid, R.T., Aziz, S.B., Kadir, M.F., Replacing synthetic polymer electrolytes in energy storage with flexible biodegradable alternatives: Sustainable green biopolymer blend electrolyte for supercapacitor device. *Mater. Today Sustain.*, 23, 100472–100488, 2023, https://doi.org/10.1016/j.mtsust.2023.100472.
8. Zhu, X., Jiang, G., Wang, G., Zhu, Y., Cheng, W., Zeng, S., Zhou, J., Xu, G., Zhao, D., Cellulose-based functional gels and applications in flexible supercapacitors. *Resour. Chem. Mater.*, 2, 177–188, 2023, https://doi.org/10.1016/j.recm.2023.03.004.
9. Ojeda, L., Oliva, J., Oliva, A.I., Garcia, C.R., Recycling expired pharmaceutical drugs as redox materials for efficient and sustainable flexible supercapacitors. *New J. Chem.*, 47, 21, 10090–10104, 2023, https://doi.org/10.1039/d3nj00497j.
10. Ding, Z., Yang, X., Tang, Y., Nanocellulose-based electrodes and separator toward sustainable and flexible all-solid-state supercapacitor. *Int. J. Biol. Macromol.*, 228, 467–477, 2023, https://doi.org/10.1016/j.ijbiomac.2022.12.224.
11. Awasthi, H., Renuka, H., Srivastava, A.K., Goel, S., Laser-induced graphene-based flexible interdigital electrode realizing micro supercapacitor sustainable in different temperatures. *Energy Storage*, 5, 3, 405–420, 2023, https://doi.org/10.1002/est2.405.
12. Heng, Y., Qin, X., Fang, H., Teng, G., Zhao, D., Hu, D., A pseudocapacitive molecule-induced strategy to construct flexible high-performance asymmetric supercapacitors. *Front. Chem. Sci. Eng.*, 17, 9, 1–13, 2023, https://doi.org/10.1007/s11705-023-2304-1.
13. Kim, J., Wi, S.M., Ahn, J.G., Son, S., Lim, H., Park, Y., Eun, H.J., Park, J.B., Lim, H., Pak, S., Jang, A.R., Engineering geometric electrodes for electric field-enhanced high-performance flexible in-plane micro-supercapacitors. *Energy Environ.*, 6, 4, 12581–12587, 2023, https://doi.org/10.1002/eem2.12581.

14. Carvalho, J.T., Cunha, I., Coelho, J., Fortunato, E., Martins, R., Pereira, L., Carbon-yarn-based supercapacitors with *in situ* regenerated cellulose hydrogel for sustainable wearable electronics. *ACS Appl. Energy Mater.*, 5, 10, 11987–11996, 2022, https://doi.org/10.1021/acsaem.2c01222.
15. Kasprzak, D., Mayorga-Martinez, C.C., Pumera, M., Sustainable and flexible energy storage devices: a review. *Energy Fuels*, 37, 1, 74–97, 2022, https://doi.org/10.1021/acs.energyfuels.2c03217.
16. Liang, Y., Wei, Z., Wang, H.E., Flores, M., Wang, R., Zhang, X., Flexible and freestanding PANI: PSS/CNF nanopaper electrodes with enhanced electrochemical performance for supercapacitors. *J. Power Sources*, 548, 232071–232079, 2022, https://doi.org/10.1016/j.jpowsour.2022.232071.
17. Sheng, H., Zhu, A., Zhang, L., Huang, J., Yang, T., Qin, S., Zhang, F., Xu, Q., Xie, H., Use of an [EMIM][OAc]/GVL-based organic electrolyte solvent to engineer chitosan into a nanocomposite organic ionogel electrolyte for flexible supercapacitors. *Green Chem.*, 25, 8, 3046–3056, 2023, https://doi.org/10.1039/d2gc04019k.
18. Zhang, R., Wang, M., Wan, Z., Wu, Z., Xiao, X., Laser direct writing based flexible solar energy harvester. *Results Eng.*, 19, 101314–101317, 2023, https://doi.org/10.1016/j.rineng.2023.101314.
19. Hao, R., Ma, R., Cao, Y., Zhou, W., Chai, H., Facile fabrication of N-self-doped porous carbons from green solid waste for supercapacitors with high cycling stability and flexibility. *Mater. Today Commun.*, 33, 104911–104921, 2022, https://doi.org/10.1016/j.mtcomm.2022.104911.
20. Zheng, J., Wang, Y., Liu, G., Yu, Z., Sun, S., Fu, Y., Chen, D., Cai, W., Cui, J., Wang, G., Zhou, H., $BaTiO_3$/PVDF-TrFE skeleton separators with superior ion conductivity and mechanical property integrated for flexible seconds self-charging supercapacitors. *J. Power Sources*, 528, 231213–231219, 2022, https://doi.org/10.1016/j.jpowsour.2022.231213.
21. Abbas, S.C., Lin, C., Hua, Z., Deng, Q., Huang, H., Ni, Y., Cao, S., Ma, X., Bamboo-derived carbon material inherently doped with SiC and nitrogen for flexible supercapacitors. *J. Chem. Eng.*, 433, 133738–133746, 2022, https://doi.org/10.1016/j.cej.2021.133738.
22. Tadesse, M.G. and Lübben, J.F., Review on hydrogel-based flexible supercapacitors for wearable applications. *Gels*, 9, 2, 106, 2023, https://doi.org/10.3390/gels9020106.
23. Wang, P., Liu, K., Wang, X., Meng, Z., Xin, Z., Cui, C., Quan, F., Zhang, K., Xia, Y., Interface engineering of calligraphic ink mediated conformal polymer fibers for advanced flexible supercapacitors. *J. Mater. Chem. A.*, 10, 29, 15776–15784, 2022, https://doi.org/10.1039/d2ta03467k.
24. Han, X., Sun, J., Li, Q., He, X., Dang, L., Liu, Z., Lei, Z., Highly flexible PEDOT film assembled with solution-processed nanowires for high-rate and long-life solid-state supercapacitors. *ACS Sustain. Chem. Eng.*, 11, 7, 2938–2948, 2023, https://doi.org/10.1021/acssuschemeng.2c06423.

25. Zhang, Y., Bai, F.Q., Xie, Y., Zhu, M., Zhao, L., An, D., Xue, D., Berda, E.B., Wang, C., Lu, G., Jia, X., A conjugated polymer with electron-withdrawing cyano group enables for flexible asymmetric electrochromic supercapacitors. *J. Chem. Eng.*, 450, 138386–138396, 2022, https://doi.org/10.1016/j.cej.2022.138386.
26. Lv, X., Xu, H., Yang, Y., Ouyang, M., Xia, M., Liu, C., Wright, D.S., Zhang, C., Flexible laterally-configured electrochromic supercapacitor with feasible patterned display. *J. Chem. Eng.*, 458, 141453–141460, 2023, https://doi.org/10.1016/j.cej.2023.141453.
27. Xin, Q., Chu, X., Yang, G., Liang, S., Lin, J., An anti-freezing and conductive glycerol-Mo-based organohydrogel electrolyte for flexible supercapacitor. *Ionics (Kiel)*, 29, 10, 4275–4283, 2023, https://doi.org/10.1007/s11581-023-05140-6.
28. Cui, S., Lv, Y., Jia, Y., Hu, Q., Hou, W., Miao, W., Shi, X., Sun, K., Peng, H., Ma, G., A flexible and hydrophilic hydrogel film based all-in-one supercapacitor. *J. Energy Storage*, 68, 107877–107884, 2023, https://doi.org/10.1016/j.est.2023.107877.
29. Suh, S., Kim, K., Park, J., Kim, W., Ultrafast flexible PEDOT:PSS supercapacitor with outstanding volumetric capacitance for ac line filtering. *J. Chem. Eng.*, 463, 142377–142386, 2023, https://doi.org/10.1016/J.CEJ.2023.142377.
30. Dharmasiri, B., Stojcevski, F., Usman, K.A.S., Alex Qin, S., Razal, J.M., Doeven, E.H., Francis, P.S., Connell, T.U., Yin, Y., Andersson, G.G., Borkar, A., Henderson, L.C., Flexible carbon fiber based structural supercapacitor composites with solvate ionic liquid-epoxy solid electrolyte. *J. Chem. Eng.*, 455, 140778–140790, 2023, https://doi.org/10.1016/J.CEJ.2022.140778.
31. Zheng, H., Du, X.J., Liu, Q., Ou, K.T., Cao, Y., Fang, X., Fu, Q., Sun, Y., Self-healing and wide temperature tolerant flexible supercapacitor based on ternary-network organo-hydrogel electrolyte. *Int. J. Hydrogen Energy*, 48, 35, 13264–13275, 2023, https://doi.org/10.1016/J.IJHYDENE.2022.12.223.
32. Kazari, H., Pajootan, E., Sowa, M., Coulombe, S., Hubert, P., Plasma-enhanced atomic layer deposition of ruthenium metal on free-standing carbon nanotube forest for 3D flexible binder-less supercapacitor electrodes. *J. Energy Storage*, 64, 107049–107059, 2023, https://doi.org/10.1016/J.EST.2023.107049.
33. Guo, J., Zhao, H., Yang, Z., Wang, Y., Liu, X., Wang, L., Zhao, Z., Wang, A., Ding, L., Liu, H., Yu, X., Hierarchical porous 3D Ni3N-CoN/NC heterojunction nanosheets with nitrogen vacancies for high-performance flexible supercapacitor. *Nano Energy*, 116, 108763–108774, 2023, https://doi.org/10.1016/J.NANOEN.2023.108763.
34. Demirel, S., Topkaya, R., Cicek, K., Stretchable and self-healable PVA–nickel–borax electrodes for supercapacitor applications. *J. Mater. Sci. Mater. Electron.*, 34, 1, 1–10, 2023, https://doi.org/10.1007/S10854-022-09392-2/FIGURES/7.

35. Lyu, S., Chang, H., Zhang, L., Wang, S., Li, S., Lu, Y., Li, S., High specific surface area MXene/SWCNT/cellulose nanofiber aerogel film as an electrode for flexible supercapacitors. *Compos. B. Eng.*, 264, 110888–110897, 2023, https://doi.org/10.1016/J.COMPOSITESB.2023.110888.
36. Bhattarai, R.M., Chhetri, K., Saud, S., Teke, S., Kim, S.J., Mok, Y.S., Eco-friendly synthesis of cobalt molybdenum hydroxide 3D nanostructures on carbon fabric coupled with cherry flower waste-derived activated carbon for quasi-solid-state flexible asymmetric supercapacitors. *ACS Appl. Nano Mater.*, 5, 1, 160–175, 2021, https://doi.org/10.1021/acsanm.1c02354.
37. Siddiqa, A., DH, N., Padaki, M., High-energy-density asymmetric supercapacitor based on layered-double-hydroxide-derived CoNi2S4 and eco-friendly biomass-derived activated carbon. *Energy Fuels*, 36, 21, 13286–13295, 2022, https://doi.org/10.1021/acs.energyfuels.2c02681.
38. Wang, J., Zhang, Q., Deng, M., Eco-Friendly preparation of biomass-derived porous carbon and its electrochemical properties. *ACS Omega*, 7, 26, 22689–22697, 2022, https://doi.org/10.1021/acsomega.2c02140.
39. Ren, X., Yuan, Z., Ma, Y., Zhang, C., Qin, C., Jiang, X., Nitrogen-/boron-doped carbon from poplar powder and carbon nanotube composite as electrode material for supercapacitors. *Energy Fuels*, 36, 5, 2841–2850, 2022, https://doi.org/10.1021/acs.energyfuels.1c04241.
40. Hou, S.P., Liao, M.D., Peng, C., Min, D., Li, X.Y., Chen, J., Lin, J.H., Honeycomb-like hierarchical porous activated carbons from biomass waste with ultrahigh specific surface area for high-rate electrochemical capacitors. *Energy Fuels*, 35, 20, 16860–16869, 2021, https://doi.org/10.1021/acs.energyfuels.1c01690.
41. Kasprzak, D., Mayorga-Martinez, C.C., Pumera, M., Sustainable and flexible energy storage devices: a review. *Energy Fuels*, 37, 1, 74–97, 2022, https://doi.org/10.1021/acs.energyfuels.2c03217.
42. Yamada, S., A transient supercapacitor with a water-dissolvable ionic gel for sustainable electronics. *ACS Appl. Mater. Interfaces*, 14, 23, 26595–26603, 2022, https://doi.org/10.1021/acsami.2c00915.
43. Qin, S., Liu, P., Wang, J., Liu, C., Zhang, S., Tian, Y., Zhang, S., *In situ* N, O Co-doped nanoporous carbon derived from mixed egg and rice waste as green supercapacitor. *Mol.*, 28, 18, 6543–6557, 2023, https://doi.org/10.3390/molecules28186543.
44. Inal, I. I. G., Koyuncu, F., Güzel, F., Investigating the surface properties of red pepper industrial waste-based activated carbons for use as reversible supercapacitor electrodes. *Diam. Relat. Mater.*, 138, 110212–110220, 2023, https://doi.org/10.1016/j.diamond.2023.110212.
45. Tatrari, G., Tewari, C., Pathak, M., Bhatt, D., Solanki, M., Shah, F.U., Sahoo, N.G., Coconut-husk derived graphene for supercapacitor applications: comparative analysis of polymer gel and aqueous electrolytes. *Mater. Adv.*, 4, 15, 3310–3322, 2023, https://doi.org/10.1039/D3MA00126A.

46. Luo, M., Chen, J., Li, Q., Wang, Y., Cotton-based activated carbon fiber with high specific surface area prepared by low-temperature hydrothermal carbonization with urea enhancement. *Ind. Eng. Chem. Res.*, 62, 22, 8744–8753, 2023, https://doi.org/10.1021/acs.iecr.3c00807.
47. Gogoi, D., Karmur, R.S., Das, M.R., Ghosh, N.N., Spent tea-waste-derived porous carbon-supported truncated octahedral $Cu_2O$ for highly efficient energy storage devices. *Energy Fuels*, 37, 18, 14350–14364, 2023, https://doi.org/10.1021/acs.energyfuels.3c02375.
48. Guo, T., Liu, Y., Xu, G., Ding, Y., Fan, B., Liu, D., Three-dimensional honeycomb-like hierarchically structured carbon nanosheets from resin for high-performance supercapacitors. *New J. Chem.*, 47, 11996–12006, 2023, https://doi.org/10.1039/D3NJ01825C.
49. Hepsiba, P., Rajkumar, S., Elanthamilan, E., Wang, S.F., Merlin, J.P., Biomass-derived porous activated carbon from anacardium occidentale shell as electrode material for supercapacitors. *New J. Chem.*, 46, 18, 8863–8873, 2022, https://doi.org/10.1039/D2NJ01041K.
50. Dhavale, S.B., Patil, V.L., Patil, S.S., Patil, D.S., Dhavale, R.P., Park, H.H., Patil, P.S., Porous carbon derived from Terminalia catappa leaves for energy storage application. *Sustain. Energy Fuels*, 7, 5, 1284–1301, 2023, https://doi.org/10.1039/D2SE01643E.
51. Dubey, P., Shrivastav, V., Jain, M., Pant, K.K., Maheshwari, P.H., Sundriyal, S., Facile synthesis of pineapple waste-derived carbon and polyaniline composite for high-energy-density asymmetric supercapacitors. *Energy Fuels*, 37, 12, 8659–8671, 2023, https://doi.org/10.1021/acs.energyfuels.3c00995.
52. Lu, K.L., Wei, X.Y., Mao, F.Z., Zhu, Z.J., Li, Z., Yin, F., Zong, Z.M., Biomass-derived nitrogen and sulfur co-doped sheet-like porous carbon for high-performance supercapacitors. *New J. Chem.*, 47, 3, 1247–1255, 2023, https://doi.org/10.1039/D2NJ05366G.
53. Zheng, J., Song, Q., Qi, Y., Leng, H., Zhou, W., Li, S., Qiu, J., N, O co-doped porous carbon derived from pine needles for zinc-ion hybrid supercapacitors. *New J. Chem.*, 47, 20, 9692–9700, 2023. https://doi.org/10.1039/D3NJ01678A.
54. Zhao, R., Yang, X., She, Z., Wu, Q., Shi, K., Xie, Q., Ruan, Y., Ultrahigh-surface-area and N, O Co-doping porous carbon derived from biomass waste for high-performance symmetric supercapacitors. *Energy Fuels*, 37, 4, 3110–3120, 2023, https://doi.org/10.1021/acs.energyfuels.2c02916.
55. Brahma, S., Ramanujam, K., Gardas, R.L., Nitrogen-doped high surface area porous carbon material derived from biomass and ionic liquid for high-performance supercapacitors. *Ind. Eng. Chem. Res.*, 61, 33, 12073–12082, 2022, https://doi.org/10.1021/acs.iecr.2c00195.
56. Benoy, S.M., Bhattacharjya, D., Bora, M., Saikia, B.K., Blowing agent-induced hierarchical porous carbon from low-quality coal for high-performance supercapacitor. *ACS Appl. Electron. Mater.*, 4, 12, 6322–6334, 2022, https://doi.org/10.1021/acsaelm.2c01464.

57. Zhu, J., Qiu, X., Sun, S., Huang, T., Huang, Z., Zhao, L., Zhang, W., Combined sustainable production technology of calcium chloride and lignin-derived porous carbon electrode materials. *J. Clean. Prod.*, 419, 138201–138210, 2023, https://doi.org/10.1016/j.jclepro.2023.138201.
58. Kitamoto, Y., Cao, K.L.A., Le, P.H., Abdillah, O.B., Iskandar, F., Ogi, T., A sustainable approach for preparing porous carbon spheres derived from Kraft lignin and sodium hydroxide as highly packed thin film electrode materials. *Langmuir.*, 38, 11, 3540–3552, 2022, https://doi.org/10.1021/acs.langmuir.1c03489.
59. Wang, S., Liu, Y., Wang, H., Wang, X., Li, L., Wang, T., Efficient Fabrication of hierarchical porous carbon fibers with tunable mesopores from discarded aramid for supercapacitors. *ACS Appl. Energy Mater.*, 6, 18, 9371–9380, 2023, https://doi.org/10.1021/acsaem.3c01314.
60. Mehdi, R., Naqvi, S.R., Khoja, A.H., Hussain, R., Biomass derived activated carbon by chemical surface modification as a source of clean energy for supercapacitor application. *Fuel*, 348, 128529–128538, 2023, https://doi.org/10.1016/j.fuel.2023.128529.
61. Narayanan, A., Siddiqa, A., Kodihalli, N.K., Hegde, G., Nagaraju, D.H., Padaki, M., Designing of a free-standing flexible symmetric electrode material for capacitive deionization and solid-state supercapacitors. *ACS Sustain. Chem. Eng.*, 11, 9, 3750–3759, 2023, https://doi.org/10.1021/acssuschemeng.2c06817.
62. Shaku, B., Mofokeng, T.P., Coville, N.J., Ozoemena, K., II, Maubane-Nkadimeng, M.S., Biomass valorisation of marula nutshell waste into nitrogen-doped activated carbon for use in high performance supercapacitors. *Electrochim. Acta*, 442, 141828–141840, 2023, https://doi.org/10.1016/j.electacta.2023.141828.
63. Arjunan, A. and Kim, S.K., Bioinspired sustainable sheetlike porous carbon derived from Cassia fistula flower petal as an electrode for high-performance supercapacitors. *Energy Fuels*, 36, 16, 9337–9346, 2022, https://doi.org/10.1021/acs.energyfuels.2c02228.
64. Zhu, H., Wu, D., Zhang, G., Li, B., Wang, A., Sun, K., Oaks-derived activated carbon by trace alkali-induced catalytic steam activation for electrochemical capacitor applications. *J. Energy Storage*, 53, 105090–105097, 2022, https://doi.org/10.1016/j.est.2022.105090.
65. Wei, X., Qiu, B., Xu, L., Qin, Q., Zhang, W., Liu, Z., Lv, Y., High performance hierarchical porous carbon derived from waste shrimp shell for supercapacitor electrodes. *J. Energy Storage*, 62, 106900–106909, 2023, https://doi.org/10.1016/j.est.2023.106900.
66. Qiu, G., Guo, Y., Zhang, Y., Zhao, X., Xu, J., Guo, S., Wu, J., Construction of N, O Co doped petal-like hierarchical porous carbon with an ultrahigh specific surface from waste bamboo for high-performance supercapacitors. *Ind. Eng. Chem. Res.*, 61, 43, 16034–16049, 2022, https://doi.org/10.1021/acs.iecr.2c02815.

67. Cevik, E., Gunday, S.T., Bozkurt, A., Iqbal, A., Asiri, S.M., Alqarni, A.N., Almofleh, A., Scalable, quasi-solid-state bio-polymer hydrogel electrolytes for high-performance supercapacitor applications. *ACS Sustain. Chem. Eng.*, 10, 33, 10839–10848, 2022, https://doi.org/10.1021/acssuschemeng.2c02281.
68. Zhang, Y., Wang, C., Chen, X., Dong, X., Meng, C., Huang, C., Bamboo leaves as sustainable sources for the preparation of amorphous carbon/iron silicate anode and nickel–cobalt silicate cathode materials for hybrid Supercapacitors. *ACS Appl. Energy Mater.*, 4, 9, 9328–9340, 2021, https://doi.org/10.1021/acsaem.1c01540.

# 2

# Role of Electrolytes in Sustainable Supercapacitors

**Soumya Jha and R. Prasanth***

*Madanjeet School of Green Energy Technology, Pondicherry Central University, Pondicherry, India*

## *Abstract*

Renewable energy sources and technologies with zero emissions are given top priority worldwide, due to which energy storage (ES) materials with efficient performance are rigorously developed. In this context, ES devices like supercapacitors and batteries are paramount choices because of their superior performance, stability, and green credentials. Supercapacitors, also known as electrochemical capacitors, are the favored option for applications requiring high power due to their higher specific power (PS > 10 kW $kg^{-1}$), ability to charge and discharge quickly (within seconds), and superior cycling stability (>105 cycles) in comparison with batteries. Sustainable supercapacitors are still an increasingly popular research topic, with ongoing efforts to boost their performance, long-term viability, and economical value. These advances are critical to achieving a more efficient and sustainable energy storage scenario. Sustainable supercapacitors are energy storage devices that are designed to enable high-capacity, quick energy storage and release while minimizing environmental effect. They are intended to be more environmentally friendly than traditional supercapacitors by utilizing eco-friendly materials, lowering consumption of resources, and focusing on end-of-life disposal concerns. Electrolytes are crucial components of electrochemical devices, and their morphological and electrochemical properties influence device performance such as rate performance, cyclability, capacity, power density, and safety. The interaction between electrodes and electrolytes is fundamental to supercapacitor performance, which is outlined in this chapter. The chapter also discusses factors such as mobility, diffusion coefficient, ionic conductivity, ion solvation, thermal stability, viscosity, ionic size, electrochemical stability, dielectric constant, and dispersion interaction that affect the overall performance of the

**Corresponding author*: prasanth.ravindran@gmail.com

Basheer Ahamed and Chaudhery Mustansar Hussain (eds.) Sustainable Supercapacitors: Next-Generation of Green Energy Storage Devices, (33–72) © 2025 Scrivener Publishing LLC

devices. Given the requirements for developing superior electrolytes, along with the hurdles that must be addressed, this research push extends beyond fundamental understanding, aiming to translate those insights into practical improvements for supercapacitors that can compete in the energy storage market.

***Keywords*:** Electrolytes, supercapacitors, pore size, rate performance, energy density, power density

## 2.1 Introduction

The swift expansion of the global economy requires much bigger energy resources and, as an outcome, more efficient energy harvesting and storing techniques have become a necessity. Such rapid worldwide economic expansion, urbanization, along with substantial increases in living standards has caused some major issues around the world, such as rapid depletion of non-renewable energy reservoirs and potentially catastrophic environmental threats such as air, water, and noise pollution, rising emission of greenhouse gases, etc. [1–3]. Therefore, renewable energy and zero-emission technologies are given the utmost importance on a global scale. The pressing demand to develop and raise feasible, sustainable energy sources and technologies associated to it is acknowledged globally [4–7]. The majority of renewable energy sources are clean, which rely heavily on both the regional weather conditions and specific time of the day. To efficiently capture these intermittent energy sources, corresponding techniques for conversion of energy and energy storage systems must be developed [8–10]. In recent years of energy scarcity, energy harvesting from sources of renewable energy has been a key research goal. The technological advances in the field of energy harvesting research have the principal objective to reduce the emissions of greenhouse gases caused by using non-renewable energy resources [11–13]. Experts are also working to devise approaches to storing this energy as electricity. The necessity of efficient energy storage and sustainable energy solutions is one of the most pressing obstacles in the increasingly globalized era [14–17].

Extensive research is underway for the development of highly efficient electrochemical energy storage (EES) materials and associated devices. Due to their superior stability, remarkable efficiency, and minimal impact on environment, EES devices like batteries and supercapacitors are regarded as the "first choice" in this regard [18–21]. Supercapacitors, additionally referred to as ultracapacitors, provide an appealing alternative to batteries in situations requiring high power delivery. This is due to their exceptional combination of properties: superior specific power exceeding 10 kW $kg^{-1}$, allowing for rapid energy release relative to their weight; fast charge and discharge capabilities, allowing them to store and release energy in seconds; and remarkable cycling stability exceeding 100,000 cycles, resulting in a long lifespan when used repeatedly. These qualities make them excellent for applications such as electric vehicles, satellites, and robots, which require short bursts of power and frequent charge/discharge cycles [22–26]. Supercapacitor devices (SCs) can be classified into three categories based on the manner in which they store charge: (i) electric double-layer capacitors (EDLCs), (ii) pseudocapacitors, and (iii) hybrid electrodes. Figure 2.1 depicts classifications of supercapacitors and mechanism to store energy in these capacitive devices [27]. In comparison to both pseudocapacitors and hybrid capacitors, EDLCs account for the vast majority of commercially available SCs, owing to their technical maturity.

Research to improve the efficiency of supercapacitive devices to match that of batteries made tremendous progress, with an $E_S$ as high as 100W h $kg^{-1}$ reported for them. As a result, extensive research has been conducted in order to achieve high capacitance (*C*); researchers concentrate on developing electrode materials that have superior capacitive characteristics, while also attempting to synthesize electrolytes with low electrochemical potential in order to boost voltage window [28–30]. The majority of the research was focused on developing EDLC electrodes, which includes

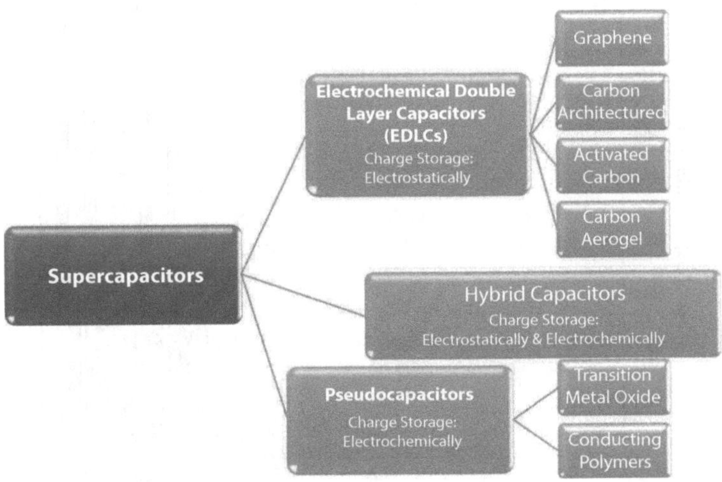


**Figure 2.1** Classifications of supercapacitors and mechanism to store energy in these capacitive devices.

enhancing high conductivity, high chemical stability, polarizability, highly developed inner surface area, corrosion resistance, and cost-effectiveness [31, 32]. In addition, scientists have paid attention to enhance electrode–electrolyte interaction by enhancing the surface area of electrode material on top of that, reducing the hydrophobicity of material that prevents enough electrode penetration by electrolyte [33].

Electrolytes are vital for assessing the electrochemical properties of supercapacitors, and electrolyte research is the next significant effort. Various electrolytes can be employed in supercapacitors composed of porous electrode materials such as aqueous solution of acid, bases, and salt, and solution of organic salt in aprotic solvent and ionic liquids [34–37]. The majority of commercially available supercapacitors contain organic electrolytes, allowing them to function over a wide potential range (2.7–3.0 V). Contrary to aqueous electrolytes, these electrolytes are unfortunately more expensive, entail a more involved assembly process, and represent a risk to the environment. Despite ionic liquids providing the highest working voltages (up to 3.5 V), these electrolytes have low ionic conductivity and high viscosity, restricting the achievement of high specific power levels [38–42]. Figure 2.2 depicts the data trends of electrode and electrolyte publications (source: Scopus). It can be observed that research in the field of enhancing the properties of electrolyte was modestly increased.

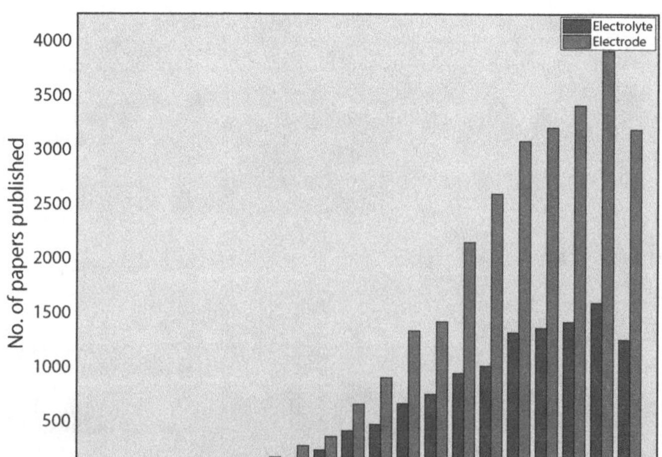


**Figure 2.2** Data trends of electrode and electrolyte publication devices from year 2004 to 2024; search from Scopus; search date: 14 September 2023.

The primary goal of this chapter is to thoroughly understand the interaction between electrodes and electrolytes, as well as the aspects that influence the overall performance of supercapacitive devices. These include the ionic diffusion, ionic conductivity and ionic mobility, solvation, thermal stability, hydrated sphere radius, electrochemical stability, surface interaction at the electrode–electrolyte interface, and dielectric constant. Furthermore, some established specifications for a suitable electrolyte for SCs are described. This chapter thoroughly investigates the various electrolyte types used in supercapacitors (aqueous, organic, ionic liquids, solid, and redox electrolytes). It additionally explores the roles of anions and cations in charge storage across various electrolytes, as well as the conductivity of cutting-edge electrolytes for SCs. It is finally summarized by discussing the difficulties associated with electrochemical supercapacitor electrolytes and is then finished by giving a quick review and an outlook for forthcoming endeavors.

## 2.2 Parameters Characterizing Sustainable Supercapacitors and Their Interactions with Electrolytes

The electrolyte serves as the primary ionic conductor between the two electrodes in SCs, supplying ionic current to enable cellular functions that have a substantial effect on the electrochemical characteristics of the devices, specifically their voltage window, speed capacity, and cycle stability [43]. Electrolytes are primarily distinguished by their ionic conductivity, electrochemical window, optimal potential window, operating temperature range, stability, and safety. The electrochemical performance of SCs can be influenced by a variety of circumstances, as shown in Figure 2.3. In this section, we will be discussing attributes of SCs and how they behave with various electrolytes.

### 2.2.1 Capacitance

The entire SC can be treated as two capacitors in series, as both electrode ($C_1$)/electrolyte ($C_2$) interfaces represent a capacitor. The overall capacitance ($C_T$) can be expressed as:

$$\frac{1}{C_T} = \frac{1}{C_1} + \frac{1}{C_2} \tag{2.1}$$

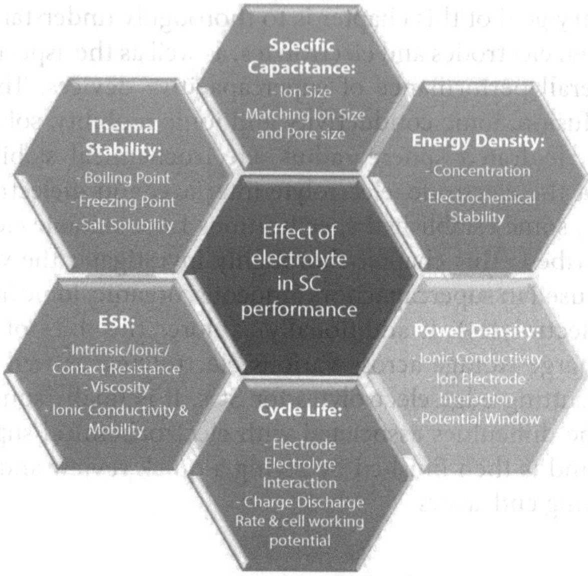


**Figure 2.3** Effect of electrolytes in supercapacitor performance.

In symmetric supercapacitor devices, the overall capacitance ($C_T$) is measured by deducing the capacitance of either half electrode since $C_1 = C_2$. However, in asymmetric supercapacitors, $C_T$ is predominantly influenced by the electrode whose capacitance is smallest among either of them. While assessing the overall capacitance of electrode material, it can be articulated by mass-specific capacitance ($C_S$) or gravimetric capacitance ($C_V$), which are conjointly known as specific capacitance. Out of these two mass-specific capacitance ($C_S$), calculated as Faraday per gram ($Fg^{-1}$), is the most often used metric for defining electrode materials. In SCs, total specific capacitance ($C_{TS}$) is defined as the ratio of overall capacitance ($C_T$) to the device's weight ($W_{TM}$), encompassing the mass of all constituent components, such as the electrodes (which includes both anode and cathode), separators, electrolyte, and current collectors.

$$C_{TS} = \frac{C_T}{W_{TM}} \tag{2.2}$$

Because electrolyte solution weight adds to the total weight of the device, it might negatively affect the capacitance of supercapacitors. Therefore, electrolyte should be designed in such a way for the device which have reduced volume and lighter weight as it is one of the most attractive features. A significant aspect of boosting supercapacitor (SC) performance is the development of an electrolyte that penetrates the pore spaces through the electrode material layer.

### 2.2.2 Power Density and Energy Density

The energy density and power density of a supercapacitor are determined by the cell voltage ($V$) that builds during charging, as expressed by the following formulas described in Equations 2.3 and 2.4 [20].

$$E = \frac{1}{2} C_{TS} V^2 \tag{2.3}$$

$$P = \frac{V^2}{4 W_{TM} R_{cell}} \tag{2.4}$$

Four important variables determine a performance of SC: $V$, $C_{TS}$, $W_{TS}$, and $R_{cell}$. From the above equations, it is evident that optimizing energy and power densities of supercapacitors entails increasing $V$ and $C_{TS}$ while minimizing $W_{TS}$ and $R_{cell}$. An electrochemically stable potential window is one of the important parameters for attaining the maximum operating voltage in SCs.

### 2.2.3 Equivalent Series Resistance

One of the pivotal roles in determining the power density of a solid-state battery and supercapacitors (SC) as outlined in Equation 2.4 is equivalent series resistance ($R_{cell}$). A high $R_{cell}$ inhibits rate capability, which leads to low power density, thus limiting charging/discharging rate [43]. The $R_{cell}$ value is even more critical than the capacitance value, for a pulsed power system, as it requires rapid intense delivery of energy over a short duration of time. To achieve high power density, an electrolyte with high ionic conductivity is needed. However, balance must be maintained between range of potential in which electrolytes can operate its ionic conductivity. Therefore, developing electrolytes having a wider operating potential and a diminutive $R_{cell}$ is essential [44].

### 2.2.4 Cycle Life

Cycle life is a crucial parameter for assessing the stability of a supercapacitor device (SC). It is measured by repeated cycles of charge–discharge cycling of electrodes in a certain electrolyte. Carbon electrodes typically have excellent cycling stability; however, the presence of pseudocapacitive reactions can jeopardize this stability over multiple cycles. Factors that impact cycle life due to the interactions among the electrode materials and ions of electrolyte that leads to non-ideal electrochemical reversibility include electrode material, electrolyte, rapid charging/discharging capability, operating potential window, cell type, and temperature [45].

### 2.2.5 Self-Discharge Rate

Self-discharge rates in SCs can limit their use for applications requiring long-term energy retention. Self-discharge is the potential discharge of charge from an charged electrode through extended periods of storage. Current leakage causes the cell voltage to drop during the self-discharge process, and mechanisms of self-discharge depends on electrolyte type, impurities, and residual gases formed [46].

### 2.2.6 Thermal Stability

SCs have potential applications in the 30°C to 70°C temperature range, and expanding their working temperature range could broaden their range. Space avionics necessitate operation at frigid temperatures down to −55°C, whereas fuel cell vehicles function across a wider range, including higher temperatures. The type of electrolyte, the concentration of electrolyte, and the solvent characteristics like boiling point, freezing point, and viscosity have an impact on SC performance [47].

## 2.3 Different Types of Electrolytes Used in Sustainable Supercapacitors

One of the most important components of SCs is the electrolyte, which is a combination of salt and solvent, which provides conductivity in ions and hence facilitates adjustment of charge on every single electrode present in the cell. Aside from its critical role in establishing the electric double layer (EDL) in electric double-layer capacitors (EDLCs), it also facilitates reversible redox processes that have a major impact charge storage mechanism in pseudocapacitors; the electrolyte in an SC also affects how well it functions [49]. Given our understanding of capacitance due to EDL formation and pseudocapacitance, it is clear that the properties of electrolyte have an enormous effect on the cycle life and the specific energy (which includes energy and power densities) of supercapacitors, which depends on (a) the ionic size and types of ion used; (b) the ionic concentration and solvent in the electrolyte; (c) the electrolyte and electrode material interaction; (d) the potential window; and (e) the ion and solvent interaction [48]. It is also noted that Equations 2.3 and 2.4 show that the electrochemical stability potential window (ESPW) of the electrolyte dictates the maximum achievable cell voltage of a supercapacitor, thereby governing its energy storage capacity, by affecting both the energy and power densities. The power density can be significantly impacted by the $R_{cell}$ of an SC, which is strongly correlated with the ion conductivity of the electrolyte. Furthermore, the lifespan and self-discharge characteristics of supercapacitors can be influenced by the ion and solvent interactions, as well as interactions in the electrolyte and electrode interface.

In a nutshell, up until this point, numerous varieties of electrolytes have been synthesized and documented in the literature. Based on the type of solvent used, these electrolytes are primarily classified as liquid electrolytes and solid/quasi-solid-state electrolytes [50]. The above-discussed electrolytes can be described as traditional electrolytes, but recently, development of novel types of electrolytes, known as hybrid electrolytes, has been under consideration, which is largely based on the fusion of classic electrolytes. The classification of electrolytes is described in Figure 2.4.

To date, no perfect electrolyte has been developed that meets all the previously listed parameters. Each electrolyte possesses its own set of properties, which have certain merits and demerits. For instance, the electrochemical capacitance of aqueous electrolytes demonstrates excellent capacitance and superior conductivity, but due to the narrow decomposition voltage of aqueous electrolyte, they suffer from limited operational potential value [52]. In case of ILs and organic electrolytes, they function at a wider range of potential window, but their ionic conductivity is typically substantially lower. Similarly, solid-state electrolytes exhibit low ionic conductivity but mitigate the issue of potential leakage, which is a major disadvantage of liquid electrolytes [51–53]. To overcome the limitations of particular electrolytes, extensive research has been undertaken to develop innovative electrolyte materials aimed at enhancing the overall performance of electrochemical capacitance [54, 55]. As mentioned earlier, liquid electrolytes fall into categories such as aqueous electrolytes, organic electrolytes, and ionic liquids (ILs), while organic and inorganic

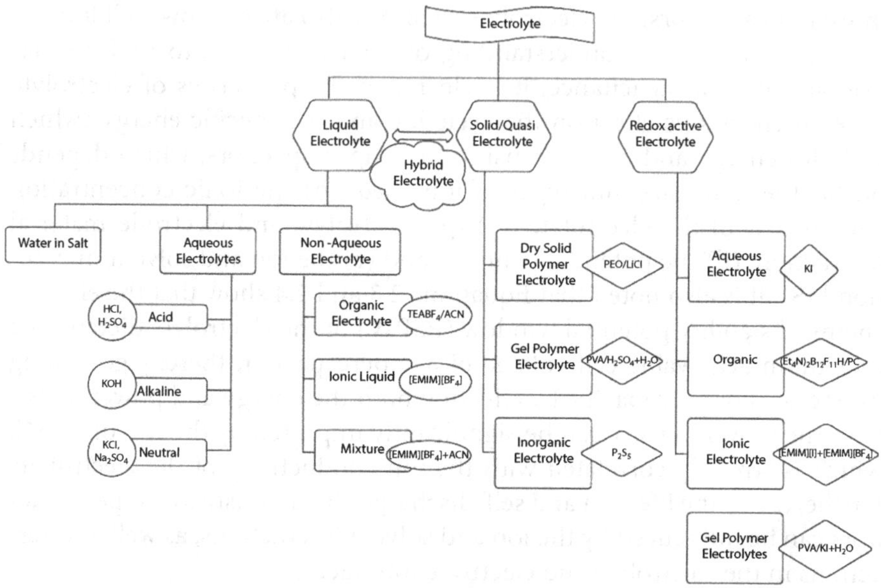


**Figure 2.4** Classification of electrolytes.

electrolytes can be classified into solid or quasi-solid state electrolytes. Several approaches have been taken to enhance the performance of electrolytes, which include the following:

(a) Advancement of novel and highly efficient electrolytes that have a large working temperature range with enhanced ionic conductivity/viscosity, expandable operating potential window, and so on.
(b) Research into the beneficial influence of given electrolytes on electrochemical parameters, such as charge storage capacity, temperature threshold, energy density, power density, self-discharge, high-speed energy delivery, and other related factors, is underway.
(c) Development of a fundamental knowledge of the electrolyte's effect on electrochemical performance using advanced experimental, modeling, and simulation methodologies.

### 2.3.1 Aqueous Electrolytes

Aqueous electrolytes have found their extensive utility in research and development endeavors because they are easier to handle in the laboratory than organic electrolytes and ILs, which need purifying procedures. Aqueous electrolytes have a higher conductivity than organic and ionic electrolytes, which helps to minimize the equivalent series resistance and enhance the high-speed energy delivery of SCs. In general, because of their small voltage windows, aqueous electrolytes are a poor choice for commercial SC devices in terms of energy density [56]. The superior energy density achievable with organic electrolytes is a major factor driving their prevalence in commercially available supercapacitors, as opposed to aqueous alternatives. Organic electrolytes and ILs, on the other hand, typically necessitate complex purifying operations in a highly regulated environment to avoid moisture introduction. Aqueous electrolytes are classified into three types: acidic, alkaline, and neutral solutions. Among aqueous electrolytes, the most frequently employed options are $H_2SO_4$, $Na_2SO_4$, $Li_2SO_4$, $K_2SO_4$, $(NH_4)_2SO_4$, $MgSO_4$, LiOH, NaOH, KOH, HCl, LiCl, NaCl, KCl, $LiNO_3$, $KNO_3$, $Li(CH_3\text{-}COO)$, $Na(CH_3COO)$, $Mg(CH_3COO)_2$, $Na_2HPO_4$, $NaHCO_3$, $Na_2B4O_7$, and so on [52].

The hydrogen and oxygen evolution play an important role in maintaining the potential window of aqueous electrolyte. Hydrogen evolution typically takes place at a negative electrode potential of around 0 V vs. SHE, whereas oxygen evolution occurs at a positive electrode potential of approximately 1.23 V, resulting in a cell voltage of SC of approximately 1.23 V. The gas evolution could potentially cause the SC cells to burst, compromising safety and lowering performance. In order to inhibit gas formation, supercapacitors employing aqueous electrolytes typically maintain a cell voltage of approximately 1.0 V. Thus, aqueous electrolytes have a narrow potential window due to water decomposition, which can cause gas evolution, potentially causing cell rupture and degrading its functionality. To avoid gas formation, SCs with aqueous electrolytes typically adhere to a cell voltage limit of approximately 1.0 V. Acidic and alkaline electrolytes generally operate within potential window below 1.3 V, while neutral electrolytes have the highest cell voltage at 2.2 V. While using aqueous electrolyte in SCs, special attention should be given to maintain the optimal temperature range for its smooth function; i.e., it must function between water melting point and its boiling point. This narrow potential window can pose safety and performance concerns.

***(a) Acidic Electrolytes***

Acidic electrolytes are commonly used in electrochemical studies, with $H_2SO_4$ being the most commonly used due to its high ionic conductivity (0.8 S $cm^{-1}$ for 1 M $H_2SO_4$ at 25°C). The ionic conductivity of the electrolyte depends on the concentration, and optimum concentrations have been extensively studied to achieve maximum conductivities at specific temperatures. At 25°C, 1.0 M of $H_2SO_4$ electrolyte concentration yields the maximum ionic conductivity. For supercapacitors that employ carbon-based electrode materials, this makes the 1.0 M $H_2SO_4$ electrolyte solution the most suitable option.

***Acidic Electrolytes in EDLCs.*** The specific capacitances of electrical double-layer capacitors (EDLCs) obtained in acid electrolytes like $H_2SO_4$ are generally higher than those in neutral electrolytes. This is attributed to the fact that supercapacitors employing $H_2SO_4$ as an electrolyte exhibit a lower $R_{cell}$, primarily because of the electrolyte's superior ionic conductivity compared to neutral electrolytes [57, 58]. Higher electrolyte conductivity leads to a corresponding enhancement in the specific capacitance of activated carbons, which directly impacts the ion mobility. Research consistently shows that EDLCs that use a strong acid electrolyte like $H_2SO_4$ typically achieve specific capacitances that range between 100 and 300 $Fg^{-1}$, which are often higher than those achieved in organic electrolytes. Electrode materials based on carbon contribute significantly to the specific capacitance, and the electrolyte contribution should be less than that of electrode materials [59, 60]. Electric double-layer capacitors (EDLCs) based on the $H_2SO_4$ electrolyte generally exhibit larger specific capacitances compared to those based on organic electrolytes, even when utilizing identical electrode materials.

***Acidic Electrolytes in Pseudocapacitors.*** The low energy density of EDLCs has led to the exploration of other types of supercapacitive materials, like pseudocapacitors in which the observed specific capacitance of carbon-based electrode materials in aqueous $H_2SO_4$ electrolytes suggests contributions beyond pure double-layer capacitance, potentially involving pseudocapacitive mechanism caused by fast redox reactions on surface functionalities which happens due to the presence of oxygenated carbon molecules [61]. This can be further enhanced by the addition of specific functional group or heteroatoms on the surface of the carbon material. The proper selection of the electrolyte has demonstrably affected the pseudocapacitive behavior of carbon-based materials, as a functional group attached to the surface exhibits distinct behaviors in various electrolytes [62, 63].

The quinone-type functional group present on the surface could produce pseudocapacitive properties in acidic aqueous electrolytes, while this effect was rarely observed in alkaline electrolytes. Thus, the strategic selection of electrolytes offers a means to manipulate the surface functionalities of carbon-based electrode materials, enabling the optimization of desired characteristics [64, 65].

Despite their higher energy density, pseudocapacitors often have a shorter cycle life than EDLCs. The primary cause of this is the functional part of the electrode material deteriorating in the presence of aqueous electrolytes. In contrast, organic electrolytes facilitate longer cycle life in supercapacitors through assisting with the formation of stable EDLs by electrostatic separation of charges. One approach to improving electrode material stability in aqueous electrolytes is incorporating phosphorus groups, which are surface functional groups, onto the carbon surface Additionally, in aqueous electrolytes, pseudocapacitive materials like electrically conductive polymers (ECPs), metal oxides, and sulfides offer theoretical capacitances that are higher compared to carbon-based materials [66, 68]. However, the efficacy of tailoring surface functionalities is compromised by the inherent instability of these materials in acidic aqueous electrolytes, which stems from their susceptibility to variations in both electrolyte composition and proton concentration (pH).

***Acidic Electrolytes in Hybrid Supercapacitors.*** Hybrid supercapacitors have been fabricated to enhance the energy as well as power densities of aqueous electrolytes. Symmetric SCs have similar anode and cathode material in aqueous electrolytes, limiting the maximum cell voltage due to gas evolution reactions. However, if aqueous electrolytes are used in asymmetric configurations (anode materials differ from cathode materials), it can result in a broader operating potential window [69]. Leveraging asymmetric electrodes in a supercapacitor can exploit beneficial voltage ranges from each, leading to a wider operating window for aqueous electrolytes. Several types of asymmetric SCs, such as carbon//$PbO_2$, carbon//$RuO_2$, carbon//ECPs [70–74], and carbon with different masses or properties in each electrode, have been tested in strong acid electrolytes and demonstrated feasibility for application. For example, typical energy densities for a carbon//$PbO_2$ hybrid SC with an $H_2SO_4$ electrolyte were reported to be 25–30 Wh $kg^{-1}$, which notably exceeds those of symmetric carbon-based electric double-layer capacitors (EDLCs) employing the same $H_2SO_4$ electrolyte. Nonetheless, it was discovered that there was cause for concern over its resistance to the chemical durability of $PbO_2$ in the $H_2SO_4$ electrolyte environment. To tackle this issue, an electrolyte composed of lead methanesulfonate and methanesulfonic acid ($CH_3SO_3H$) was explored as an alternative electrolyte to address this issue. Hybrid SCs with a diverse combination of pseudocapacitive materials with ELDC or battery-type material, such as carbon//ECP, carbon//surface functionalized carbon, AQ-modified carbon//$RuO_2$, and so on, were also further explored [75]. Other types of acid electrolytes, such as perchloric acid, hexafluorosilicic acid, and tetrafluoroboric acid, have not been studied for SC application due to safety concerns and self-discharge in concentrated electrolytes.

***(b) Alkaline Electrolyte***

Potassium hydroxide (KOH) reigns supreme as the most commonly used type of alkaline electrolyte; this is attributed to its exceptional ionic conductivity. These electrolytes are suitable for application in carbon-based EDLCs, pseudocapacitive materials like $Co_3O_4$ and $Ni(OH)_2$, as well as in hybrid SCs. Affordable metal materials such as nickel (Ni) can serve as cost-effective current collectors for supercapacitors. Additional basic electrolytes that are subjects of investigation are NaOH and LiOH.

***Alkaline Electrolyte in EDLCs.*** Literature indicates that energy densities and specific capacitances of electric double-layer capacitors (EDLCs) with an aqueous alkaline electrolyte like KOH are comparable to those with an acidic electrolyte like $H_2SO_4$. However, there has been significant effort to improve energy densities of SC by using basic electrolytes for enhancing charge storage capacity or broadening the operating potential window [76–78]. This includes augmenting the capacitance of carbon-based electrode materials by leveraging the pseudocapacitance mechanism, developing high-specific capacitance pseudocapacitive materials, exploring composite materials, combining EDLC materials with polymers and metal oxides, and by using basic electrolytes and thus further enhancing cell voltage through asymmetric SC design [79].

***Alkaline Electrolyte in Pseudocapacitors.*** The ability of carbon-based electrode materials to exhibit pseudocapacitance is heavily influenced by the specific functional groups present on their surfaces. The intricate interplay between surface functional groups and ions in the electrolyte governs the faradaic interactions at the electrode surface, ultimately impacting the pseudocapacitive behavior. Research has revealed that the KOH electrolyte has benefits for carbon electrode materials doped with non-metallic atoms like nitrogen and phosphorus, indicating that the associated pseudocapacitance has a significant impact on electrolyte properties such as type and pH. Co-doping porous carbons with phosphorus and nitrogen offers a compelling strategy to enhance supercapacitor performance. This approach can simultaneously deliver increased specific capacitance, a wider operating voltage window, and improved stability compared to materials lacking phosphorus, hence improving supercapacitor performance even further [80–82].

Due to their high theoretical capacitances, transition metal oxides, hydroxides, sulfides, and nitrides have been explored in alkaline electrolytes. The interaction between ions in the electrolytes and the electrode materials plays an important role in the pseudocapacitive behavior of these materials [83]. Charge storage mechanisms in pseudocapacitive electrode materials are primarily based on reversible ion transfer processes at the electrode interface. These processes can include adsorption/desorption of ions present in electrolyte to the electrode surface as well as their insertion/extraction within the surface. Normally, the electrolyte properties, such as ion type, concentration, and operating temperature, can affect the performance of SCs. For example, the alkaline electrolyte concentration can affect the value of $R_{cell}$, specific capacitance, and the oxygen evolution reaction. A major setback associated with utilization of concentrated electrolytes introduces a potential vulnerability to the electrode substrate in the form of accelerated corrosion. The potential for corrosion at the electrode/substrate interface, exacerbated by concentrated electrolytes, necessitates a judicious selection of electrolyte concentration. This optimization process strives to achieve a balance between maximizing supercapacitor performance and preserving electrode structural integrity. The impact of electrolyte ion type on supercapacitor (SC) performance is somewhat complex. Research has demonstrated a superior specific capacitance for $MnO_2$ electrodes when employed in conjunction with a LiOH electrolyte, compared to counterparts utilizing KOH or NaOH electrolytes. However, other comparative studies with different alkaline electrolytes for electrode materials such as $Co_2P_2O_7$, $MnFe_2O_4$, and $Bi_2WO_6$ showed that higher specific capacitances could be achieved in KOH electrolytes than those in the NaOH or LiOH electrolyte [82–86].

Pseudocapacitive materials generally demonstrate lower cycling stability when compared to non-pseudocapacitive counterparts. The observed instability especially with respect to electrode materials could be attributed to the repetitive processes of ion intercalation/deintercalation within a basic electrolyte. Furthermore, the disintegration of electrode materials in alkaline electrolytes poses a substantial challenge in achieving long-term stability for these devices, which may contribute to an overall decrease in capacitive performance during prolonged charging–discharging cycles. A primary focus of contemporary research in supercapacitors revolves around expanding their operational voltage range, which is aimed at substantially enhancing energy density while using alkaline electrolytes.

***Alkaline Electrolyte in Hybrid supercapacitors.*** Alkaline electrolytes have been developed for hybrid capacitors to enhance energy density, with positive electrodes distinguished from negative electrodes. The positive electrode can either be a battery-type or pseudocapacitive electrode, where faradaic reactions govern charge storage mechanism, while the carbon-based material constitute a negative electrode, where storing charge primarily occurs through formation of the electric double-layer mechanism. These SCs can achieve higher energy densities ranging from 20 to 40 W h $kg^{-1}$, and occasionally, few of the electrode materials reach their energy density of 140 W h $kg^{-1}$, which is analogous to rechargeable lithium-ion batteries. Nevertheless, due to the utilization of faradaic-type electrodes in these asymmetric supercapacitors (SCs), their cycle stabilities are frequently lower than those of EDLCs. Additionally, these SCs usually suffer from a slower charging–discharging process and a few-second response compared to the EDLCs.

***(c) Neutral Electrolytes***

Neutral electrolytes, such as $Na_2SO_4$, are widely studied for enhanced security electrochemical systems due to their merits such as less corrosion, wider potential window, and greater safety. These electrolytes are typically composed of Li, Na, K, Ca, and Mg salts. Within the spectrum of neutral electrolytes, the most employed neutral electrolyte is $Na_2SO_4$, especially metal oxide-based ($MnO_2$-) electrode material, which is pseudocapacitive material. These are predominantly employed in pseudocapacitors and hybrid supercapacitors, although certain investigations have also been dedicated to electric double-layer capacitors (EDLCs).

***Neutral Electrolytes in EDLCs.*** Neutral electrolytes have been found to be more effective than acidic and alkaline electrolytes in electrical double-layer capacitors (EDLCs). These electrolytes have lower specific capacitances and higher $R_{cell}$ values compared to $H_2SO_4$ or KOH electrolytes [87, 88]. However, carbon-based SCs with neutral electrolytes can provide larger operating voltages due to increased electrolyte stable potential windows (ESPWs) compared to both acidic and alkaline aqueous electrolytes. In comparison to acidic and alkaline electrolytes, neutral electrolytes have lower quantities of H+ and OH- ions, which increases the overpotential for processes involving the evolution of hydrogen and oxygen. According to study results, carbon-based supercapacitors with neutral electrolytes can operate reliably up to a cell voltage of 1.5 V, which is lower than what other studies have reported. The operating voltages achieved for supercapacitors (SCs) based on neutral electrolytes are notably higher than those utilizing KOH and $H_2SO_4$ electrolytes, and further, neutral electrolytes offer a significant advantage over their strongly acidic and alkaline counterparts: reduced corrosiveness. Even within the category of neutral electrolytes, the performance of carbon-based supercapacitors can vary depending on the specific electrolyte chosen. Comparative studies have been conducted to investigate the impact of variation of ions on the efficiency of carbon-based SCs with different salts. However, some controversial results have been observed, such as the order of SC-specific capacitance values for alkali metal sulfate electrolytes, such as $Na_2SO_4$, $Li_2SO_4$, and $K_2SO_4$ [34, 89]. Other factors, such as method of fabrication of electrode material and electrochemical measurement parameters, may also affect the results. Regarding $R_{cell}$, it was noticed that an increase in series resistance mirrored the sequence of bulk electrolyte resistivity, therefore resulting in a decreasing pattern in both power density and rate capability performance: $K_2SO_4 > Na_2SO_4 > Li_2SO_4$ [91]. Additionally, Chae *et al.* found that because of the smaller ionic size of $Cl^-$ ions than $SO_4^{2-}$ ions, switching electrolytes having anions from $SO_4^{2-}$ to $Cl^-$ ions has the potential to boost specific capacitance [90, 91]. Recently, lithium, sodium, and potassium salts of silicotungstic acid (Li–SiW, Na–SiW, and K–SiW) were explored as aqueous neutral electrolytes for EDLCs, exhibiting much higher ionic conductivities compared to their counterparts with $Cl^{2-}$, $SO_4^{2-}$, or $NO^{3-}$ anions. EDLCs made up of carbon-based electrodes have achieved a cell voltage of 1.5 V utilizing neutral electrolyte.

***Neutral Electrolytes in Pseudocapacitors.*** For electrochemical storage devices that extensively use neutral electrolytes, pseudocapacitive electrode materials based on $MnO_2$ and $V_2O_5$ have demonstrated a positive response. $MnO_2$ is a pseudocapacitive material that has been the focus of extensive research in neutral electrolytes, as it allows for the variation of the Mn oxidation state during the charging–discharging process. The mechanism behind charge storage in these SCs involves the interplay between two processes: the accumulation and depletion of electrolyte cations (K+, Na+, and Li+) on the electrode surface, and the movement of these cations and potentially protons (H+) into and out of the electrode material itself. The nature of the neutral electrolytes, such as type of cation and anion species, additives, salt concentrations, ionic size, pH, and temperatures of electrolyte solution, has been found to significantly influence the SC performance [92–94]. The correlation between specific capacitance and cation species of electrolyte is not thoroughly understood, primarily due to variations in fabrication processes and modification in electrode material architectures. While many studies utilize specific capacitance to assess supercapacitor performance in different electrolytes, it is essential to recognize that the value of specific capacitance is also influenced by parameters such as morphological and structural characteristics, route of synthesis, and amount of electrode material. Relying on specific capacitances for understanding the unique effect of cationic species poses a challenge for the comparison of supercapacitor behavior. Different fabrication procedures of similar electrode material can provide varying morphological features with different pore shapes that affect the diffusion path of electrolyte ions and complicating the whole process. Further research using electrode materials with clearly established surfaces might assist in revealing how cation species affect the inherent capacitance of electrode materials. Ion intercalation/deintercalation has been proposed to be the crucial stage in the charging–discharging process of SC, and a decreased scan rate or charging/discharging rate could facilitate a better understanding of the intercalation–deintercalation process. The intercalation–deintercalation process is directly correlated to the bare cation size, which may be crucial in the present scenario. For instance, the presence of $Li^+$ ions in electrolyte would be beneficial to SC performance because the smaller $Li^+$ ion size might favor the intercalation–deintercalation process. In addition to alkaline metal ions, the effect of alkaline-earth cations on SC performance has also been studied. For example, by substituting bivalent cations for monovalent cations, the specific capacitance of $MnO_2$ was doubled. Boisset *et al.* conducted research to study the influence of anions on the electrochemical performance in neutral aqueous electrolytes containing various lithium salts; they used Birnessite-type and Cryptomelane-type $MnO_2$ electrode materials [94]. They examined ten different lithium salts composed of mineral and organic anions. Their findings suggested that the basicity and volume of the anion in each aqueous lithium salt electrolyte influenced the electrochemical performance of $MnO_2$ electrodes. Further study shows that $MnO_2$ electrode materials undergo structural changes at elevated temperatures during the rapid charging–discharging process, which is a potential cause for decreased cycling stability. Hence, it is crucial to consider the impact of electrolyte temperature on pseudocapacitive behavior, especially concerning the cycling stability of $MnO_2$-based supercapacitors (SCs). Approaches such as augmenting neutral electrolytes with additives have been explored to improve supercapacitor performance. For instance, Komaba *et al.* reported that adding a small amount of additives such as $NaHCO_3$, $Na_2HPO_4$, or $Na_2B_4O_7$ into the $Na_2SO_4$ electrolyte increased the specific capacitance of the birnessite electrode and significantly improved cyclability (41,000 cycles) [98]. The SC cycle life in neutral electrolytes is also considered an important issue, as the charge storage process of most pseudocapacitive materials involves intercalation/deintercalation of ions from the electrolyte. The electrolyte properties have a strong impact on the electrode material used, for example, $RuO_2$-based SCs with a neutral aqueous electrolyte achieving a high operating voltage and relatively high energy density.

***Neutral Electrolytes in Hybrid Capacitors.*** Neutral electrolytes are widely used in the development of hybrid supercapacitors as this arises from their capacity to offer a substantially larger functional potential window, resulting in enhanced energy density. However, most pseudocapacitive materials like $MnO_2$ have high specific capacitances but limited potential windows, restricting the limit of the cell voltage and charge storage capacity of symmetrical SCs. $MnO_2$-based supercapacitors have a distinctive cell voltage window of approximately 1 V [95]. A significant advantage can be achieved by interchanging the anode with alternative materials like activated carbons. This substitution allows for a substantial increase in the cell voltage. Hybrid supercapacitors, which uses a battery-type electrode material such as cathode in strong acidic or alkaline electrolytes when compared with asymmetric AC//$MnO_2$ SCs in neutral electrolytes, have shown a prolonged long cycle life due to the pseudocapacitive behavior of $MnO_2$. For initial studies, neutral electrolytes were used for a better understanding of behaviors of the charge storage mechanism of asymmetric AC//$MnO_2$ SCs, and efforts have focused on developing neutral electrolyte-based asymmetric SCs. Various types of anode and cathode electrode materials have been researched for hybrid SCs by employing neutral electrolytes, and these efforts have yielded devices boasting a remarkable maximum operating voltage window of approximately 1.8–2.0 V [92–98]. Most reported energy densities can achieve high values above 20 Wh $kg^{-1}$, and some even as high as 50 Wh $kg^{-1}$. The advantages of neutral electrolytes in supercapacitors (SCs) not only solve corrosion concerns, but also provide an economical and sustainable way to boost operating voltage and energy density. However, further advances in SC performance using neutral electrolytes are required to obtain substantial increases in energy density and cycle life.

### 2.3.2 Organic Electrolytes

Organic electrolytes for SCs are a popular choice due to their high voltage window and high resistivity compared to aqueous electrolytes. The properties of solvents and salts of organic solvent, such as ionic size, interaction between ion and solvent, viscosity, and ionic conductivity, significantly impact the electrochemical behavior of organic electrolyte-based SCs. Studies have demonstrated a significant correlation between the size of cations within small quaternary ammonium tetrafluoroborate ($BF_4$) salts and the performance of EDLCs [99]. For instance, trimethylethylammonium $BF_4$ and trimethylpropylammonium $BF_4$ showed 10% higher capacitance than tetraethylammonium $BF_4$ using microporous activated carbon electrodes. The capacitance of the electrode exhibits a greater dependence on the size of the cations compared to the $BF_4$ anion due to the significantly larger size of the quaternary ammonium cations relative to the $BF_4$ anion [100].

Organic electrolytes are usually used with metal oxides, conductive polymers material, and its composite, which are combinedly known as pseudocapacitive material. The presence of lithium ions is prevalent in many organic electrolytes used for pseudocapacitors owing to their small bare ion size and the ease of ion intercalation/deintercalation [101]. Some typically used solvents in organic electrolytes include tetraethylammonium tetrafluoroborate ($TEABF_4$), acetonitrile, dimethyl sulfoxide (DMSO), propylene carbonate, and a mixture of these solvents.

Asymmetric electrochemical supercapacitors employing organic electrolytes have garnered interest for their ability to further boost energy density. Numerous organic electrolytes have been explored in conjunction with asymmetric electrochemical supercapacitors, such as carbon//$TiO_2$, graphite//AC, carbon//$V_2O_5$, carbon//ECP, and carbon//$Li_4Ti_5O_{12}$, which have been reported by Amatucci *et al.* In comparison to aqueous-based asymmetric supercapacitors, these electrolytes have the potential for higher energy densities due to their substantially broader voltage range, which typically ranges from 3 to 4 V [102].

Maeshima *et al.* used *ab initio* molecular orbital theory to investigate the operational potential window of seven organic electrolytes used in EDLCs. To examine the impact of intermolecular interactions, four different models were developed: ion-pair *in vacuo* model, ion-in-solvent model, and single-ion *in vacuo* model. The study revealed that the specific capacitance of an EDLC is dependent not only on the surface area of electrode material but also on the distribution of pores of the carbon components [103]. The properties of organic electrolytes, such as ion–solvent interactions, anionic size, and cationic size, have a strong influence on the porous nature of working electrodes. Carbon-based working electrode materials with small pore sizes can enhance a specific surface area but may impede the accessibility of electrolyte ions, limiting their transport between the surfaces of the electrodes. Hence, it is crucial to ensure that the size of electrolyte ions matches the pore size of carbon materials to optimize specific capacitance. Furthermore, it is crucial to examine the dimensions of both bare and solvated ions. In conclusion, organic electrolytes for SCs play a crucial role in improving the performance of devices based on organic electrolytes.

### 2.3.3 Ionic Liquid Electrolytes

Ionic liquids are molten salts that become liquid at low temperatures, usually below 100°C. They consist completely of ions, frequently organic cations combined with organic or inorganic anions, and feature unique qualities such as excellent minimal vapor pressure and large ESPW, and their physical and chemical properties can be configured. They have gained significant interest as alternative electrolytes for electrochemical storage liquids due to their potential advantages such as high thermal, chemical, and electrochemical stability, inflammability, and negligible volatility [104, 105]. ILs, based on their composition, are divided into aprotic, protic, and zwitterionic types [109]. So far, the ILs used in ESs constitute a very small subset of the broad family of ILs whose chemical structure is discussed in Figure 2.5 [107, 108].

ILs are commonly based on pyrrolidinium, sulfuronium, phosphonium cations, ammonium, and so on [106]. The anions of ILs include dicyanamide (DCA), bis(fluorosulfonyl)imide (FSI), bis(trifluoromethanesulfonyl)imide (TFSI), tetrafluoroborate ($BF_4$), and hexafluorophosphate ($PF_6$). The proper balancing act between ionic conductivity of IL and electrochemical stable potential window exists, with IL-electrolyte-based SCs potentially providing operative cell voltages above 3 V.

However, ILs have several main drawbacks, such as low ionic conductivity, high viscosity, and costly, which are major setbacks for their practical use in SCs [110]. For instance, the $TEABF_4$/ACN electrolyte (59.9 mS cm$^{-1}$ at 25°C) is more conductive than the [EMIM][$BF_4$] electrolyte. Additionally, ILs exhibit a marked increase in viscosity relative to organic electrolytes (e.g., ACN organic electrolytes have a viscosity of 0.3 cP, whereas [EMIM][$BF_4$] and [BMIM][$BF_4$] ionic electrolytes have a viscosity of 41 cP and 219 cP, respectively).

Optimizing the performance of supercapacitors utilizing ILs has been a multi-pronged approach, leveraging both experimental and theoretical

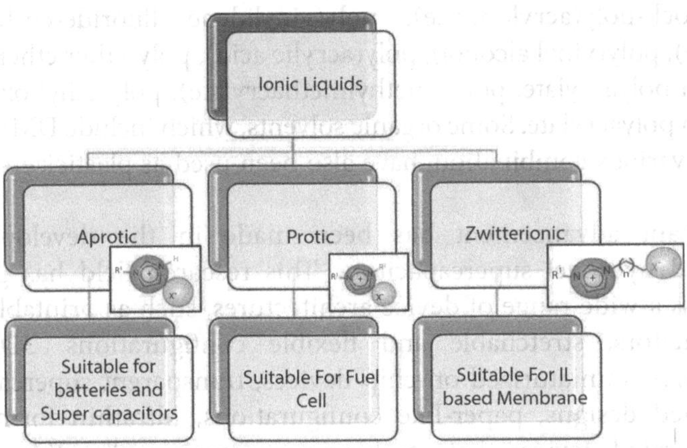


**Figure 2.5** Basic types of ionic liquid.

investigations. This effort encompasses the strategic selection of IL composition and the design of the electrochemical cell itself. Recent achievements have been made in understanding the morphological structure, chemical composition, electrochemical characteristics, and capacitive behavior corresponding to it, in the as-formed electric double-layer (EDL) of the IL/electrode interface [110–112].

Researchers are currently investigating the fundamental properties of the EDL produced at the electrode surface in IL-based electrolytes. This knowledge is critical for optimizing capacitance, which is the primary performance metric for supercapacitors. In conclusion, ILs have shown potential as alternative electrolytes for ESs due to their unique properties; high chemical, thermal, and electrochemical stability; low volatility; and being inflammable in nature. However, there are still several drawbacks, such as reduced ionic conductivity, highly viscous solution, and exorbitant price, which are setbacks in their practical application in SCs.

### 2.3.4 Solid and Quasi-Solid-State Electrolytes

Solid-state electrolyte-based solid-state capacitors (SCs) have gained significant interest due to their advantages such as economical packaging, elevated ionic conductivity, and no liquid spillage. properties. These electrolytes are classified into three types: solid polymer electrolytes (SPEs), gel polymer electrolytes (GPEs), and polyelectrolytes. Among quasi-solid-state electrolytes, gel polymer electrolytes exhibit superior ionic conductivity, positioning them as the preeminent choice for solid-state supercapacitors [113, 114]. Development of gel polymer electrolytes has involved exploration of diverse polymer hosts, including poly(acrylonitrile)-block poly(ethylene glycol)-block-poly(acrylonitrile), poly(vinylidene fluoride-co-hexauoropropylene), poly(vinyl alcohol), poly(acrylic acid), poly(ether ether ketone), potassium polyacrylate, poly-(methylmethacrylate), poly(ethyl oxide), and potassium polyacrylate. Some organic solvents, which include DMF, PC, EC, and their various combination, have also been used as plasticizers in GPEs [115].

Significant advancement has been made in the development of PVA-based hydrogel supercapacitors. This research field has grown to encompass a wide range of device architectures, such as printable micro-supercapacitors, stretchable and flexible configurations, 3D micro-architectures, miniaturized on-chip devices, transparent supercapacitors, fiber-shaped designs, paper-like configurations, ultrathin formats, and even integrated devices that combine supercapacitor functionality with other complementary functions. Poly (vinyl alcohol) has been extensively studied as a polymer matrix with various aqueous solutions such as alkaline (KOH), neutral, and strong acid ($H_2SO_4$ and $H_3PO_3$) electrolytes to prepare hydrogels [116–118]. The thickness of the electrode layer for PVA/$H_3PO_3$ and $H_2SO_4$-based supercapacitors plays a crucial role in achieving high capacitance. Electrodes with various material structures, such as CNTs grown on carbon cloth porous graphene ribbons, 3D graphene networks, graphene/porous carbon aerogels, and activated carbon cloth, have been investigated using PVA-based hydrogel electrolytes to simplify the penetration of a hydrogel electrolyte into the porous electrode and improve the performance of SCs [116].

The selection of electrolyte solution for polyvinyl alcohol (PVA) must be done meticulously to guarantee the optimal operation of the device. Chen *et al.* used a variety of electrolytes to discover six distinct PVA-based hydrogels for graphene-based supercapacitors, thus finding that the PVA–$H_3PO_4$ hydrogel electrolyte displayed the best capacitive performance among other PVA-based hydrogels [119]. The higher capacitance in the $H_3PO_4$–PVA electrolyte was due to the smaller ionic radius of $H^+$ ions compared to other ions in the gel electrolyte, allowing $H^+$ ions to diffuse quickly and easily between the graphene layers.

Acrylic hydrogel electrolytes, such as poly(2-hydroxyethyl methacrylate) (PHEMA), potassium polyacrylate (PAAK), poly(acrylamide) (PAAm), poly (2-acrylamido-2- poly (acrylic acid) (PAA), and methyl-1-propanesulfonic acid) (PAMPS), have been used for supercapacitors due to their ability to promote proton conduction. Studies have shown that the specific capacitance of these electrolytes increases in the order of PAMPS/$H_2SO_4$ < PAAK/$H_2SO_4$ < PAA/$H_2SO_4$ < 1 M $H_2SO_4$ < PAMPS/$H_2O$. The interaction between the acrylic polymer and $H_2SO_4$ solvent molecule is more advantageous than the $H_2O$-based electrolyte due to the higher specific capacitance (400–500 $Fg^{-1}$) even at high scan rates.

Indeed, organic solvents have been utilized in gel polymer electrolytes (GPEs) for supercapacitors with the aim of enhancing the working cell voltage of the device. Various polymer hosts, such as poly(methyl methacrylate) (PMMA), poly(acrylonitrile) (PAN), polyvinylpyrrolidone (PVP), poly(vinylidene fluoride) (PVDF), poly(ethylene oxide) (PEO), and polyether ether ketone (PEEK), have been reported as potential materials within the realm of organogel electrolytes in scientific literature. Recent research has showed potential in using block copolymers with different monomeric units, such as poly(acrylonitrile)-block-poly(ethylene glycol)-block-poly(acrylonitrile) (PAN-b-PEG-b-PAN), as polymeric scaffolds within organogel electrolytes for applications involving supercapacitors.

Research efforts have been focused on improving heat resistance and ESPW in ionogel electrolytes. A variety of polymeric materials, including PMMA, poly(ethylene glycol) diacrylate, PEO, PVA, PVP, PVDF-HFP, and PEEK, have been explored as potential host structures for ionogel electrolytes. Tamilarasan *et al.* created a transparent and flexible ionogel using PMMA/[BMIM][TFSI], while Yamagata *et al.* demonstrated that employing a chitosan host enhances the performance of ionogel-based EDLCs. Integrating nanoscale inorganic fillers into the polymer matrix of solid polymer electrolytes to create composite solid polymer electrolytes is considered one of the most effective strategies for enhancing ionic conductivity.

### 2.3.5 Redox Active Electrolytes

Researchers have explored a new technique for increasing SC capacitance by obtaining the pseudocapacitive properties from redox-active electrolytes and harnessing diffusion happening due to oxidation and reduction reactions occurring within the electrolyte to store and release charge, hence increasing the overall capacitance of the electrochemical device. Electrolytes that exhibit faradaic responses can enhance the overall capacitance of SCs. The attribution of pseudocapacitive behavior in this system transcends the electrode material alone. The electrolyte demonstrably contributes through faradaic redox reactions involving its dissolved redox mediator. Supercapacitors can increase their capacitance by inclusion of redox-active electrolytes that assist in enhancing pseudocapacitive behavior. For instance, high capacitance (1840 F $g^{-1}$) is observed in carbon-based SCs when an $I^-/I_3^-$ redox pair is dissolved in redox active aqueous electrolyte [120]. When nanoscale organic fillers are inserted in the polymer matrix of SPEs, a composite SPE is formed which is one of the most effective approaches to improving ionic conductivity. However, capacitance value increased only at the cathode region with a narrow potential range. A boost in the van der Waals radius of alkali metals within the cathode results in a proportional rise in capacitance, indicating an association that is beneficial: LiI (300 F $g^{-1}$) < NaI (492 F $g^{-1}$) < KI (1078 F $g^{-1}$) < RbI (2272 F $g^{-1}$) [121].

Silicotungstic acid and phosphotungstic acid, sometimes known as heteropoly acids, have been studied as redox mediators for redox-active electrolytes. These compounds have several rapid electron transfer redox processes and high proton conductivities, making them interesting candidates for improving the performance of energy storage devices [122, 123].

The phosphotungstic acid-based electrolyte, along with enhancing pseudocapacitance for carbon-based supercapacitors, increases the potential window up to 1.6 V. $Cu^{2+}$ reduction by the surface oxygen group on the carbon electrode enhances the specific capacitance.

Organic redox mediators as pseudocapacitive sources have been studied, including humic acid, hydroquinone, methylene blue, m-phenylenediamine, p-phenylenediamine, sulfonated polyaniline , and indigo carmine lignosulfonates. Incorporating redox-active polyfluorodecaborate cluster ions into an organic PC–DMC mixture solvent has been shown to effectively enhance the pseudocapacitance component of carbon-based supercapacitors, thereby contributing to the overall capacitance of the device.

## 2.4 Difficulties Associated with Electrolytes in a Sustainable Supercapacitor

As previously discussed, electrolytes have been identified as a critical component in electrochemical supercapacitor performance. The key hurdles in the development of sustainable supercapacitors include increasing cycle life, energy, and power densities, and addressing safety concerns. Operating temperature range, optimal potential window, ion mobility, ion conductivity, and electrochemical and thermal stabilities of the electrolytes are the primary parameters that determine both efficiency and practical applications of sustainable supercapacitors [125]. Although significant progress has been achieved in utilizing electrolytes in SCs, a few obstacles remain that are impeding the technology's practical application maturity. Some difficulties are described below:

The energy and power densities of SCs are dependent on the operating cell voltage of the electrolytes. The energy density of supercapacitors can be further enhanced by increasing the cell voltage of the capacitors using electrolytes that possess a more optimal potential window. However, increasing cell voltage frequently compromises other electrolyte properties. For instance, ionic liquids and organic electrolytes exhibit relatively low ionic conductivity but substantially higher cell voltages (up to 4 V) compared to aqueous electrolytes [126]. Furthermore, when compared to aqueous electrolytes, these nonaqueous electrolytes have lesser specific capacitance. As a result, developing electrochemical supercapacitors with better energy densities while maintaining other features is a significant problem.

Most of the publications have not addressed in detail the mechanisms of rapid charge/discharge capability occurring at electrode and electrolyte interfaces of recently developed materials. To improve the performance of sustainable supercapacitors, it is critical to assure compatibility or optimize the interaction between newly developed electrolytes and electrode materials. Despite significant advances in experimental techniques and theoretical investigations, there are still a few differences in capacitance at the interface between ionic liquid electrolytes and electrodes, the impact of organic solvent modifications on the electric double layer (EDL) structure, and the effects of temperature. A complete comprehension of the ion transport and charging dynamics of electrolyte ions is necessary. There is still a lack of understanding regarding the behavior of more complex electrode structures (such as hybrid nanostructures) during rapid charging and discharging processes.

Improving the energy density of sustainable supercapacitors using organic electrolytes and ionic liquid as their electrolyte solution is one viable option. Nonetheless, they often exhibit lower ionic conductivity than aqueous solutions, culminating in higher equivalent series resistance. Accordingly, using electrolytes with a larger ESPW of SCs opens up the possibility of greater $R_{cell}$ values, which, in turn, lower device performance. Because of this reason, all-solid-state supercapacitors and flexible supercapacitors have poor physical properties and inferior electrochemical performance.

There is an immediate necessity to develop standardized methodologies that can characterize the performance of numerous new electrolytes designed for electrochemical supercapacitors. It is also difficult to pick appropriate electrolytes from various published data because performance measurements are typically acquired under diverse situations. Furthermore, gravimetric capacitance, specific capacitance, and the energy density and power density of SCs are often stated dependent on the mass of active electrode materials. However, the contribution of electrolyte to total mass is significant, which should not be overlooked.

## 2.5 Potential Research Avenues for Resolving the Problems with Electrolytes

Aqueous electrolytes, organic electrolytes, ionic liquids (ILs), solid-state or quasi-solid-state electrolytes, and redox-active electrolytes are among the various kinds of electrolytes that have been studied and discussed in the literature. These types of electrolytes can be categorized and summarized in the following ways. Various electrolytic properties such as ionic conductivity, radius of bare and hydrated ions, solvent dielectric constant, mobility, viscosity, and thermal stability on electrochemical supercapacitor performance such as rate performance, specific capacitance, cycling stability, and energy and power densities are thoroughly discussed. Despite significant advances in this field, several major challenges remain, including low energy density, suboptimal compatibility between recently synthesized electrolytes and electrode materials, absence of standardized techniques for robust evaluation of electrolytes performance in SCs, the exorbitant price of some electrolytes, and a lack of basic understanding regarding charge storage mechanisms for some advanced electrolytes with more realistic and complex electrode structures [127–129]. To address these issues, the following future research prospects are proposed:

### 2.5.1 Improving the Potential Window of Electrolyte Values to Boost the Energy Density of the SCs

The electrochemical stabilized potential window can be improved by experimenting with new solvents, conducting salts, or optimizing/modifying frequently used electrolytes. It is worth noting that various additional electrolyte characteristics, such as ion mobility and viscosity, may degrade as a result of an increase in ESPWs. To address this challenge, it is necessary to comprehensively assess the entire electrolyte composition. For example, an effective approach involves the strategic combination of a salt with improved ionic conductivity or solubility with organic solvents possessing a broader potential window. This strategy has the potential to provide a viable solution that overcomes the drawbacks associated with existing electrolytes. Organic electrolytes are known to have the highest ESPW due to the existence of cation and anion in conducting salt and the incorporation of a single or a combination of solvents in organic solutions. As an instance, sulfone-based organic solvents have outperformed frequently used PC solvent-based electrolytes. On top of that, employing ILs as electrolytes in SCs is advantageous for boosting the potential window and consequently enhancing capacitance energy. The research on ILs for SCs is still in its early stages, and many ILs have yet to be investigated. ILs with wider ESPWs, as well as appropriate ionic conductivities and viscosities, can be created by carefully designing cation and anion structures. or employing an electrolyte mixture. On the other hand, meeting all the parameters (high ESPW, excellent ionic conductivity, robust thermal stability, low viscosity, low cost, and environmental compatibility) at the same time is tough. In order to solve actual challenges, certain compromises may be necessary. We believe that the compromises are heavily influenced by the interaction of electrode with electrolyte material, as well as the various need of SCs in different application perspectives.

### 2.5.2 Increasing the Purity of Electrolytes

It is extremely desirable to employ a feasible cleansing procedure to reduce the number of impurities in electrolytes, as residual impurities might have a detrimental effect on the electrolyte-specific potential and cause a noteworthy increase in self-discharge. Advanced lithium-ion battery research has given us a better understanding to develop various methods of purifying the organic solvents that could potentially be acquired through the lithium-ion battery industry. Special consideration should be given to ILs containing hydrophobic anions while developing IL-based electrolytes. This might make room to develop more stable electrolytes by reducing water content in the IL.

### 2.5.3 Enhancing the Compatibility of the Electrode Materials and Electrolyte to Improve Overall Performance

It is important to match SCs based on their power density as well as their specific capacitance or energy density. More basic research is needed to completely grasp the correlation between pore size and capacitance of electrochemically active normalized surface area with respect to specific capacitance, as varying results have been reported in this regard. This would serve as the foundation for optimizing the porosity and size of existing pores of carbon-based material, and to tackle this challenge, it is necessary to develop more precise methods for estimating the exact surface area of electrode materials. Additionally, further research is needed to fabricate electrodes with distinct porous structures, such as multilayered porous networks, which can offer a high surface area to achieve enhanced capacitance while ensuring effective ion transport for high power capability. Additionally, consideration of the size of electrolyte ions is important in designing electrode materials.

### 2.5.4 Effect of Ionic Radii at the Electrode–Electrolyte Interface to Enhance Overall Supercapacitive Performance

A key factor in defining the nonclassical behavior at the electrode–electrolyte interface and dynamic behavior is the ionic radius, which also has an impact on the capacitance, energy density, and ion distributions under the electric double layer [129]. In electrochemical systems, the effect of ionic radius is a crucial factor in influencing ion distributions and electric double layer (EDL) capacitance. It gives insight into how the curvature of micropores influences the separation of positive and negative ions, leading to higher capacitance. Thus, it is important to find more reliable methods to understand the magnitude of ion segregation, and capacitance increase depends on the curvature and the applied potential. In addition, the research might look at the effects of ionic radius on impurity molecules and additives, as well as their impact on the capacitance and energy density of the EDL [130]. Furthermore, the study could seek to comprehend the significance of ionic radius in phase transitions inside the EDL, particularly in organic electrolytes and ionic liquids. Investigating the impact of ionic radius on charging kinetics and ion transport in nanopores could potentially be a worthwhile endeavor. Overall, a thorough examination of the effect of ionic radius on numerous elements of electrochemical systems should yield useful insights for the development of new energy storage technologies.

### 2.5.5 Extend Basic Comprehension via Theoretical and Experimental Research

Enhancing fundamental knowledge through both experimental and theoretical studies is essential for the development and formulation of new electrolytes, as well as for the down-selection of electrolytes with respect to maintaining temperature threshold, enhancing electrochemical stable potential windows, and decreasing viscosity and ionic conduction. A basic understanding of ion dynamics, charging dynamics, charge storage mechanisms, and solvation/desolvation is necessary during the charging–discharging process for more convincing electrode shape. This can be accomplished through the use of both theoretical modeling (at the molecular/electronic level) and practical approaches, particularly *in situ* characterization methods such as NMR and FTIR. Special consideration should be given to recently discovered electrolytes comprising multivalent ions, IL-organic solvent mixtures, and solid-state electrolytes. Furthermore, fundamental research on the interaction of the electrolyte and electrode materials for a better understanding of the charging mechanism of SC is required for the creation of desired electrode–electrolyte systems and hybrid SCs. This fundamental understanding will not only guide the selection or development of novel electrolytes but also be able to accommodate specific types of electrolytes that could be used during fabrication and characterization of electrode material. Understanding the electrolyte aging and failure types is also required to mitigate electrolyte degradation. Both experimental and theoretical modeling approaches can be used to achieve this. Characterizing the electrolytes during lifespan tests, for instance, can be done using a range of experimental analysis techniques, particularly *in situ* or online characterization techniques (such online DEMS and Raman microscopy).

### 2.5.6 Expanding the Temperature Range Where the SC Functions

Depending on the kind of electrolyte, alternative strategies for expanding the working temperature range of SCs may be investigated. In the circumstance of an aqueous electrolyte, different additives such as conductive salt (KCl, $TEABF_4$, $LiClO_4$, etc.), pH buffer, surfactants, and redox mediators can be used to reduce the temperature threshold. Regarding organic electrolytes, it is feasible to create an electrolyte with a broader temperature range of operation by creating a novel organic solvent mixture as opposed to using just one solvent. Regarding ILs, the temperature range can be effectively widened by developing electrode materials, specifically in close proximity to the cooler temperature limit of IL-based SCs and thorough investigation into a eutectic mixture of ILs.

### 2.5.7 Standardization of Technique for Evaluating Electrolyte Performance

With the recent advancement in this field to understand the role of electrolytes, it is crucial to establish and standardize methodologies for assessing, understanding, and comparing the performance of various electrolytes and their related SCs. When characterizing the performance of SCs, the mass and volume of the electrolyte utilized must be known. Moreover, a standardized specification and methodology are essential for assessing the physical characteristics and electrochemical performance of flexible supercapacitors (SCs) utilizing solid-state electrolytes under applied stress. Furthermore, highly efficient and rapid methods of screening for evaluating electrolyte performance are required to accelerate the development of novel electrolytes. When employing a three-electrode setup for evaluating the effectiveness of electrolytes, it is advisable in some cases to utilize electrodes identical to those used in actual supercapacitors (SCs). This ensures more accurate performance metrics for SC electrolytes.

## 2.6 Conclusion

SCs are evolving as revolutionary micro energy storage devices with a wide range of technical applications, including electronic skins, medical devices, wearable electronics, and biosensors. Recent trends in electrolytes for supercapacitors, highlighting their potential as energy storage technologies due to their short charge/discharge time, long life cycle, and good temperature behavior, have been discussed thoroughly. This study investigated current developments in charge storage strategies by different kinds of electrolytes. Each electrolyte comes with its own set of properties. When selecting an electrolyte, the operating voltage and ionic conductivity are important variables to take into account. The energy density and power density of the SCs increase with increasing operating voltage and ionic conductivity. Overall, the chapter presents a detailed overview of recent advancements in supercapacitor electrolytes, emphasizing the importance of high-performance electrolytes in enhancing supercapacitor capabilities and enabling their wide-ranging applications in diverse electronic devices.

## References


1. Reddy, M.V. and Subba Rao, G.V., and Chowdari, B.V.R., Metal oxides and oxysalts as anode materials for Li ion batteries. *Chem. Rev., 113*, 7, 5364–5457, 2013.
2. Zhou, C., Bag, S., Thangadurai, V., Engineering materials for progressive all-solid-state Na batteries. *ACS Energy Lett.*, *3*, 9, 2181–2198, 2018.
3. Burke, A., R&D considerations for the performance and application of electrochemical capacitors. *Electrochim. Acta*, *53*, 3, 1083–1091, 2007.
4. Palakkathodi Kammampata, S. and Thangadurai, V., Cruising in ceramics—discovering new structures for all-solid-state batteries—fundamentals, materials, and performances. *Ionics*, *24*, 639–660, 2018.
5. Aravindan, V., Chuiling, W., Reddy, M.V., Rao, G.S., Chowdari, B.V.R., Madhavi, S., Carbon coated nano-LiTi 2 (PO 4) 3 electrodes for non-aqueous hybrid supercapacitors. *PCCP*, *14*, 16, 5808–5814, 2012.
6. Gogotsi, Y. and Simon, P., True performance metrics in electrochemical energy storage. *Science*, *334*, 6058, 917–918, 2011.
7. Chmiola, J., Yushin, G., Gogotsi, Y., Portet, C., Simon, P., Taberna, P.L., Anomalous increase in carbon capacitance at pore sizes less than 1 nanometer. *Science*, *313*, 5794, 1760–1763, 2006.
8. Jha, S. and Prasanth, R., FNM-Based Supercapacitor in Futuristic Application, in: *Functionalized Nanomaterials Based Supercapacitor*, p. 679.

9. Simon, P. and Gogotsi, Y., Materials for electrochemical capacitors. *Nat. Mater.*, *7*, 11, 845–854, 2008.
10. Andreas, H.A. and Conway, B.E., Examination of the double-layer capacitance of an high specific-area C-cloth electrode as titrated from acidic to alkaline pHs. *Electrochim. Acta*, *51*, 28, 6510–6520, 2006.
11. Fic, K., Lota, G., Meller, M., Frackowiak, E., Novel insight into neutral medium as electrolyte for high-voltage supercapacitors. *Energy Environ. Sci.*, *5*, 2, 5842–5850, 2012.
12. Brandt, A., Pohlmann, S., Varzi, A., Balducci, A., Passerini, S., Ionic liquids in supercapacitors. *MRS Bull.*, *38*, 7, 554–559, 2013.
13. Choudhury, N.A., Sampath, S., Shukla, A.K., Hydrogel-polymer electrolytes for electrochemical capacitors: an overview. *Energy Environ. Sci.*, *2*, 1, 55–67, 2009.
14. Yang, Z., Zhang, J., Kintner-Meyer, M.C., Lu, X., Choi, D., Lemmon, J.P., Liu, J., Electrochemical energy storage for green grid. *Chem. Rev.*, *111*, 5, 3577–3613, 2011.
15. Thackeray, M.M., Wolverton, C., Isaacs, E.D., Electrical energy storage for transportation—approaching the limits of, and going beyond, lithium-ion batteries. *Energy Environ. Sci.*, *5*, 7, 7854–7863, 2012.
16. Gao, H. and Lian, K., 2014. Proton-conducting polymer electrolytes and their applications in solid supercapacitors: a review. *RSC Adv.*, *4*, 62, 33091–33113, 2014.
17. Lota, G. and Frackowiak, E., Striking capacitance of carbon/iodide interface. *Electrochem. Commun.*, *11*, 1, 87–90, 2009.
18. Wang, J.Y.Q., Wei, T., Fan, Z., Recent advances in design and fabrication of electrochemical supercapacitors with high energy densities. *Adv. Energy Mater.*, *4*, 1300816–1300859, 2014.
19. Béguin, F., Presser, V., Balducci, A., Frackowiak, E., Carbons and electrolytes for advanced supercapacitors. *Adv. Mater.*, *26*, 14, 2219–2251, 2014.
20. Zhang, L.L. and Zhao, X.S., Carbon-based materials as supercapacitor electrodes. *Chem. Soc. Rev.*, *38*, 9, 2520–2531, 2009.
21. Jiang, H., Lee, P.S., Li, C., 3D carbon based nanostructures for advanced supercapacitors. *Energy Environ. Sci.*, *6*, 1, 41–53, 2013.
22. Shao, Y., Li, J., Li, Y., Wang, H., Zhang, Q., Kaner, R.B., Flexible quasi-solid-state planar micro-supercapacitor based on cellular graphene films. *Mater. Horiz.*, *4*, 6, 1145–1150, 2017.
23. Thiagarajan, K., Theerthagiri, J., Senthil, R.A., Arunachalam, P., Madhavan, J., Ghanem, M.A., Synthesis of Ni 3 V 2 O 8@ graphene oxide nanocomposite as an efficient electrode material for supercapacitor applications. *J. Solid State Electrochem.*, *22*, 527–536, 2018.
24. Singu, B.S., Goda, E.S., Yoon, K.R., Carbon Nanotube–Manganese oxide nanorods hybrid composites for high-performance supercapacitor materials. *J. Ind. Eng. Chem.*, *97*, 239–249, 2021.

25. Miller, J.R., Outlaw, R.A., Holloway, B.C., Graphene double-layer capacitor with ac line-filtering performance. *Science*, *329*, 5999, 1637–1639, 2010.
26. Long, J.W., Bélanger, D., Brousse, T., Sugimoto, W., Sassin, M.B., Crosnier, O., Asymmetric electrochemical capacitors—Stretching the limits of aqueous electrolytes. *MRS Bull.*, *36*, 7, 513–522, 2011.
27. Abdah, M.A.A.M., Azman, N.H.N., Kulandaivalu, S., Sulaiman, Y., Review of the use of transition-metal-oxide and conducting polymer-based fibres for high-performance supercapacitors. *Mater. Des.*, *186*, 108199, 2020.
28. Wang, Y., Song, Y., Xia, Y., Electrochemical capacitors: mechanism, materials, systems, characterization and applications. *Chem. Soc. Rev.*, *45*, 21, 5925–5950, 2016.
29. Zhong, C., Deng, Y., Hu, W., Qiao, J., Zhang, L., Zhang, J., A review of electrolyte materials and compositions for electrochemical supercapacitors. *Chem. Soc. Rev.*, *44*, 21, 7484–7539, 2015.
30. Wang, Y., Song, Y., Xia, Y., Electrochemical capacitors: mechanism, materials, systems, characterization and applications. *Chem. Soc. Rev.*, *45*, 21, 5925–5950, 2016.
31. Jiang, H., Lee, P.S., Li, C., 3D carbon based nanostructures for advanced supercapacitors. *Energy Environ. Sci.*, *6*, 1, 41–53, 2013.
32. Chen, D., Wang, Q., Wang, R., Shen, G., Ternary oxide nanostructured materials for supercapacitors: a review. *J. Mater. Chem. A*, *3*, 19, 10158–10173, 2015.
33. Dubal, D.P., Chodankar, N.R., Kim, D.H., Gomez-Romero, P., Towards flexible solid-state supercapacitors for smart and wearable electronics. *Chem. Soc. Rev.*, *47*, 6, 2065–2129, 2018.
34. Zhong, C., Deng, Y., Hu, W., Qiao, J., Zhang, L., Zhang, J., A review of electrolyte materials and compositions for electrochemical supercapacitors. *Chem. Soc. Rev.*, *44*, 21, 7484–7539, 2015.
35. Qin, W., Zhou, N., Wu, C., Xie, M., Sun, H., Guo, Y., Pan, L., Mini-review on the redox additives in aqueous electrolyte for high performance supercapacitors. *ACS Omega*, *5*, 8, 3801–3808, 2020.
36. Zhu, Y. and Fontaine, O., Most Modern Supercapacitor Designs Advanced Electrolyte and Interface, in: *Supercapacitors for the Next Generation*, 2021.
37. Fu, L., Qu, Q., Holze, R., Kondratiev, V.V., Wu, Y., Composites of metal oxides and intrinsically conducting polymers as supercapacitor electrode materials: the best of both worlds? *J. Mater. Chem. A*, *7*, 25, 14937–14970, 2019.
38. Bigdeloo, M., Kowsari, E., Ehsani, A., Chinnappan, A., Ramakrishna, S., AliAkbari, R., Review on innovative sustainable nanomaterials to enhance the performance of supercapacitors. *J. Energy Storage*, *37*, 102474, 2021.
39. Gogotsi, Y. and Simon, P., True performance metrics in electrochemical energy storage. *Science*, *334*, 6058, 917–918, 2011.
40. Samui, A.B. and Sivaraman, P., Solid polymer electrolytes for supercapacitors, in: *Polymer Electrolytes*, pp. 431–470, 2010.

41. Reddy, M.V., Subba Rao, G.V., Chowdari, B.V.R., Metal oxides and oxysalts as anode materials for Li ion batteries. *Chem. Rev., 113*, 7, 5364–5457, 2013.
42. Kurzweil, P. and Chwistek, M., Electrochemical stability of organic electrolytes in supercapacitors: Spectroscopy and gas analysis of decomposition products. *J. Power Sources, 176*, 2, 555–567, 2008.
43. Zhong, C., Deng, Y., Hu, W., Qiao, J., Zhang, L., Zhang, J., A review of electrolyte materials and compositions for electrochemical supercapacitors. *Chem. Soc. Rev., 44*, 21, 7484–7539, 2015.
44. Ue, M., Ida, K., Mori, S., Electrochemical properties of organic liquid electrolytes based on quaternary onium salts for electrical double-layer capacitors. *J. Electrochem. Soc., 141*, 11, 2989, 1994.
45. Lu, M., *Supercapacitors: materials, systems, and applications*, John Wiley & Sons, 2013.
46. Conway, B.E., *Electrochemical supercapacitors: scientific fundamentals and technological applications*, Springer Science & Business Media, 2013.
47. Yu, A., Chabot, V., Zhang, J., *Electrochemical supercapacitors for energy storage and delivery: fundamentals and applications*, p. 383, Taylor & Francis, 2013.
48. Burke, A. and Miller, M., The power capability of ultracapacitors and lithium batteries for electric and hybrid vehicle applications. *J. Power Sources, 196*, 1, 514–522, 2011.
49. Burke, A., Ultracapacitors: why, how, and where is the technology. *J. Power Sources, 91*, 1, 37–50, 2000.
50. Fang, X. and Yao, D., November. An overview of solid-like electrolytes for supercapacitors, in: *ASME International Mechanical Engineering Congress and Exposition*, vol. 56284, American Society of Mechanical Engineers, p. V06AT07A071, 2013.
51. Galiński, M., Lewandowski, A., Stępniak, I., Ionic liquids as electrolytes. *Electrochim. Acta, 51*, 26, 5567–5580, 2006.
52. Jabeen, N., Hussain, A., Xia, Q., Sun, S., Zhu, J., Xia, H., High-performance 2.6 V aqueous asymmetric supercapacitors based on *in situ* formed Na0. 5MnO2 nanosheet assembled nanowall arrays. *Adv. Mater., 29*, 32, 1700804, 2017.
53. Iqbal, J., Numan, A., Rafique, S., Jafer, R., Mohamad, S., Ramesh, K., Ramesh, S., High performance supercapattery incorporating ternary nanocomposite of multiwalled carbon nanotubes decorated with Co3O4 nanograins and silver nanoparticles as electrode material. *Electrochim. Acta, 278*, 72–82, 2018.
54. Kim, H., Hong, J., Park, K.Y., Kim, H., Kim, S.W., Kang, K., Aqueous rechargeable Li and Na ion batteries. *Chem. Rev., 114*, 23, 11788–11827, 2014.
55. Zhao, L., Qiu, Y., Yu, J., Deng, X., Dai, C., Bai, X., Carbon nanofibers with radially grown graphene sheets derived from electrospinning for aqueous supercapacitors with high working voltage and energy density. *Nanoscale, 5*, 11, 4902–4909, 2013.

56. Xiong, T., Tan, T.L., Lu, L., Lee, W.S.V., Xue, J., Harmonizing energy and power density toward 2.7 V asymmetric aqueous supercapacitor. *Adv. Energy Mater.*, *8*, 14, 1702630, 2018.
57. Jiménez-Cordero, D., Heras, F., Gilarranz, M.A., Raymundo-Piñero, E., Grape seed carbons for studying the influence of texture on supercapacitor behaviour in aqueous electrolytes. *Carbon*, *71*, 127–138, 2014.
58. Wu, H., Wang, X., Jiang, L., Wu, C., Zhao, Q., Liu, X., Yi, L., The effects of electrolyte on the supercapacitive performance of activated calcium carbide-derived carbon. *J. Power Sources*, *226*, 202–209, 2013.
59. Jin, Z., Yan, X., Yu, Y., Zhao, G., Sustainable activated carbon fibers from liquefied wood with controllable porosity for high-performance supercapacitors. *J. Mater. Chem. A*, *2*, 30, 11706–11715, 2014.
60. Chen, Q., Hu, Y., Hu, C., Cheng, H., Zhang, Z., Shao, H., Qu, L., Graphene quantum dots–three-dimensional graphene composites for high-performance supercapacitors. *PCCP*, *16*, 36, 19307–19313, 2014.
61. Andreas, H.A. and Conway, B.E., Examination of the double-layer capacitance of an high specific-area C-cloth electrode as titrated from acidic to alkaline pHs. *Electrochim. Acta*, *51*, 28, 6510–6520, 2006.
62. Oda, H., Yamashita, A., Minoura, S., Okamoto, M., Morimoto, T., Modification of the oxygen-containing functional group on activated carbon fiber in electrodes of an electric double-layer capacitor. *J. Power Sources*, *158*, 2, 1510–1516, 2006.
63. Yan, X., Yu, Y., Yang, X., Effects of electrolytes on the capacitive behavior of nitrogen/phosphorus co-doped nonporous carbon nanofibers: an insight into the role of phosphorus groups. *RSC Adv.*, *4*, 48, 24986–24990, 2014.
64. Hulicova, D., Kodama, M., Hatori, H., Electrochemical performance of nitrogen-enriched carbons in aqueous and non-aqueous supercapacitors. *Chem. Mater.*, *18*, 9, 2318–2326, 2006.
65. Pognon, G., Brousse, T., Demarconnay, L., Bélanger, D., Performance and stability of electrochemical capacitor based on anthraquinone modified activated carbon. *J. Power Sources*, *196*, 8, 4117–4122, 2011.
66. Ruiz, V., Blanco, C., Granda, M., Santamaría, R., Enhanced life-cycle supercapacitors by thermal treatment of mesophase-derived activated carbons. *Electrochim. Acta*, *54*, 2, 305–310, 2008.
67. Liu, X. and Pickup, P.G., Ru oxide/carbon fabric composites for supercapacitors. *J. Solid State Electrochem.*, *14*, 231–240, 2010.
68. Mendoza-Sánchez, B., Brousse, T., Ramirez-Castro, C., Nicolosi, V., Grant, P.S., An investigation of nanostructured thin film α-MoO3 based supercapacitor electrodes in an aqueous electrolyte. *Electrochim. Acta*, *91*, 253–260, 2013.
69. Long, J.W., Bélanger, D., Brousse, T., Sugimoto, W., Sassin, M.B., Crosnier, O., Asymmetric electrochemical capacitors—Stretching the limits of aqueous electrolytes. *MRS Bull.*, *36*, 7, 513–522, 2011.

70. Perret, P., Khani, Z., Brousse, T., Bélanger, D., Guay, D., Carbon/PbO2 asymmetric electrochemical capacitor based on methanesulfonic acid electrolyte. *Electrochim. Acta, 56*, 24, 8122–8128, 2011.
71. Algharaibeh, Z., Liu, X., Pickup, P.G., An asymmetric anthraquinone-modified carbon/ruthenium oxide supercapacitor. *J. Power Sources, 187*, 2, 640–643, 2009.
72. Salinas-Torres, D., Sieben, J.M., Lozano-Castello, D., Cazorla-Amorós, D., Morallon, E., Asymmetric hybrid capacitors based on activated carbon and activated carbon fibre–PANI electrodes. *Electrochim. Acta, 89*, 326–333, 2013.
73. Shen, J., Yang, C., Li, X., Wang, G., High-performance asymmetric supercapacitor based on nanoarchitectured polyaniline/graphene/carbon nanotube and activated graphene electrodes. *ACS Appl. Mater. Interfaces, 5*, 17, 8467–8476, 2013.
74. Oickle, A.M. and Andreas, H.A., Examination of water electrolysis and oxygen reduction as self-discharge mechanisms for carbon-based, aqueous electrolyte electrochemical capacitors. *J. Phys. Chem. C, 115*, 10, 4283–4288, 2011.
75. Lang, J.W., Yan, X.B., Liu, W.W., Wang, R.T., Xue, Q.J., Influence of nitric acid modification of ordered mesoporous carbon materials on their capacitive performances in different aqueous electrolytes. *J. Power Sources, 204*, 220–229, 2012.
76. Wang, D.W., Li, F., Yin, L.C., Lu, X., Chen, Z.G., Gentle, I.R., Lu, G.Q., Cheng, H.M., Nitrogen-doped carbon monolith for alkaline supercapacitors and understanding nitrogen-induced redox transitions. *Chem. Eur. J., 18*, 17, 5345–5351, 2012.
77. Wang, C., Sun, L., Zhou, Y., Wan, P., Zhang, X., Qiu, J., P/N co-doped microporous carbons from H3PO4-doped polyaniline by *in situ* activation for supercapacitors. *Carbon, 59*, 537–546, 2013.
78. Nam, K.W. and Kim, K.B., A study of the preparation of NiO x electrode via electrochemical route for supercapacitor applications and their charge storage mechanism. *J. Electrochem. Soc., 149*, 3, A346, 2002.
79. Yuan, A. and Zhang, Q., A novel hybrid manganese dioxide/activated carbon supercapacitor using lithium hydroxide electrolyte. *Electrochem. Commun.*, 8, 7, 1173–1178, 2006.
80. Dong, X., Guo, Z., Song, Y., Hou, M., Wang, J., Wang, Y., Xia, Y., Flexible and wire-shaped micro-supercapacitor based on Ni (OH) 2-nanowire and ordered mesoporous carbon electrodes. *Adv. Funct. Mater., 24*, 22, 3405–3412, 2014.
81. Rakhi, R.B., Alhebshi, N.A., Anjum, D.H., Alshareef, H.N., Nanostructured cobalt sulfide-on-fiber with tunable morphology as electrodes for asymmetric hybrid supercapacitors. *J. Mater. Chem. A, 2*, 38, 16190–16198, 2014.

82. Choi, D., Blomgren, G.E., Kumta, P.N., Fast and reversible surface redox reaction in nanocrystalline vanadium nitride supercapacitors. *Adv. Mater.*, *18*, 9, 1178–1182, 2006.
83. Hou, L., Yuan, C., Li, D., Yang, L., Shen, L., Zhang, F., Zhang, X., Electrochemically induced transformation of NiS nanoparticles into Ni (OH) 2 in KOH aqueous solution toward electrochemical capacitors. *Electrochim. Acta*, *56*, 22, 7454–7459, 2011.
84. Wang, Y.G., Wang, Z.D., Xia, Y.Y., An asymmetric supercapacitor using RuO2/TiO2 nanotube composite and activated carbon electrodes. *Electrochim. Acta*, *50*, 28, 5641–5646, 2005.
85. Kong, L.B., Lang, J.W., Liu, M., Luo, Y.C., Kang, L., Facile approach to prepare loose-packed cobalt hydroxide nano-flakes materials for electrochemical capacitors. *J. Power Sources*, *194*, 2, 1194–1201, 2009.
86. Vidyadharan, B., Abd Aziz, R., Misnon, I.I., Kumar, G.M.A., Ismail, J., Yusoff, M.M., Jose, R., High energy and power density asymmetric supercapacitors using electrospun cobalt oxide nanowire anode. *J. Power Sources*, *270*, 526–535, 2014.
87. Lee, H.Y. and Goodenough, J.B., Supercapacitor behavior with KCl electrolyte. *J. Solid State Chem.*, *144*, 1, 220–223, 1999.
88. Kuo, S.L. and Wu, N.L., Investigation of pseudocapacitive charge-storage reaction of MnO2· nH2O supercapacitors in aqueous electrolytes. *J. Electrochem. Soc.*, *153*, 7, A1317, 2006.
89. Xu, C., Li, B., Du, H., Kang, F., Zeng, Y., Supercapacitive studies on amorphous MnO2 in mild solutions. *J. Power Sources*, *184*, 2, 691–694, 2008.
90. Komaba, S., Tsuchikawa, T., Tomita, M., Yabuuchi, N., Ogata, A., Efficient electrolyte additives of phosphate, carbonate, and borate to improve redox capacitor performance of manganese oxide electrodes. *J. Electrochem. Soc.*, *160*, 11, A1952, 2013.
91. Chae, J.H. and Chen, G.Z., Influences of ions and temperature on performance of carbon nano-particulates in supercapacitors with neutral aqueous electrolytes. *Particuology*, *15*, 9–17, 2014.
92. Shao, J., Li, X., Qu, Q., Wu, Y., Study on different power and cycling performance of crystalline KxMnO2· nH2O as cathode material for supercapacitors in Li2SO4, Na2SO4, and K2SO4 aqueous electrolytes. *J. Power Sources*, *223*, 56–61, 2013.
93. Wu, M., Zhang, L., Gao, J., Xiao, C., Wang, D., Chen, A., Zhang, S., Effects of thickness and electrolytes on the capacitive characteristics of anodically deposited hydrous manganese oxide coatings. *J. Electroanal. Chem.*, *613*, 2, 125–130, 2008.
94. Boisset, A., Athouel, L., Jacquemin, J., Porion, P., Brousse, T., Anouti, M., Comparative performances of birnessite and cryptomelane MnO2 as electrode material in neutral aqueous lithium salt for supercapacitor application. *J. Phys. Chem. C*, *117*, 15, 7408–7422, 2013.

95. Qu, Q.T., Liu, L.L., Wu, Y.P., Holze, R., Electrochemical behavior of V2O5·0.6 H2O nanoribbons in neutral aqueous electrolyte solution. *Electrochim. Acta, 96*, 8–12, 2013.
96. Yang, Q., Pang, S.K., Yung, K.C., Study of PEDOT–PSS in carbon nanotube/conducting polymer composites as supercapacitor electrodes in aqueous solution. *J. Electroanal. Chem., 728*, 140–147, 2014.
97. Kim, M. and Kim, J., Development of high power and energy density microsphere silicon carbide–MnO 2 nanoneedles and thermally oxidized activated carbon asymmetric electrochemical supercapacitors. *PCCP, 16*, 23, 11323–11336, 2014.
98. Komaba, S., Tsuchikawa, T., Tomita, M., Yabuuchi, N., Ogata, A., Efficient electrolyte additives of phosphate, carbonate, and borate to improve redox capacitor performance of manganese oxide electrodes. *J. Electrochem. Soc., 160*, 11, A1952, 2013.
99. Koh, A.R., Hwang, B., Roh, K.C., Kim, K., The effect of the ionic size of small quaternary ammonium BF 4 salts on electrochemical double layer capacitors. *PCCP, 16*, 29, 15146–15151, 2014.
100. McDonough, J.K., Frolov, A.I., Presser, V., Niu, J., Miller, C.H., Ubieto, T., Fedorov, M.V., Gogotsi, Y., Influence of the structure of carbon onions on their electrochemical performance in supercapacitor electrodes. *Carbon, 50*, 9, 3298–3309, 2012.
101. Krishnan, S.G., Reddy, M.V., Harilal, M., Vidyadharan, B., Misnon, I.I., Ab Rahim, M.H., Ismail, J., Jose, R., Characterization of MgCo2O4 as an electrode for high performance supercapacitors. *Electrochim. Acta, 161*, 312–321, 2015.
102. Amatucci, G.G., Badway, F., Du Pasquier, A., Zheng, T., An asymmetric hybrid nonaqueous energy storage cell. *J. Electrochem. Soc., 148*, 8, A930, 2001.
103. Maeshima, H., Moriwake, H., Kuwabara, A., Fisher, C.A., Quantitative evaluation of electrochemical potential windows of electrolytes for electric double-layer capacitors using ab initio calculations. *J. Electrochem. Soc., 157*, 6, A696, 2010.
104. Liew, C.W., Arifin, K.H., Kawamura, J., Iwai, Y., Ramesh, S., Arof, A.K., Effect of halide anions in ionic liquid added poly (vinyl alcohol)-based ion conductors for electrical double layer capacitors. *J. Non-Cryst. Solids, 458*, 97–106, 2017.
105. Shi, M., Kou, S., Yan, X., Engineering the Electrochemical Capacitive Properties of Graphene Sheets in Ionic-Liquid Electrolytes by Correct Selection of Anions. *ChemSusChem, 7*, 11, 3053–3062, 2014.
106. Dou, Q., Lian, C., Lei, S., Chen, J., Liu, H., Yan, X., Silica-grafted ionic liquid for maximizing the operational voltage of electrical double-layer capacitors. *Energy Storage Mater., 18*, 253–259, 2019.

107. Armand, M., Endres, F., MacFarlane, D.R., Ohno, H., Scrosati, B., Ionic-liquid materials for the electrochemical challenges of the future. *Nat. Mater.*, *8*, 8, 621–629, 2009.
108. Dou, Q., Lian, C., Lei, S., Chen, J., Liu, H., Yan, X., Silica-grafted ionic liquid for maximizing the operational voltage of electrical double-layer capacitors. *Energy Storage Mater.*, *18*, 253–259, 2019.
109. Wasserscheid, P. and Welton, T. (Eds.), *Ionic liquids in synthesis*, vol. 1, p. 145, Wiley-Vch, Weinheim, 2008.
110. Demarconnay, L., Calvo, E.G., Timperman, L., Anouti, M., Lemordant, D., Raymundo-Piñero, E., Arenillas, A., Menéndez, J.A., Béguin, F., Optimizing the performance of supercapacitors based on carbon electrodes and protic ionic liquids as electrolytes. *Electrochim. Acta*, 108, 361–368, 2013.
111. Timperman, L., Skowron, P., Boisset, A., Galiano, H., Lemordant, D., Frackowiak, E., Béguin, F., Anouti, M., Triethylammonium bis (tetrafluoromethylsulfonyl) amide protic ionic liquid as an electrolyte for electrical double-layer capacitors. *PCCP*, *14*, 22, 8199–8207, 2012.
112. Timperman, L., Béguin, F., Frackowiak, E., Anouti, M., Comparative study of two protic ionic liquids as electrolyte for electrical double-layer capacitors. *J. Electrochem. Soc.*, *161*, 3, A228, 2013.
113. Murugan, R., Thangadurai, V., Weppner, W., Fast lithium ion conduction in garnet-type Li7La3Zr2O12. *Angew. Chem. Int. Ed.*, *46*, 41, 7778–7781, 2007.
114. Chong, M.Y., Numan, A., Liew, C.W., Ng, H.M., Ramesh, K., Ramesh, S., Enhancing the performance of green solid-state electric double-layer capacitor incorporated with fumed silica nanoparticles. *J. Phys. Chem. Solids*, *117*, 194–203, 2018.
115. Ramzy, A. and Thangadurai, V., Tailor-made development of fast Li ion conducting garnet-like solid electrolytes. *ACS Appl. Mater. Interfaces*, *2*, 2, 385–390, 2010.
116. Francisco, B.E., Jones, C.M., Lee, S.H., Stoldt, C.R., Nanostructured all-solid-state supercapacitor based on Li2S-P2S5 glass-ceramic electrolyte. *Appl. Phys. Lett.*, *100*, 10, 103902, 2012.
117. Pal, B., Yasin, A., Kunwar, R., Yang, S., Yusoff, M.M., Jose, R., Polymer versus cation of gel polymer electrolytes in the charge storage of asymmetric supercapacitors. *Ind. Eng. Chem. Res.*, *58*, 2, 654–664, 2018.
118. Ramasamy, C., Palma del vel, J., Anderson, M., An activated carbon supercapacitor analysis by using a gel electrolyte of sodium salt-polyethylene oxide in an organic mixture solvent. *J. Solid State Electrochem.*, *18*, 2217–2223, 2014.
119. Chen, Q., Li, X., Zang, X., Cao, Y., He, Y., Li, P., Wang, K., Wei, J., Wu, D., Zhu, H., Effect of different gel electrolytes on graphene-based solid-state supercapacitors. *RSC Adv.*, *4*, 68, 36253–36256, 2014.
120. Lota, G. and Frackowiak, E., Striking capacitance of carbon/iodide interface. *Electrochem. Commun.*, *11*, 1, 87–90, 2009.

121. Lota, G., Fic, K., Frackowiak, E., Alkali metal iodide/carbon interface as a source of pseudocapacitance. *Electrochem. Commun.*, *13*, 1, 38–41, 2011.
122. Frackowiak, E., Fic, K., Meller, M., Lota, G., Electrochemistry serving people and nature: High-energy ecocapacitors based on redox-active electrolytes. *ChemSusChem*, *5*, 7, 1181–1185, 2012.
123. Lian, K. and Li, C.M., Asymmetrical electrochemical capacitors using heteropoly acid electrolytes. *Electrochem. Solid-State Lett.*, *12*, 1, A10, 2008.
124. Mao, B.S., Wen, Z., Bo, Z., Chang, J., Huang, X., Chen, J., Hierarchical nanohybrids with porous CNT-networks decorated crumpled graphene balls for supercapacitors. *ACS Appl. Mater. Interfaces*, *6*, 12, 9881–9889, 2014.
125. Pal, B., Yang, S., Ramesh, S., Thangadurai, V., Jose, R., Electrolyte selection for supercapacitive devices: a critical review. *Nanoscale Adv.*, *1*, 10, 3807–3835, 2019.
126. Largeot, C., Portet, C., Chmiola, J., Taberna, P.L., Gogotsi, Y., Simon, P., Relation between the ion size and pore size for an electric double-layer capacitor. *J. Am. Chem. Soc.*, *130*, 9, 2730–2731, 2008.
127. Centeno, T.A., Sereda, O., Stoeckli, F., Capacitance in carbon pores of 0.7 to 15 nm: a regular pattern. *PCCP*, *13*, 27, 12403–12406, 2011.
128. Miller, J.R. and Burke, A., Electrochemical capacitors: challenges and opportunities for real-world applications. *Electrochem. Soc. Interface*, *17*, 1, 53, 2008.
129. Wu, J., Understanding the electric double-layer structure, capacitance, and charging dynamics. *Chem. Rev.*, *122*, 12, 10821–10859, 2022.
130. Liu, K., Lian, C., Henderson, D., Wu, J., Impurity effects on ionic-liquid-based supercapacitors. *Mol. Phys.*, 1154, 454–464, 2016.

# 3

# Green Supercapacitors

**Priya R. and S. Sonia***

*Department of Physics, Holy Cross College (Autonomous), Nagercoil, Tamil Nadu, India*

***Abstract***

Recent research on green renewable materials for energy storage provides humanity with a dependable hope for a pollutant-free green world free of conventional fossil fuels. Because of their good properties such as rapid charge/discharge rate, environmentally safe, long life span, and low cost, supercapacitors are emerging as a viable option among existing energy storage devices. This overview focuses on the general information of supercapacitors, supercapacitor applications, the future of supercapacitors, the importance of supercapacitors in our lives, the role of metal–organic framework in supercapacitors, and the most recent trend in the field of supercapacitors, which shows a new perspective into all green supercapacitors charge storing devices.

***Keywords:*** Bio-sourced materials, pollutant free, porous structure, porous electrode, longer lifespan, redox process, charge adsorption, power source

## 3.1 Introduction

In recent decades, the usage of nonrenewable fossil fuels for energy has risen, putting human life in grave danger. Therefore, it is essential to create ecologically safe, dependable, and sustainable energy storage systems. New electrochemical sources that are efficient, effective, and sustainable have therefore garnered more interest. Supercapacitors are considered recently as intriguing electrochemical energy conversion and energy storing technologies, owing to their higher power density, outstanding cyclic

**Corresponding author*: sonia.s@holycrossngl.edu.in

Basheer Ahamed and Chaudhery Mustansar Hussain (eds.) Sustainable Supercapacitors: Next-Generation of Green Energy Storage Devices, (73–96) © 2025 Scrivener Publishing LLC

stability with rapid charge, and discharge operation. To address the energy dilemma, effective energy storage and conversion systems must be built.

Supercapacitors (SCs), fuel cells, and batteries are frequently used energy storage devices in industrial settings and daily life. SCs are also known as ultracapacitors (UCs) or electrochemical capacitors (ECs). It should be noted that supercapacitors are capable of rapid recharging with significant amounts of released power, overcoming the gap between capacitors and batteries, although having electrical storage that is many times more than that of batteries and fuel cells. This capability attracted the intense attention of scientists working in the energy storing field.

The creation of an effective and environmentally friendly energy storage technology is crucial to meeting the portable demand of all types of devices. Because of its higher power density and fast charge–discharge speed, supercapacitors (SCs), a form of energy storage device, have gained a lot of interest [1–6]. The material used for the electrodes has a significant impact on the performance of supercapacitors; thus, it is vital to innovate and develop new, superior electrode materials. Due to its extremely high porosity, tunable pore size distribution, excellent structural adaptability, etc., MOFs are becoming increasingly used as supercapacitor electrode materials. MOFs have sparked a lot of interest in recent years for developments in SCs because of their unique features and simple synthesis processes. Additionally, MOFs can be combined with a variety of other substances, including polyoxometalates, metal nanoparticles, metal oxides, and members of the carbon family. It can fix the issue of the low conductivity of the pure MOFs without harming their wonderful qualities of high surface area and pore structure.

## 3.2 History of Supercapacitors

Supercapacitors have a relatively short history, having been created in the middle of the nineteenth century. To begin, engineers at General Electric (GE) created capacitors between the early 1950s and 1957. Becker [7] invented the first supercapacitor that did not have a recognized double layer. In 1966, the SOHIO company patented another version of the SC as an electrolytic capacitor [8, 9]. In 1982, the Pinnacle Research Institute (PRI) developed the first SCs for military applications, known as PRI ultracapacitors. Charge and discharge currents increased as capacitance values increased, while equivalent series resistance (ESR) levels decreased. In 1992, Maxwell Laboratories took over this development and called them "Boost Caps" to emphasize their power uses [10]. Evans constructed a high voltage tantalum electrolytic capacitor in 1994 that has the mixed properties of electrolytic and electrochemical capacitors; nonetheless, it was only used in a few military applications [11, 12]. The FDK group business recently developed hybrid lithium-ion capacitors, which is the combination of an electrostatic carbon electrode with an electrochemical electrode to improve capacitance [13, 14]. Nowadays, researchers aim to increase the supercapacitors' characteristics, performance, and the production costs [15, 16].

## 3.3 Supercapacitors

The world's rising population and quicker economic growth necessitate the use of clean and renewable energy sources, as well as sophisticated energy storing technology such as batteries and supercapacitors. For the current advancement of electric vehicles, energy storing systems with high energy and power densities and long cycle lifetimes are required. Despite the battery's high energy density, its low power density and limited life span need research into other technologies such as supercapacitors. Because of its desirable qualities, such as high-power density, quick charge–discharge, a wider working temperature range, and improved cyclic life, supercapacitors have piqued the scientific community's interest in research.

The supercapacitor is a potentially valuable electrochemical energy storing device. It recently gained a lot of attention because of its excellent performance, simplicity, and stability. In comparison to batteries, it has a higher power density and cycle stability. The reason for this is mainly due to their charge-storing mechanism, wherein the ions are deposited on the electrode surface during charging and liberated during discharging. This mechanism comes from batteries, where continuous redox processes result in weak stability and weak power density. A supercapacitor works similarly to a capacitor, with two conductors separated by a dielectric substance. Capacitance is caused by charge separation at the conductor surface. The conductors in a supercapacitor have been replaced with a porous electrode, which provides efficient surface areas for ion adsorption. Furthermore, the gap between two opposite charges is in the nanoscale range, contributing to higher capacitance than conventional capacitors.

### 3.3.1 Mechanism

SCs are principally powered by two types of charge storing process: (i) charge adsorption and (ii) redox processes linked to chemical changes.

Carbon materials, such as graphene, activated carbon, and carbon nanotubes, frequently function as SC electrode materials in the first mechanism, producing high power densities and extended cycling lives but flat energy densities. The second way is commonly utilized by transition metal oxides, hydroxides, and conducting polymers, which have greater energy density but poor cycling stability due to microstructure deformation caused by continuous redox reactions [17–20].

As a result, developing novel SC electrode materials is critical, particularly in tackling the concerns of low energy density, weak cycling stability, and chasing higher capacitance.

MOFs (metal-organic frameworks) and materials derived from them such as carbon, metal oxides, sulfides, etc. exhibit diverse properties and applications due to their tunable porosity, and versatile chemical functionality.

### 3.3.2 Supercapacitor Specifications

1. Supercapacitors act as a link between regular capacitors and rechargeable batteries.
2. Supercapacitors are capable of storing an enormous amount of energy.
3. Supercapacitors have high capacitances of up to 2 kF.
4. A supercapacitor takes 1–10 s to charge.
5. These components are capable of storing electricity via electrostatic charge absorption/desorption.

### 3.3.3 Characteristics of a Supercapacitor

#### *3.3.3.1 Charging Time*

Supercapacitors have the same charge and discharge times as conventional capacitors. Because of its low internal resistance, it can support high charge and discharge currents. For example, a mobile phone battery can take many hours to fully charge, whereas supercapacitors can reach the same charge state in less than two minutes.

#### *3.3.3.2 Specific Performance of SCs*

The maximum power output divided by the device total mass is used to compare various technologies that utilize the specific power of a battery or supercapacitor. Supercapacitors have five to ten times the specific power of batteries. For example, a typical supercapacitor has a specific power of about ten kilowatts/kilogram, whereas lithium-ion batteries have a specific power of one to three kilowatts/kilogram.

#### 3.3.3.3 *Supercapacitor Life Cycle*

Batteries have a maximum life span of five hundred times; however, supercapacitors may be charged and discharged millions of times and have a nearly limitless life cycle. As a result, supercapacitors are particularly useful in applications that demand the storage and release of energy on a regular basis.

#### 3.3.3.4 *Safety of Supercapacitors*

Supercapacitor batteries are safer than conventional batteries when overused. Supercapacitors do not heat up as much as batteries do because of their low internal resistance, however batteries have been known to explode due to excessive heating when short-circuited.

## 3.4 Advantages of Supercapacitors

1. In comparison to a standard capacitor, supercapacitors have a high energy and power density.
2. The range of capacitance is high in the case of a supercapacitor.
3. Supercapacitors have the ability to charge quickly.
4. They have excellent low-temperature performance (-40°C to 70°C).
5. They have a longer service life and a longer lifespan (for approximately ten to fifteen years against five to ten years for lithium-ion batteries). They have an almost infinite cycle life and can be cycled millions of times.
6. They improve performance reliability.
7. They reduce the battery's size, weight, and, as a result, cost.
8. Supercapacitors are environmentally friendly.

## 3.5 Disadvantages of Supercapacitors

1. They are releasing themselves more frequently. This is significantly more than a battery.
2. The voltages in individual cells are low. As a result, series connections are required in order to attain higher voltages.

3. There is far less energy stored per unit weight than in an electrochemical battery. This translates to three to five Wh/kg for an ultracapacitor and thirty to forty watts/kg for a battery.
4. They have a low energy density as compared to batteries. which is equivalent to one-fifth to one-tenth of the energy stored in the battery.
5. It cannot be utilized in alternating current or other higher-frequency circuits.

## 3.6 Applications of Supercapacitors

1. Power source for laptops
2. Backup power systems in missiles
3. MP3 player
4. Supercapacitors can be used as an energy storage for a cordless electric tool.
5. Due to very fast charging and discharging, supercapacitors make a great tool for regenerative breaking.

Figure 3.1 showcases the innovative implementation of a supercapacitor hybrid bus, representing a cutting-edge advancement in transportation technology. This pioneering design integrates supercapacitors into the bus's power system, offering enhanced energy storage capabilities and promoting greater efficiency and sustainability in urban transit. Figure 3.2 illustrates

**Figure 3.1** Supercapacitor hybrid bus.

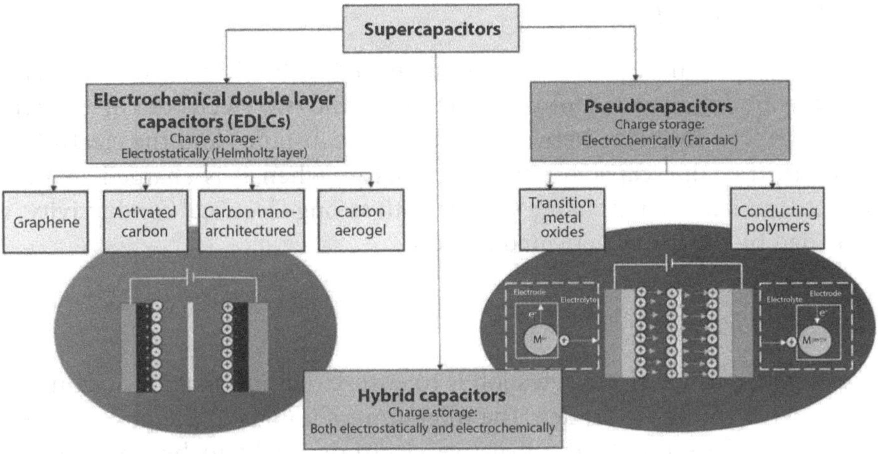


**Figure 3.2** Classification of Supercapacitors [21].

the comprehensive classification of supercapacitors, presenting a taxonomy that categorizes these energy storage devices based on various characteristics and operational principles. This classification scheme encompasses a diverse range of supercapacitor types, including but not limited to electrochemical double-layer capacitors (EDLCs), pseudo capacitors, and hybrid capacitors. Each category is delineated based on its underlying electrochemical mechanisms, material composition, structural design, and performance attributes. By systematically categorizing supercapacitors, this classification framework facilitates a deeper understanding of their fundamental principles and technological capabilities, thereby guiding research, development, and application efforts in the burgeoning field of energy storage and conversion.

## 3.7 Classification of Supercapacitors

The three types of supercapacitors are as follows:

1. Electrostatic double-layer capacitor
2. Pseudo-capacitors
3. Hybrid capacitors

### 3.7.1 Electrostatic Double-Layer Capacitors

The two electrodes of a supercapacitor are separated by a separator that is submerged in an electrolyte.

#### 3.7.1.1 Electrodes

A porous substance, usually carbon (activated carbon), is deposited over a thin metal film, usually aluminum, to create the two electrodes of a supercapacitor. The interface between the porous substance and the electrolyte is where the supercapacitor stores its charge when it is charged. A very large active surface for charge storage with strong electrical conductivity is provided by the use of activated carbon.

#### 3.7.1.2 Electrolyte

The supercapacitor electrolyte's job is to ensure that internal ions can move freely between the electrodes. The migration of anions to the positive electrode and of cations to the negative electrode ought to be unrestricted. Although they can be solid, electrolytes are usually liquid. Ionic conductivity and supercapacitor cell voltage are frequently compromised in the selection of the electrolyte, and the goal of maximizing ionic conductivity may result in a breakdown voltage of the electrolyte as low as 1 V. The operating voltage of the supercapacitor must be restricted (2.5 3V) since the redox reaction will cause irreversible reactions during the charge and discharge phases.

#### 3.7.1.3 Separator

The separator of a supercapacitor is often a piece of paper, which serves as insulation and prevents any conductive contact between the electrodes. It must, however, be able to dissolve in the electrolyte without impairing the electrolyte's ionic conductivity.

The EDLC system is frequently used as an energy storage method. When compared to traditional electrolytic capacitors, this device has a higher energy density and a high-power density when compared to re-chargeable battery. Because the electrodes or electrolyte contact electrostatically retains the charge without any redox reaction, the charging–discharging process in EDLC can occur quickly. Its enhanced energy-storage capacity is primarily attributable to the high conductivity of the electrolyte and the electrodes' wide surface areas. In the early stages of electrostatic double-layer capacitors' energy device research, liquid-based electrolytes were used because they could produce outstanding and realistic electrochemical properties. However, a number of issues have impeded the development of liquid electrolytes (LEs), such as volatility, leakage problems, electrode corrosion, toxicity, thermal instability, and flammability.

#### *3.7.1.4 Carbon Nanotubes*

Carbon nanotubes are frequently employed as a high-power electrode material due to their high electrical conductivity and wide surface area. Furthermore, due to their great mechanical robustness and open tubular network, they are an excellent foundation for active materials. Carbon nanotubes' discovery has substantially expanded carbon material research and engineering. The overall resistance of the components of a supercapacitor is the most important factor in determining power density. Carbon nanotubes have received a lot of attention for supercapacitor electrode applications due to their unique pore structure, exceptional electrical characteristics, and good mechanical and thermal durability. They are classified as either single-walled carbon nanotubes or multiwalled carbon nanotubes, and they both have received considerable attention as electrode materials for energy storage.

Other carbon structures researched for supercapacitor applications include activated carbon fibers, carbon aerogels, and carbon onions. The general principles for supercapacitor electrode material selection are a large specific surface area with strong electrical conductivity. ACFs have high surface areas, up to three thousand $m^2\ g^{-1}$, and a controlled pore size distribution. Carbon aerogels are yet another intriguing material that could be used as an electrode for supercapacitors. They are ultra-light, extremely porous materials, primarily with mesopores, and can be used without the use of binding chemicals. Despite the difficulties in preparation, a novel form of carbon nanotube aerogel electrode material has shown promising capacitive capabilities.

### 3.7.2 Pseudo-Capacitors

Pseudo-capacitors, otherwise known as faradaic supercapacitors, are devices that differ from EDLCs. This capacitor's electrodes contain redox-active materials that store electrical energy via a different method than EDLCs. In reality, the EDLC only accounts for a portion of the charge, but faradaic mechanisms such as redox reactions, electro sorption, and intercalation can account for a considerable portion of the energy transfer and storage. When an external voltage is supplied to this capacitor, fast and reversible redox processes on the electrode occur, including the passage of charges between electrode and electrolyte. The charge and discharge operation of this capacitor is similar to that of an electric battery.

Pseudo-capacitors are a hybrid of a battery and an electrostatic double-layer capacitor. An electrolyte separates the two electrodes in this capacitor. The most popular techniques of charge storage are chemical and electrostatic reactions. Transfer rates are superior when the charge transfer is comparable to those of a battery due to thinner redox material over the electrode, since fewer ions permeate from the electrolyte into the structure would otherwise occur. Because of the various methods that perform charge storage, the pseudo-capacitors capacitance values are higher.

#### 3.7.2.1 *Working Principle of a Pseudo-Capacitor*

A pseudo-capacitor accumulates electrical energy by transferring electron charge in between the electrode and an electrolyte reduction/oxidation reaction, electro sorption, and intercalation processes known as pseudo capacitance. A pseudo-capacitor is an integral component of an electrochemical capacitor that, along with an electrostatic double-layer capacitor, creates a supercapacitor. Metal sulfides, metal oxides, metal hydroxides, metal nitrides, and conducting polymers are common constituents of pseudocapacitive. Metal oxides such as ruthenium dioxide, nickel oxide, $MnO_2$, and conducting polymers such as polyaniline and polypyrrole are some of the examples of pseudo-capacitor materials.

Throughout the faradaic reactions, energy can be stored in pseudo-capacitors. Thus, they store charge electrostatically, where charge can be transferred between electrode and an electrolyte. When a voltage is supplied to the pseudo-capacitor, the electrode materials undergo both reduction and oxidation process. The faradic method used in these capacitors, as compared to electrostatic double-layer capacitors, has the ability to speed electrochemical processes, resulting in higher specific capacitance and energy densities. Pseudo-capacitor materials improve energy density, allowing for greater energy storage density in the volume of electrode materials at their surface. The main characteristics of these capacitor materials are that they are electrically conductive and have two oxidation states within a specific voltage window. When compared to conventional capacitors, these capacitors have the highest capacitance density due to their unique storing charge principles. Hence, the result is that the total amount of electric charge stored in a pseudo capacitance is proportional to the voltage supplied to them.

#### 3.7.2.2 *Classifications of Pseudo-Capacitors*

Pseudo-capacitors are subdivided into two categories, depending on the electrode materials used to store the charge within them.

- Metal oxide
- Conducting polymers

#### 3.7.2.2.1 Metal Oxide

Metal oxide is a type of pseudo capacitive substance that undergoes reversible and quick redox reactions on the electrode materials' surfaces. This electrode material is ideal for creating high-power supercapacitors because of their low resistance and high specific capacitance. Metal oxides that are often used as supercapacitor electrodes include $MnO_2$, $RuO_2$, NiO, $SnO_2$, $IrO_2$, $Fe_3O_4$, $V_2O_5$, and MoO. Because of its superior electrical conductivity and specific capacitance, Ruthenium (II) oxide is the one of the most suitable electrode materials among various metal oxide electrode materials in the application of supercapacitors. However, the planet's scarcity of "Ru" has limited its practical application. Given this fundamental difficulty, it is necessary to identify low-cost pseudo capacitance materials.

Metal oxides have varied oxidation states at different potentials and crystalline morphology that allow for maximum conductivity and charge distribution in their network. Metal oxides can change their oxidation state during oxidation and reduction surface processes, and protons can be introduced to and removed from the oxide lattice.

#### 3.7.2.2.2 Conducting Polymers

Conducting polymers are utilized in redox pseudo-capacitors due to their rapid and reversible oxidation or reduction processes, low cost, and high electrical conductivity. The most often used conducting polymers are polypyrrole, polyaniline, polythiophene, p-poly p-phenylene vinylene, and p-poly e-ethylene dioxythiophene. These materials are typically formed through electrochemical or chemical oxidation of the monomer and are rendered conductive through the use of a conjugated link system and the polymer backbone.

Conducting polymers provide capacitance, higher conductivity, and lower equal series resistance as compared to carbon-based electrode materials. Once the ions have been oxidized and reduced, they are simply monitored or floated from the electrolyte to the conducting polymers before being returned to the electrolyte. Because there is no phase transition, these polymers exhibit extremely reversible effects, resulting in greater cycling stability. As a result of redox activities that promote greater conductivity, they become positively or negatively charged.

### *3.7.2.3 Advantages of Pseudo-Capacitors*

a. The power density of these capacitors is higher.
b. They have a long life span.
c. Pseudo-capacitors can charge and discharge significantly more quickly than lithium-ion batteries.

d. The electrode materials used in pseudo-capacitors boost energy density and allow for energy density storage both within and on the surface of the electrode materials.

#### 3.7.2.4 *Disadvantages of Pseudo-Capacitors*

(a) Because of their poor energy density, these capacitors cannot be used to replace batteries in energy storage applications.
(b) As long-term energy storage solutions, they are ineffective.
(c) These capacitors' output voltage does not rise linearly with charge.

#### 3.7.2.5 *Applications of Pseudo-Capacitors*

(a) Pseudo-capacitors store electrical energy through a faradaic reaction.
(b) It is an electrochemical capacitor component that has the combination of an electrostatic double-layer capacitor with an electrochemical capacitor to generate a supercapacitor.
(c) These are used in consumer electronics.
(d) In wearable or flexible electronics
(e) Regenerative braking within automobile applications
(f) Kinetic energy (K.E) recovery systems like cranes, elevators, wind turbines, etc.

Thus, this is a brief overview of pseudo-capacitors and how they interact with applications. EDLC materials cannot improve energy density significantly; pseudo-capacitor materials, on the other hand, boost energy density and allow for energy storage density within and on the outermost layers of electrode materials. The major advantages of pseudo-capacitors include its low cost, lightweight, eco-friendliness, comfort, flexibility, increased safety, tuneable electrochemical characteristics, and so on.

### 3.7.3 Hybrid Capacitors

Hybrid capacitors are the result of combining an electrostatic double-layer capacitor and a pseudo-capacitor technique. Various electrodes with differing properties are utilized in these components. One electrode shows electrostatic capacitance while the other electrode shows electrochemical capacitance. Lithium-ion battery capacitors are an example of a hybrid capacitor.

## 3.8 Importance of Supercapacitors in Our Everyday Life

Supercapacitors are electronic devices that have an incredibly higher energy density when compared to regular capacitors. Because of their unlimited lifespan, they are most frequently utilized in backup power applications. These energy storage systems, according to some researchers, will be the most effective replacement for gas-powered cars. The endless energy lifespan of supercapacitors makes them valuable. Batteries and traditional power sources, which are necessary for our personal electronic devices and automobiles, do not have a lot of energy storage capacity. Supercapacitors are increasingly being used in mobile phones, palm pilots, and other portable electronic gadgets because of their great energy storage capacity.

## 3.9 The Future of Supercapacitors

Although there are now just a few uses for supercapacitors, future design improvements may increase their usefulness. For instance, scientists are still working on and experimenting with novel dielectric materials, such as barium titanate, polypyrrole, and carbon nanotubes, which may increase capacitance and energy density. Within the contemporary "green" movement, the idea of using supercapacitors in conjunction with alternative energy sources to replace automobile batteries has gained popularity. As a result, some public transportation agencies have established test programs for capacitor-powered buses and trains. The electric double-layer capacitor may become more functional and play a bigger role in the energy sector if these and other advances are successful.

## 3.10 Comparison of Supercapacitor versus Battery

Commonly used batteries with a certain volume and weight have a higher energy density. Supercapacitors are capacitors with a high capacity and power density. A supercapacitor, as opposed to a battery, has a higher charging–discharging capacity, the ability to withstand low–high temperatures, high reliability, and low impedance. The battery is inexpensive, but the supercapacitor is not. Self-discharge is a property of supercapacitors. The charging and discharging modes are determined by the battery's operational voltage. The dielectric material used between the plates determines the allowed voltage in a supercapacitor. Capacitance may be increased by the electrolyte in the capacitor.

There are lead-acid batteries, Ni-MH batteries, Li-Po batteries, lithium-ion batteries, LMP batteries, and more types of batteries available. Supercapacitors can be made of organic electrolytes, aqueous electrolytes, ionic liquids, hybrids, or pseudo supercapacitors. Supercapacitors give great power density while batteries store vast amounts of energy.

## 3.11 Role of Metal–Organic Framework in Supercapacitors

Metal–organic frameworks are organic, inorganic hybrid materials having crystalline pores composed of a predictable arrangement of positively charged metal ions ringed by organic "linker" molecules. The metal ions serve as nodes, linking the arms of the linkers to form a repeating, cage-like structure. Due to their hollow nature, metal–organic frameworks have a massive internal surface area. Researchers developed a metal–organic framework with a surface area per gram of around 7,800 $m^2$. To put this into perspective, one teaspoon of this substance, or around one gram of solid, has enough surface area to fill an entire soccer field.

Three categories can be used to classify the most recent MOF research for supercapacitor applications [22–25]:

1. Metal–organic framework physicochemical properties can be controlled by using them as building blocks or templates to develop metal–organic framework-based products such as porous carbon, metal oxides, metal hydroxides, metal sulfides, and a range of other derivatives. In this category, the synthetic process utilized to make precursor electrodes would significantly increase the shape, size, and chemical composition of the created electrode materials when compared to pristine metal–organic framework.
2. Mixing metal–organic framework with conductive polymers like polyaniline or polypyrrole or carbon substances like activated carbon, graphene, and carbon nanotubes to create nanocomposites. This strategy is essential for enhancing the MOF-derived electrodes' structural stability, conductivity, and cycling stability.
3. The third category entails creating innovative MOFs that are only employed as electrodes for supercapacitors. Even if they are brand new, stability and conductivity are crucial factors to take into account before using pristine MOFs as supercapacitor electrodes [26, 27].

Porous coordinated polymers, otherwise known as metal–organic frameworks (MOFs), are a new materials science technology that is crucial in energy storage, particularly supercapacitors. Because of their broad functionality, structural richness, and extensive redox chemistry between metal species and organic linkers, metal–organic framework-based supercapacitors perform well. As developing sophisticated functional materials, metal–organic frameworks and metal–organic framework-derived materials are identified as excellent candidate materials for electrochemical energy storage. They not only contain porosity to improve ion, electron transfer and to tolerate electrode volume changes during cycling, but they also have design qualities that allow for facile functionalization to produce the requisite electrochemical properties.

Recent breakthroughs in MOF-based derived materials have demonstrated outstanding efficiency in energy conversion and storage by overcoming many of the obstacles frequently encountered in the fabrication of supercapacitor electrodes. They have active charge storage sites, as well as a high surface area, porosity, varying redox activity, chemical and mechanical stability, and a variety of crystalline structures and scaffold topologies. Acid base and metal ligand reactivity can be used to tune MOFs, with the addition of an acidic ligand permitting pore size modulation. This improves the speed of electron and electrolytic ion transport in supercapacitors, surpassing the limitations of commercially available materials. A hierarchical porous nickel-based metal–organic framework, for example, built with nanosheets morphology, demonstrated exceptional specific capacitance of 1,057 at a current density of 1.0 Ag1 and delivered high specific capacitance of 649 at a current density of thirty Ag1, indicating that it has good rate capability of 63.4% even up to 30 Ag1. This is because the shorter route length improves charge transfer mobility, redox stability, and charge accumulation on the electrode's surface, resulting in high capacitance, power density, and electrode stability.

Because of its well-coordinated structure, excellent porosity, huge surface area, and ease of tunability and usefulness, metal–organic frameworks have lately emerged as promising electrodes for supercapacitors. Poor electronic conductivity and low stability, on the other hand, are important constraints that preclude them from being used to make commercial supercapacitors. Metal–organic frameworks are either used as sacrificial templates or composited with conductive substrates such as carbon-based materials and conductive polymers while keeping the structural integrity of the parent MOFs to overcome this issue. These composite electrodes' synergistic effects give novel physicochemical properties and improve electrochemical performance. In recent years, significant progress has been achieved in the use of metal–organic framework-derived electrodes for supercapacitors. Nonetheless, more progress needs be done, particularly in terms of cyclic stability and energy density [28].

## 3.12 Eco-Friendly Supercapacitors

As a result of decreasing fossil fuel resources and greater environmental awareness, there has been an increase in demand for green and renewable energy generation and storage technologies. Energy storage devices must consider other issues in the upcoming years, such as sustainability through the use of green materials and the possibility of providing both affordable materials used in batteries right now, which account for as much as 70% of total cell costs, and powerful devices to alleviate energy poverty or meet the increasing requirement for electric vehicles. Green synthesis is popular because it is a green technique that does not involve the use of hazardous chemicals during the experimental procedure. Because of their renewability and sustainability, bio-sourced materials have gained in appeal.

Because of their renewability and sustainability, bio-sourced materials have become more important in the production of energy devices such as batteries and solar cells. Many battery types, including lithium-ion and sodium-ion, have demonstrated promising performance when biosourced materials were used as electrolytes and electrode materials.

Environmentally friendly materials, such as biopolymers derived from chlorophyll, porous carbon based on chitosan, fallen leaves, apricot shell, or even used face masks made of polypropylene, have been examined for use in green supercapacitors. Gel electrolytes could increase specific capacitance and ionic transport capabilities while avoiding downsides like higher weight or electrolyte leakages. Some extensive research has been carried out and demonstrates the benefits of environmentally friendly gel electrolytes.

For example:

1. Canan Baslak and Serkan Deminal used the hydrothermal approach to create carbon quantum dot (CQD) particles using the S. vuralii plant. The results of the CV measurements showed that the rechargeable symmetrical capacitor behaviors of the CQD electrodes were 10.42 F/g charge and 8.26 F/g discharge capacitances. It is possible to enhance the green synthesized CQDs for supercapacitor applications [29].
2. Utilizing nanostructures, such as carbon nanotubes and graphene, makes it possible to boost supercapacitor capacitance by naturally increasing the surface area, which has a

significant effect on energy power density. The expansion of the hydrogel-based electrolyte's electrochemical window has a favorable impact on the power density as well. M. Pacheco and B. Lefort synthesized a green prototype with a cycling life of 2000 cycles, which is better than some supercapacitors based on hydrogel electrolyte, but is still far from an efficient system. It also has an excellent power density of 1779.41 W $kg^{-1}$, which is typical of supercapacitors, as well as a decent energy density of 105.37 W h $kg^{-1}$. Those findings need to be refined, but they promise to create a green supercapattery by fusing the greatest qualities of supercapacitors with rechargeable batteries [30].

3. Biomass-based materials have gained a lot of attention from academics working on energy storage in recent years due to their benefits such as being ecologically benign and having naturally porous properties. Because of its high carbon content and natural heteroatom N concentration, soybean straw, for example, is an underutilized agricultural and forestry waste. Agroforestry waste of soybean straw was successfully turned into biomass porous carbon using *in situ* doping and green activation. Carbon composite materials' unique porosity and surface topologies may significantly reduce capacity attenuation in supercapacitors at high current densities [31].
4. Due to the drawbacks of synthetic polymers, researchers are now concentrating on investigating biodegradable natural polymers for the development of SPE (solid polymer electrolytes). High expenses, toxic electronic waste, non-biodegradable nature, and unfavorable high crystallinity are a few of these drawbacks. Many natural biopolymers, including cellulose derivatives, gums, agar, carrageenan, pectin, gelatine, natural rubber, and the widely used biomaterials chitosan and starch, have so far been used in the manufacture of SPE. Glycerol (GCL) is a common plasticizer because of their low molecular weight, strong polarity, and high dielectric constant = 42.5. It is also appropriate for biopolymer electrolyte systems. Furthermore, glycerol increases the interfacial properties in-between the electrode and electrolyte, which are critical in energy storing device applications.
5. One-dimensional carbon nanofibers, synthesized using carbonization of electrospun polymeric sheets, are a suitable

option among the different carbon materials investigated in the supercapacitor's energy storage field [32–34].

6. Muthu Sankar Eswaran *et al.* highlighted a cost-effective, environmentally friendly method of producing a bio-supercapacitor material utilizing nanostructured hydroxyapatite (FNHAp). The FNHAp used is created from abundant fish scale trash obtained from Garra mullya fish using an easy alkaline heat treatment procedure. They have obtained the mechanical cyclic stability over 1000 cycles at 1 mA/g current density with 84% coulombic efficiency. The discovered biocompatible material has the potential to provide a green alternative to traditional supercapacitors [35].
7. To give good electrochemical properties for usage in supercapacitors, biowaste from cooking and oven-produced carbon nanoparticles are combined into polyaniline synthesized via *in situ* chemical oxidative polymerization. WCP polyaniline composite electrodes have showed excellent charge storage performance. Furthermore, the microelectrode's energy-power performance demonstrates a nearly 5500% increase in power density with only a 1% drop in energy density. These rationally synthesized WCP polyaniline composite electrodes based on biowaste carbon nanomaterials pave the way for the development of future-oriented environmental supercapacitors with improved energy storage efficiency [36].
8. Abrar Khan *et al.* created extremely porous carbon containing tube-like structures with waste from tea utilizing simple and clean precarbonization, acid washing, and potassium hydroxide activating procedures. The new biomass porous carbon derived from tea waste displayed good electrochemical behavior, making it a potential candidate as a cathode material in the development of highly efficient and environmentally friendly supercapacitors as sophisticated devices to store energy [37].
9. Rongke Sun and colleagues used long-fiber coconut silk with a vascular bundle-like structure as a precursor and varied the potassium hydroxide activation temperature to adjust the pore structure and electrical properties of the electrode material. The specific area of the surface of the activated carbon electrode material etched with potassium

hydroxide can reach 2794 $m^2$ $g^{-1}$ at 900°C. When combined with a polyvinyl alcohol/sulfuric acid gel electrolyte, the symmetric flexible supercapacitor's large specific surface area and proper distribution of pore size provide an abundance of active sites for electrolyte ion adsorption, resulting in a good electrode specific capacitance. It highlights the great benefits and potential of coconut silk-based biomass carbon materials in the fabrication of flexible supercapacitors, and it supports the development of energy storage units for future wearable electronic gadgets [38].

10. Peimin Zhan *et al.* synthesized a cement-based electrolyte mixed with recycled steel slag powder and waste glass powder as a promising candidate for manufacturing multifunctional structure supercapacitor cells (SSCs). The proposed SSCs, which contain a cement-based electrolyte with twenty percentage steel slag and ten percentage glass powder, are used for the construction of green building energy storage devices [39].
11. Van Dien Dang *et al.* created the NiNF@TBC flower-like nickel hydroxide, tea leaf-derived biochar composite for high-performance supercapacitor applications. After 10,000 charge–discharge cycles, a high long-term cyclic stability of 94% is achieved. This outstanding efficiency has demonstrated the practicality of employing waste from agriculture as green carbon sources, which may be coupled with various metal hydroxides in order to create hybrid nanomaterials as a highly intriguing electrode material for environmentally friendly supercapacitor applications [40].
12. Boris *et al.* look into concepts for a fully functional supercapacitor built of materials that are entirely secure to dispose of as well as easy to incinerate. Material selections for the current collector, electrolyte, separator, particle binder, and packaging are all totally appropriate, with the majority of them being commonly used materials. As a current collector, they chose graphite foil, a solution of sodium acetate as an electrolyte, an ester as an acetate cellulose porous membrane, and polymers made from polyvinyl alcohol as environmentally acceptable device components. The materials presented are all obtained from straightforward and cheap source chemicals, which decreases their impact on

the environment while rendering them more practicable for integrating into industrial devices for large-scale stationary and transport energy storage uses [41].

13. Porous carbons generated from waste biomass are well-known active materials for the development of high-performance, low-cost, and environmentally acceptable electrochemical double-layer supercapacitors. However, efficient microstructure management of biomass activated carbons to obtain high capacitor performance remains a difficulty. Manman Xu *et al.* used an agricultural waste, biomass corncobs, which were simply cut and shown to be efficiently activated by soaking them in a simple solution of potassium hydroxide at room temperature prior to a one-step carbonization process. The inherent hierarchical architecture of the corncob was fully used and partially conserved by this green and energy-efficient process, and the produced porous carbon has a high degree of graphitization and effective specific surface area with increased electrochemical performance. These porous carbons are envisioned as appropriate active materials for high-performance supercapacitors [42].

Figure 3.3 showcases the utilization of agricultural corncob waste as electrode materials in supercapacitors, exemplifying a sustainable approach to energy storage technology. This innovative application repurposes a readily available biomass resource, corncobs, to fabricate electrodes for supercapacitors, thereby reducing reliance on traditional, non-renewable materials. By harnessing the unique properties of corncob-derived materials, such as high surface area and carbon content, the supercapacitor electrodes exhibit

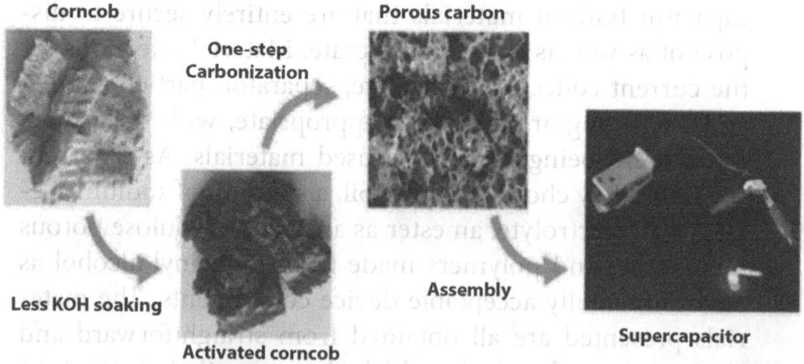


**Figure 3.3** Supercapacitor electrode materials from agricultural corncob waste [42].

enhanced electrochemical performance and stability. This eco-friendly strategy not only mitigates environmental impact by diverting agricultural waste from landfills but also contributes to the development of cost-effective and sustainable energy storage solutions. Through the integration of agricultural by-products into supercapacitor electrode materials, this research underscores the potential for synergy between renewable resources and advanced technology to address global energy challenges.

## 3.13 Conclusions

Because of their remarkable performance in power systems, supercapacitors have become a research hotspot. Recent studies have confirmed the exceptional performance of supercapacitors employing green materials such as biopolymers, porous carbon, and so on. These findings pave the way for a greener synthesis of high-quality supercapacitors that use environmentally friendly solvents rather than harmful compounds to improve supercapacitor performance.

## References


1. Dang, T., Zhang, G., Li, Q., Cao, Z., Zhang, G., Duan, H., Ultrathin hetero-nanosheets assembled hollow Ni-Co-P/C for hybrid supercapacitors with enhanced rate capability and cyclic stability. *J. Colloid Interface Sci.*, 577, 368–378, 2020.
2. Yin, X., Li, H., Han, L., Yuan, R., Lu, J., NiCo2O4 nanosheets sheathed SiC@CNTs core-shell nanowires for high-performance flexible hybrid supercapacitors. *J. Colloid Interface Sci.*, 577, 481–493, 2020.
3. Wang, J.G., Ren, L., Hou, Z., Shao, M., Flexible reduced graphene oxide/prussian blue films for hybrid supercapacitors. *Chem. Eng. J.*, 397, 125521, 2020.
4. Wang, P., Wang, R., Lang, J., Zhang, X., Chen, Z., Yan, X., Porous niobium nitride as a capacitive anode material for advanced Li-ion hybrid capacitors with superior cycling stability. *J. Mater. Chem. A*, 4, 9760–9766, 2016.
5. Lim, E., Kim, H., Jo, C., Chun, J., Ku, K., Kim, S., Lee, H.I., Nam, I.S., Yoon, S., Kang, K. *et al.*, Advanced hybrid supercapacitor based on a mesoporous niobium pentoxide/carbon as high-performance anode. *ACS Nano*, 8, 8968–8978, 2014.
6. Yan, Y., Li, A., Lu, C., Zhai, T., Lu, S., Li, W., Zhou, W., Double-layered yolk-shell microspheres with NiCo2S4 -Ni9S8 -C hetero-interfaces as advanced

battery-type electrode for hybrid supercapacitors. *Chem. Eng. J.*, 396, 125316, 2020.
7. Ho, J., Jow, R., Boggs, S., Historical Introduction to Capacitor Technologies. *IEEE Electra. Insul. Mag.*, 26, 20–25, 2010.
8. Schindall, J.G., The Change of the Ultra-Capacitors. *IEEE Spectra*, 44, 42–46, 2007.
9. D.L. Boos, Electrolytic Capacitor Having Carbon Paste Electrodes. U.S. Patent 3536963, 27 January 1970.
10. Namisnyk, A.M., *A Survey of Electrochemical Supercapacitor Technology; Technical Report*, University of Canterbury, Christchurch, New Zealand, 2003.
11. D.A. Evans, Containers with Anodes and Cathodes with Electrolytes. U.S. Patent 5369547, 29 November 1994.
12. Evans, D.A., *The Littlest Big Capacitor—An Evans Hybrid Technical Paper*, Evans Capacitor Company, East Providence, RI, USA, 2007.
13. FDK, FDK Corporate Information, *FDK History 2000s*, MDPI, Basel, Switzerland, 2023. Available online: (accessed on 29 October 2021).
14. Muzaffar, A., Ahamed, M.B., Deshmukh, K., Thirumalai, J., A review on recent advances in hybrid supercapacitors: Design, fabrication and applications. *Renew. Sustain. Energy Rev.*, 101, 123–145, 2019.
15. Naoi, K. and Simon, P., New Materials and New Configurations for Advanced Electrochemical Capacitors. *Electrochem. Soc. Interface*, 17, 34–37, 2008.
16. Zhao, J. and Burke, A., Review on supercapacitors : Technologies and performance evaluation. *J. Energy Chem.*, 59, 276–291, 2021.
17. Xu, S., Su, C., Wang, T., Ma, Y., Hu, J., Hu, J., Hu, N., Su, Y., Zhang, Y., Yang, Z., One-step electrodeposition of nickel cobalt sulfide nanosheets on Ni nanowire film for hybrid supercapacitor. *Electrochim. Acta*, 259, 617–625, 2018.
18. Xu, S., Li, X., Yang, Z., Wang, T., Xu, M., Zhang, L., Yang, C., Hu, N., He, D., Zhang, Y., A novel Ni@Ni (OH)2 coaxial core-sheath nanowire membrane for electrochemical energy storage electrodes with high volumetric capacity and excellent rate capability. *Electrochim. Acta*, 182, 464–473, 2015.
19. Xu, S., Li, X., Yang, Z., Wang, T., Jiang, W., Yang, C., Wang, S., Hu, N., Wei, H., Zhang, Y., Nano foaming to Boost the Electrochemical Performance of Ni@Ni (OH)2 Nanowires for Ultrahigh Volumetric Supercapacitors. *ACS Appl. Mater. Interfaces*, 8, 27868–27876, 2016.
20. Hong, M., Zhou, C., Xu, S., Ye, X., Yang, Z., Zhang, L., Zhou, Z., Hu, N., Zhang, Y., Bi-metal organic framework nanosheets assembled on nickel wire films for volumetric-energy-dense supercapacitors. *J. Power Sources*, 423, 80–89, 2019.
21. Abdaha, M.A.A.M., Azman, N.H.N., Kulandaivalu, S., Sulaiman, Y., Review of the use of transition-metal-oxide and conducting polymer-based fibres for high-performance supercapacitors. *Mater. Sci.*, 186, 108199, 2020.

22. Mohanty, A., Jaihindh, D., Fu, Y.P., Senanayak, S.P., Mende, L.S., Ramadoss, A., An extensive review of three-dimension architectural metal-organic frameworks towards supercapacitor application. *J. Power Sources*, 488, 229444, 2021.
23. Wang, L., Han, Y., Feng, X., Zhou, J., Qi, P., Wang, B., Metal–organic frameworks for energy storage: Batteries and supercapacitors. *Coord. Chem. Rev.*, 307, 361–381, 2016.
24. Wang, K., Bi, R., Huang, M., Lv, B., Wang, H., Li, C., Wu, H., Zhang, Q., Porous cobalt metal–organic frameworks as active elements in battery–supercapacitor hybrid devices. *Inorg. Chem.*, 59, 10, 6808–6814, 2020.
25. Zhao, X., Gong, L., Wang, C., Wang, C., Yu, K., Zhou, B., A facile grinding method for the synthesis of 3D Ag metal–organic frameworks (MOFs) containing Ag6Mo7O24 for high-performance supercapacitors. *Chem. Eur. J.*, 26, 20, 4613–4619, 2020.
26. Liu, Y., Xu, N., Chen, W., Wang, X., Sun, C., Su, Z., Supercapacitor with high cycling stability through electrochemical deposition of metal–organic frameworks/polypyrrole positive electrode. *Dalton Trans.*, 47, 38, 13472–13478, 2018.
27. Gu, M., Wu, M., Wang, S.C., Chen, C., Xiong, D., Yi, F.Y., Morphology control of nanoscale metal-organic frameworks for high-performance supercapacitors. *Electrochim. Acta*, 343, 135617, 2020.
28. Opeyam, O.K., *Rationaldesign of metal organic framework (MOF)-based Supercapacitor electrodesfor enhanced energy storage*, MDPI, Basel, Switzerland, 2021.
29. Başlak, C., Öztürk, G., Demirel, S., Kocyigit, A., Doğu, S., Yıldırım, M., Green synthesis of carbon quantum dots from *Sideritis vuralii* and its application in supercapacitors. *Inorg. Chem. Commun.*, 153, 2, 110845.
30. Pacheco, M., Lefort, B., Pacheco, J., Santana, A., Valdivi, R., *Hydrogel and carbon nanostructures-based supercapacitor: Initial steps toward green supercapatteries*, Elsevier, Dutch, 2023, Corpus ID: 260819463.
31. Jiao, S., Guo, L., Yao, Y., Zhang, J., Gao, M., Zhao, X., He, D., Chen, H., Jiang, J., Controlled porous structures of soybean straw-based carbon via green liquefication and N/S *in-situ* doping facilitate supercapacitor performance at large current density. *Ind. Crops Prod.*, 204, 117338, 2023.
32. Dahal, B., Mukhiya, T., Ojha, G., Muthurasu, A., Chae, S.H., Kim, T., Kang, D., Kim, H.Y., In-built fabrication of MOF assimilated B/N co-doped 3D porous carbon nanofiber network as a binder-free electrode for supercapacitors. *Electrochim. Acta*, 301, 209–219, 2019.
33. Tian, D., Lu, X., Zhu, Y., Li, M., Wang, C., Fabrication of two-dimensional metal-organic frameworks on electro spun nanofibers and their derived metal doped carbon nanofibers for an advanced asymmetric supercapacitor with a high energy density. *J. Power Sources*, 413, 50–58.34, 2019.

34. Mi, K., Song, L., Nie, H., Liu, T., Li, X., Layer-by-layer growth of ZIF-8 on electro spun carbon nanofiber membranes for high-performance supercapacitor electrode. *J. Energy Chem.*, 47, 221–224, 2020.
35. Eswaran, M., Sathiskumar, S., Chokkiah, B., Ragupathy, D., Bharathkumar, S., VinothKumar, P., A green and economical approach to derive nanostructured hydroxyapatite from Garra mullya fish scale waste for biocompatible energy storage applications. *Mater. Lett.*, Elsevier, 130341, 302, 2021.
36. Goswami, S., Dillip, G.R., Nandy, S., Banerjee, A.N., Pimentel, A., Joo, S.W., Martins, R., Fortunato, E., Biowaste-derived carbon black applied to polyaniline-based high-performance supercapacitor microelectrodes: Sustainable materials for renewable energy applications. *Electrochim. Acta*, 316, 202–218, 2019.
37. Khan, A., Senthil, R.A., Pan, J., Osman, S., Sun, Y., Shu, X., A new biomass derived rod-like porous carbon from tea-waste as inexpensive and sustainable energy material for advanced supercapacitor application. *Electrochim. Acta*, 335, 135588, 2020.
38. Sun, R., Zhang, X., Chan, Z., Ma, Y., Ma, L., Application of long fibrous coconut silk- supercapacitor. *J. Energy Storage*, 66, 1107410, 2023.
39. Zhan, P., Xu, J., Wang, J., Zuo, J., He, Z., Structural supercapacitor electrolytes based on cementitious composites containing recycled steel slag and waste glass powders. *Cem. Concr. Compos.*, 137, 104924, 2023.
40. Khedekar, A.P., Dang, V.D., Pandit, B., Bui, T.A.N., Tran, H.L., Doong, R., Flower-like nickel hydroxide@tea leaf-derived biochar composite for high-performanc,e supercapacitor application. *J. Colloid Interface Sci.*, 623, 845–855, 2022.
41. Dyatkin, B., Presser, V., Heon, M., Lukatskaya, M.R., Beidaghi, M., Development of a Green Supercapacitor Composed Entirely of Environmentally Friendly Materials. *Chemistry and Sustainability, Energy & Materials*, 6, 12, 2269–2280, 2013.
42. Xu, A., Huang, Q., Lu, J., Green synthesis of high-performance supercapacitor electrode materials from agricultural corncob waste by mild potassium hydroxide soaking and a one-step carbonization. *Ind. Crops Prod.*, 61, 113215, 2021.

# 4

# Materials for Sustainable Supercapacitors

**Arunima Verma[1], Vandana[2] and Tanuj Kumar[1]***

*[1]Department of Nanoscience and Materials, Central University of Jammu, Jammu, India*
*[2]Department of Physics and Astronomical Sciences, Central University of Jammu, Jammu, India*

***Abstract***
The use of sustainable materials in the development of supercapacitors—a vital part of the advancement of energy storage technologies—is examined in this research. Supercapacitors are becoming a more important component of renewable energy applications because of their high power density and quick charge times. We dive into the traditional and environmentally friendly materials used in the production of supercapacitors, and we explore the fundamental principles of their operation. Innovations in electrode materials, including metal oxides, conductive polymers, and carbon-based materials, as well as novel electrolytes, such as gel and ionic liquids, are given particular focus. The performance of supercapacitors is evaluated in terms of energy density, power density, and cycle stability, and the research highlights the significance of including sustainable materials into their design and construction. The importance that these sustainable supercapacitors will play in promoting a more sustainable future in energy storage is highlighted by the discussion of its possible uses in electric vehicles, portable devices, and renewable energy systems.

***Keywords:*** Supercapacitors, sustainable, application, technology, capacitor

## 4.1 Introduction to Supercapacitors and Sustainability

An increasing global demand for energy has intensified our reliance on carbon-based fuels. Consequently, experts have begun exploring

**Corresponding author*: tanuj.nsm@cujammu.ac.in

Basheer Ahamed and Chaudhery Mustansar Hussain (eds.) Sustainable Supercapacitors: Next-Generation of Green Energy Storage Devices, (97–122) © 2025 Scrivener Publishing LLC

alternative, safer energy sources. While renewable energy sources offer a promising solution, they face challenges such as lower efficiency, intermittent availability, and high initial setup costs. Because of these limitations, researchers have been working hard to come up with effective and environmentally friendly ways to store energy that can make the energy system stronger [1, 2]. Using electrode materials made using environmentally friendly methods important to solve the problems that are currently being faced [3]. A lot of different fields need green supercapacitors. These are the business segments: biomedicine, energy, computing, aircraft, packaging, and environmental protection. Scientists are interested in supercapacitors because they can be used in "situations where batteries are being overused" and also when a lot of power (more than 500 W/kg) is needed [4]. While supercapacitors use less energy, they are more expensive and do not work as well as other options. Once you choose electrode materials, it is important to think about both their electrical performance and their "*sustainability performance.*" The focus of this review will be on greenness in the setting of green supercapacitors. It is explained in detail how the understanding and knowledge base that are needed to make this greenness acceptance happen in new applications can be developed.

### 4.1.1 Overview of Supercapacitor Technology

Electrochemical energy storage devices called "*supercapacitors*" can be charged or discharged by connecting or eliminating a DC power source. Their use is essential in many business settings. Ultracapacitors, electrochemical double-layer capacitors (EDLCs), and supercapacitors are some of the names for these batteries. High-capacitance supercapacitors have values from a few to tens of thousands of Farads. Supercapacitors have a higher power density, a potentially infinite cycle life, and a higher charge/discharge rate than other battery technologies. Unfortunately, supercapacitors do not store energy as efficiently as batteries do. Combining batteries and supercapacitors makes for a powerful power source. Therefore, the battery's performance and cycle life can be improved since the supercapacitor can be charged at very low current from the battery. Charged supercapacitors can give electrical engines a very high current [5]. The Standard Oil Company in Cleveland, Ohio (SOHIO), USA, created the first cell in the 1960s. It has two layers of activated charcoal divided by a thin, porous barrier. The idea behind today's double-layer capacitors came from this technology, but because Standard Company was having trouble getting their idea sold, they gave it to a different company called NEC in 1971. NEC called their product a "supercapacitor" and was able to sell it

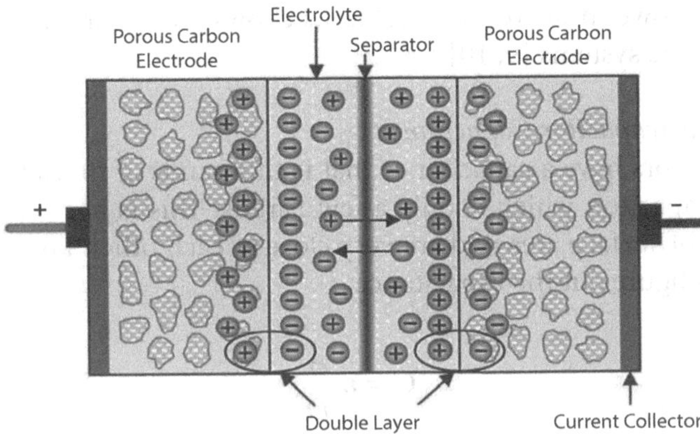


**Figure 4.1** Schematic diagrams of a supercapacitor [7].

as a backup power source for computers. For example, Panasonic made the Gold Capacitor in 1978, and ELNA made the Dynacap in 1987. All three of those types could be used in low-power situations. The first high power EDLCs used in military uses were made by the Pinnacle Research Institute (PRI) in 1982. Maxwell Company made EDLCs, which are also called Ultracapacitors, so that they can be used in electric cars [6].

There are two main types of supercapacitors: dual-layer capacitors and pseudo capacitors. These are based on how the charge builds up. The dual-layer capacitor is a capacitive-type (non-Faradic) supercapacitor, which means that when it is filled, each electrode will have a mix of positive and negative charges. The solution separates the combos, as shown in Figure 4.1 [7]. According to Faraday's laws, chemical processes happen in the active materials of the poles in a pseudo capacitor. Therefore, the type of active materials and the voltage of the wires affect the storage of charge [8].

#### 4.1.1.1 *Characteristics of Supercapacitor Technology*

Capacitance, regulated voltage, power and energy levels, and finally how it charges and discharges over time are some of the things that matter.

➢ *Supercapacitor Power and Energy Densities*

Batteries, regular capacitors, and supercapacitors all have characteristics in common, such as their energy and power levels. Although batteries can store a lot of energy, they do not deliver it very quickly. It is possible for a regular capacitor to give a lot of power, but it is not very good at keeping energy. Unlike batteries and regular capacitors, supercapacitors store energy and power more densely. Table 4.1 shows the features of the various energy storage systems [9, 10].

➤ *Supercapacitor Capacitance*

Supercapacitors have a capacitance that is thousands of times higher than a regular capacitor, which is only a few Farads. This is because the wires cover a lot of surface area and are very close to each other. From Equation 4.1, we can figure out the capacitance

$$C = \varepsilon \frac{A}{D} \tag{4.1}$$

where "ε" is the dielectric constant, "*A*" is the geometric area, and "*d*" is the distance between the formed layers.

**Table 4.1** Details on lead acid batteries, capacitors, and supercapacitors [9, 10].

| S. no. | Parameter | Battery | Supercapacitor | Capacitor |
|---|---|---|---|---|
| 1. | Charge/discharge efficiency | 50–90% | 90% | 99% |
| 2. | Specific energy (Wh/kg) | 30–280 | 0.5–0.15 | 0.01–0.3 |
| 3. | Power density (W/kg) | 100–200 | $10^6$ | $10^4$ |
| 4. | Energy density (kJ/kg) | 100–360 | 0.360 | 30 |
| 5. | Cycle life | 200–3000 | $10^6$ | Must-have, limitless, and enduring for decades |
| 6. | Life expectancy (years) | 5 | 30 | 30 |
| 7. | Specific power (W/kg) | 10–3000 | 20–$10^5$ | $10^1$–$10^5$ |

If Equation 4.1 is true and the size is thousands of square meters and the distance is nanometers, then the equation says that the capacitance will be thousands of Farads. For example, Equation 4.2 can be used to find the capacitance, which is the ratio of the stored charge to the applied voltage.

$$P = \frac{Q}{V} \tag{4.2}$$

➢ *Charging/Discharging Supercapacitor*

The way a supercapacitor charges and discharges is the same as a regular capacitor but different from how a battery does it. It charges quickly and enormously, and it drains in the same way, slowly, over a short period of time [11]. RC circuit is the only thing that limits how long it can charge. The charge time is determined by the value of each RC circuit in the supercapacitor, which is made up of many of them. Also, unlike batteries, the cycle of charging and releasing is thought to be endless because there is no chemical reaction happening during the process. The information in Table 4.1 shows how these types of storage devices vary in terms of how long they take to charge and discharge, how long their cycles last, and how well they charge and discharge.

### *4.1.1.2 Application of Supercapacitor*

Power systems, standby power systems, and automobile systems (both electric and fossil fuel) are only a few of the many industrial applications of supercapacitors. Digital cameras and telephones also make use of it. A basic explanation of the supercapacitor and its uses is as follows:

1. ***Cold starting of diesel fuelled engine:*** With the lubricating fluid temperature dropping below 0°C, increasing the viscous friction and necessitating considerable cranking torque, diesel-fueled engines are notoriously difficult to start in cold weather. Due to its internal resistance increasing with decreasing temperature, the lead acid battery's capacity to discharge current and maintain the cranking current during engine start is reduced in cold weather, which in turn impacts the cranking torque [12]. Because they are unaffected by cold temperatures, banks of supercapacitors can be used to generate the cranking current spike, thereby resolving this issue.

2. ***Hybrid electric vehicle:*** High energy density and power density storage devices are necessary for the functioning of hybrid electric vehicles because of the transient power demand that occurs during acceleration and deceleration. The battery life of a hybrid electric vehicle is compromised due to the bank of batteries' low power density and high energy density; however, by integrating a supercapacitor with the battery, the supercapacitor can supply the transient current, thereby extending the battery life.
3. ***Transmission lines***: A storage device becomes essential when there is a change in power consumption. The supercapacitor is a key component of FACTs devices, which are storage devices that increase power transmission, power quality, and frequency and voltage stability.

### 4.1.2 Importance of Sustainability in Energy Storage

Our ever-increasing need for energy may be attributed to two main factors: the ever-increasing human population and the ever-increasing quality of life throughout the globe. Fossil fuel combustion has been the primary means of meeting this need up to this point. Concerns about climate change, energy insecurity, and its uneven and finite supplies have turned the focus to renewable energy and other alternative energy systems. Renewable energy systems provide great benefits in terms of energy security and environmental impact, but they cannot be relied on as a constant source of energy. Therefore, ecologically friendly, secure, effective, and dependable methods of storing this energy are necessary for its long-term sustainability [13]. Smart energy systems for the future have been the subject of several academic investigations, and it is generally acknowledged that such systems need a range of energy storage solutions tailored to certain end-use needs. Precisely adjusting supply to suit shifting demand is the problem of traditional energy-producing and distribution systems. Common alternatives to electricity for the surplus include potential, kinetic, and chemical energy.

Electronic, chemical, mechanical, electrochemical, thermal, and a host of other energy storage technologies are detailed in the literature. Important considerations for developing (or selecting) suitable energy storage systems are shown in Figure 4.2. Some other critical factors are reaction time, black start capability, adaptable capacity, congestion control, and power quality, in addition to the requirements shown in Figure 4.2. To illustrate the point, flexible capacity may take a back seat to emergency response and

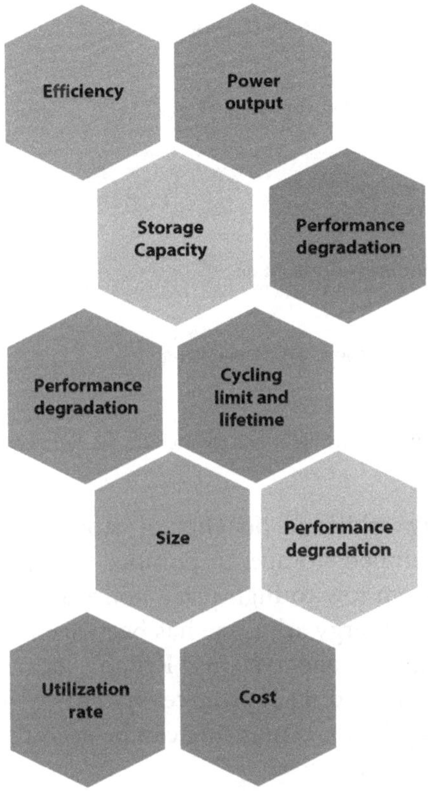


**Figure 4.2** Criteria to evaluate the performance of an energy storage system.

congestion management in times of crisis (such as earthquakes or epidemics). Choosing the right energy storage technology involves several unique considerations, as shown in Figure 4.3. Variations in demand provide a problem for energy infrastructure. Energy providers are compelled to discover supplementary energy sources in order to meet the consistent and unwavering demands of their end consumers, even during peak demand. It may be rather expensive to fulfill peak demand, which occurs relatively infrequently (just a few hours a year, for example), by adding power producing capacity. In these situations, energy storage may help with peak demand without putting additional strain on generating by providing variable capacity [14]. One benefit of energy storage devices is that they can handle peak demand. The storage system instantly fills the gap between supply and demand, and the generation side may operate in a continuous state of optimal performance because of its adjustable capacity.

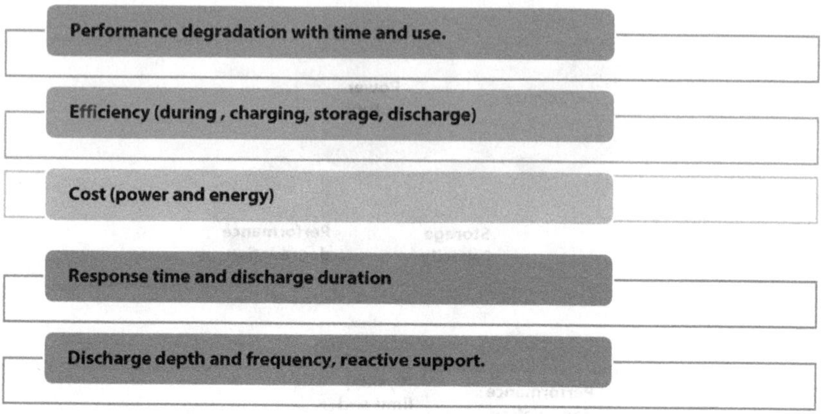


**Figure 4.3** Important unique considerations for picking the right energy storage technology.

Currently, energy arbitrage benefits of storage systems are mostly accomplished via pumped storage. In pumped storage, bulk quantities of energy are shifted from low to high price hours and vice versa. Together with flexible capacity, energy arbitrage has been seen as the key economic advantages of energy storage systems. Reliable operation of the entire energy supply chain (including 3S: source–system–service) requires effective system balancing. System balancing can be accomplished via a variety of ancillary services to ensure:

- Effective energy transmission from source to service (load)
- Reliable and safe operation
- Uninterrupted service to end users (customers)
- Balanced supply and demand
- Adequate contingency (or emergency) reserves
- Stability of the entire network

It is crucial to have a stable energy supply–demand chain, which entails being able to swiftly return to optimal functioning following a disturbance (such an outage) without causing any more harm to the system. When traditional power plants are sluggish to react to disruptions, energy storage solutions might step in and keep the lights on. This is of the utmost importance in remote areas where big, traditional facilities may take too long to respond to disruptions and get back to normal operations. As a result, the whole energy supply and demand networks cannot function without energy storage technologies. When recovering from interruptions or disturbances, the ability to "black start" is crucial for the system's stability. When a power source or generator has black start capability, it may begin operating immediately without the need for a constant source of electricity. The majority of energy producers nowadays require fixed electricity in order to begin operation. However, there is a way to reliably and effectively guarantee the dark start capability using energy storage solutions.

## 4.2 Fundamentals of Supercapacitors

### 4.2.1 Basic Principles of Supercapacitor Operation

The energy storage concept classifies superconductors into three broad groups. Ions (charged complexes) in a liquid electrolyte allow charge transmission in supercapacitors, which classify them as wet electrolytic capacitors. The so-called EDLC supercapacitors, which are electric double-layer capacitors, are the first to be considered. As far as commercially available supercapacitors go, this is by far the most prevalent kind. Electrolytes in liquid form are utilized by EDLC supercapacitors; ethylene carbonate (PC), propylene carbonate (DEC), dimethyl carbonate (DME), and ethylene carbonate (EC) are aprotic solvents used by the majority of these electrolytes. These solvents contain dissolved ions such as tetraethylammonium tetrafluoroborate ($TEABF_4$) and lithium hexafluoroarsenate ($LiAsF_6$). Some examples of ionic liquids used as electrolytes include triethylsulfonium bis(trifluoromethylsulfonyl) imide, 1-ethyl-3-methylimdazolium thiocyanate, and many more. To store energy, these EDLC supercapacitors form Helmholtz double layers at the interface of the electrolyte and electrode surfaces, to electrostatic contact. Surface energy held electrostatically at the interface of capacitor electrodes causes double-layer capacitance, which is dependent on voltage [15]. The energy is held non-radically, and there is no redox reaction or electron exchange in these supercapacitors. An exceptionally high capacity can only be achieved by utilizing electrodes with a large surface area and a thick Helmholtz layer. After millions of cycles, EDLC supercapacitors still show excellent durability. For EDLC supercapacitors, activated carbon (AC) is a popular electrode material because of its large specific surface area. A model of Helmholtz double layer is shown in Figure 4.4 [16].

The second group includes what are sometimes referred to as faradaic supercapacitors or pseudo-supercapacitors. Compared to EDLC supercapacitors, this sort of capacitor is far less common and is only sold by a small

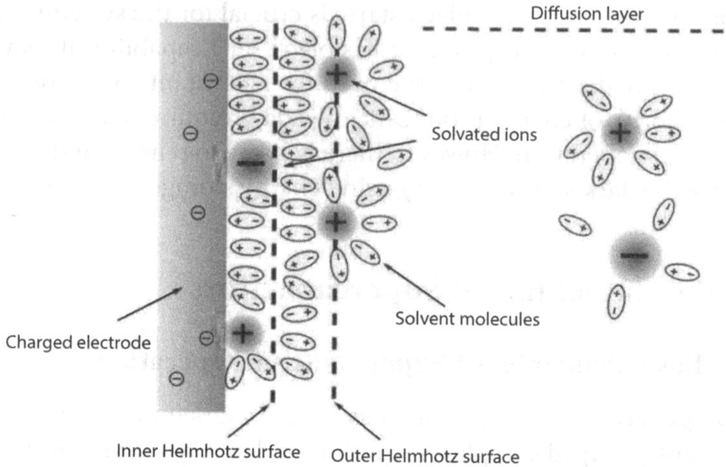


**Figure 4.4** Diagram of the Helmholtz double layer on a liquid–solid (electrolyte–electrode) interface [16].

number of manufacturers [17]. They work more like batteries than capacitors, at least in theory. When materials used as electrodes undergo redox reactions and intermediary electron transfer, a phenomenon known as pseudocapacitance occurs. Pseudocapacitance results from reactions that start at the electrode surfaces and involve the transfer of energy across the double layer, much like when a battery is charged or discharged. However, capacitance increases because of a specific relationship between the amount of energy accepted ($\Delta q$) and the change in potential ($\Delta V$), such that the derivative of $d(\Delta q)/d(\Delta V)$ or $dq/dV$ is the same as capacitance $C$. A redox reaction (resulting from the bonds in the compounds) and the transfer of energy between the electrolyte and the electrode take place during charging and discharging. The "dielectric" layer does not really store energy; rather, it is a representation of the energy in the links between molecules. The idea itself is a drawback of these systems; unlike electrostatic storage, the stress and degradation of the electrodes while charging and discharging accelerates their degradation. The process is related to making supercapacitors with a higher internal resistance. Elements such as ruthenium oxide ($RuO_2$) or manganese dioxide ($MnO_2$) are used for the electrodes of pseudo-supercapacitors. Reduced charging efficiency, longer time response, lower discharge rate, and worse stability and cyclability compare to EDLC supercapacitors [16, 18].

Hybrid supercapacitors are the third and final kind of supercapacitors. A new kind of supercapacitor has just been developed. The enhanced double-layer capacitor (EDLC) and pseudo-supercapacitors are both incorporated into this state-of-the-art supercapacitor. The capacity to provide large currents and increased volumetric and gravimetric energy densities are the key benefits. The negative electrode of a hybrid supercapacitor, usually constructed from pseudocapacitive material, undergoes a faradaic reaction, resulting in a greater energy density [19, 20]. The double layer of electrostatic energy stored by the positive electrode, which is usually activated carbon, is the electrode's defining characteristic. High currents are delivered by hybrid supercapacitors as a result of the electrostatic interaction between charge carriers and the electrode surface on the positive electrode side. From a design and operational perspective, hybrid supercapacitors are similar to lithium-ion batteries.

### 4.2.2 Types of Supercapacitors: Electrochemical Double-Layer Capacitors (EDLCs)

The idea of an electrochemical double layer, which outlined the supercapacitor's operating principle, was first presented in a US patent in 1950 [21]. Then, other models were described, which provided the performance of electrodes in an electrolyte under potentiostatic control. These models included the Helmholtz model, the Grahame model, the Gouy-Chapman model, and the Gouy-Chapman-Stein model. Electrochemical double-layer capacitors (EDLCs) typically store charge electrostatically [21]. The rapid formation of two opposing charges at the electrode–electrolyte contact occurs upon dipping the electrode into the electrolyte. Electric charge layers with two charges each stand in for these opposing charges. Figure 4.5 shows a schematic representation of this charging process [22]. A positive-polarity charge layer forms on the electrode surface, whereas a negative-polarity charge layer forms in the electrolyte solution. The solvent or water molecules act as a barrier between the two charge levels. The dielectric function is performed by this layer that separates the charges. Electrode and electrolyte charge transfer is prevented by these solvate molecules. Charging of the double-layer capacitor begins upon proper connection of the supply to the electrodes.

Immersing an electrode in an electrolyte causes two things to happen: either the ions of the electrolyte are adsorbed onto the surface of the electrode, or the atoms of the electrode dissolve into the electrolyte. However,

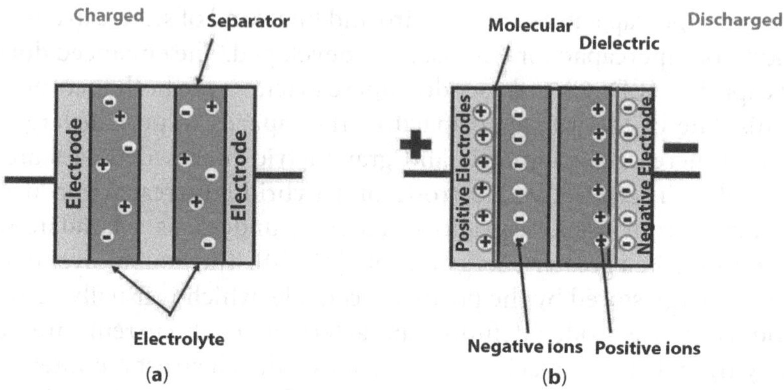


**Figure 4.5** Schematic diagrams of a capacitor showing (a) the charged (left) (b) discharged (right) states [22].

a charge separation between the electrolyte and electrode surface can occur in any of these scenarios. A charge-storage electrochemical double layer is formed. It is possible to increase the quantity of charge that is stored by raising the applied voltage. The electrolyte's oppositely charged ions will congregate around the electrode as the applied voltage rises, creating an electric field that will attract charges to the surface of the electrode. It is this capacity to hold charge that gives rise to the concept of double-layer capacitance. Supercapacitors work by storing charges in this way. In this case, the capacitance of the two layers is given by

$$\frac{1}{C_{double\ layer}} = \frac{1}{C_{Helmholtz\ capacitance}} + \frac{1}{C_{diffusion\ layer}}$$

When it comes to supercapacitors, electrodes with a large surface area and conductivity are crucial for optimal performance. A substantial amount of research has been conducted on carbon graphite electrodes [22]. Among the most promising options are carbon compounds because of their inexpensive cost, high conductivity, and excellent electrochemical stability. It is the size of the electrodes that determines the electrochemical capacitor's capacity to hold charge per unit voltage.

As the concentration of adsorbed ions increases, the double-layer electrostatic energy storage becomes linear with respect to the stored charge. Capacitance of double-layer capacitors is dependent on the restricted mobility of ions in the electrolyte and the resistive porous structure of the electrodes, as opposed to the electrons used in traditional capacitors. A limitless number of charging and discharging cycles may theoretically be achieved with electric double layers since no chemical reactions occur inside the electrode or electrolyte [23].

## 4.3 Sustainable and Eco-Friendly Materials for Supercapacitors

### 4.3.1 Substances Derived from Carbon

Activated carbon, which may be found in coconuts and other plant sources, is a popular ingredient in supercapacitors because of its high surface area and excellent conductivity. Electrodes made of single- or multi-walled CNTs are ideal for use in supercapacitors due to their high electrical conductivity. The study conducted by Kuzmenko *et al.* [24] focuses on the development of carbon nanofiber/nanotube composites derived from cellulose, which are used as electrodes in supercapacitors with a sustainable approach. Supercapacitor electrodes made from cellulose, which is the most abundant renewable resource, were developed. The findings indicate that cellulose-derived CNF/CNT composites serve as durable and effective carbonaceous electrodes for supercapacitors. The imperative task at hand is to create supercapacitors with elevated energy density and extended cycle life by utilizing sustainable materials, in order to meet the demands of future applications.

### 4.3.2 Components Found in Biomass

Sugarcane bagasse, peanut shells, and rice husks are also potential sources of carbonaceous materials for use as electrodes in supercapacitors. Chitosan and cellulose are two examples of natural polymers that can be transformed into environmentally friendly electrode materials. Lignin and cellulose-based supercapacitors are economically viable and environmentally friendly because of their renewability, sustainability, and easy accessibility. Adam *et al.* [25] study a comprehensive analysis of previously published research on electrospun lignin, cellulose, and lignin/cellulose nanofibers, specifically focusing on their potential as materials for supercapacitor electrodes. The materials suitable for next-generation supercapacitors, which are both sustainable and useful, encompass metal compounds, conducting polymers, and carbon materials.

Recently, Thangavel *et al.* [26] provide a study on utilizing watermelon rind, a by-product of watermelons, for the development of high-energy and very stable green supercapacitors that can withstand high temperatures. These supercapacitors are powered by a high-voltage ionic liquid electrolyte. Ramadoss *et al.* [27] outlines the process of creating a self-sustaining system by combining a supercapacitor with energy collecting devices like nanogenerators and solar cells. This system is designed to provide power for portable electronic devices. This study initially discusses the synthesis and electrochemical characterization of different electroactive materials used in supercapacitors. It then moves on to the fabrication of a supercapacitor device. Lin [28] also focuses on the detailed examination of nickel cobalt-based metal complexes. The green or eco-friendly nanomaterials has the characteristics of being inexpensive, lightweight, environmentally benign, and highly capable for a variety of energy-related uses [29]. Recently, Kausar [30] offers a range of methods that offer essential insights for the design and creation of versatile energy storage materials.

### 4.3.3 Porous Organic Polymers (POPs)

In order to tackle the problem of global warming and the scarcity of fossil fuels, there is an urgent requirement for environmentally acceptable and sustainable renewable energy technology. Liu *et al.* [31] examine the latest progress and design principles of porous organic polymers (POPs) as highly effective electrode materials for supercapacitors (SCs). Figure 4.6 shows the schematic representation of a process or a set of chemical structures and reactions related to porous organic polymers (POPs). It seems to illustrate the synthesis and modification of these materials, showing chemical structures, molecular models, and possibly a process for creating carbonized POPs, which involves a carbonization step. It also includes chemical structures of specific covalent organic frameworks (COFs), such as DAB-TFP COF, TaPa-Py COF, and Tz-COF-3, and shows how they might be interconverted or modified into more stable forms, hybridized with other materials, or used to produce composites such as COF-316@PPy TCE. Redox-active units are highlighted in the structures, suggesting a focus on the electronic or catalytic properties of these materials.

They analyze the synthetic approaches and the correlations between the structure and performance of POPs. The attapulgite@carbon/NiCo layered double hydroxide nanocomposites are produced using environmentally friendly calcination and hydrothermal techniques, utilizing waste adsorbents. These nanocomposites have the potential to serve as electrode

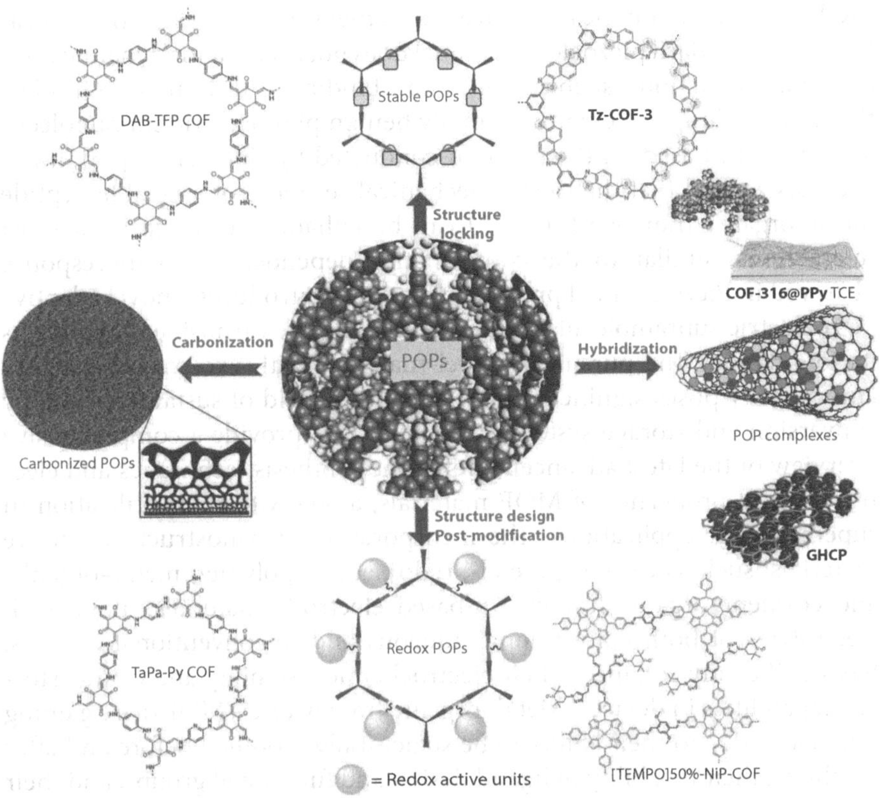


**Figure 4.6** Schematic representations of different modes of POPs as electrodes in SCs.

materials in improved supercapacitors. This approach would enhance the advancement of supercapacitors in a highly efficient and productive manner, while also offering a viable strategy for addressing tetracycline pollution and recycling waste adsorbents to promote sustainable development [32].

## 4.3.4 Metal–Organic Frameworks

Metal–organic frameworks (MOFs) are gaining more recognition as very promising materials for use as electrodes in supercapacitors (SCs) [33]. An approach was devised to decrease the overall electric resistance of MOFs by intertwining MOF crystals with polyaniline (PANI) chains that are coated on MOFs using electrochemical means [34]. Kaur *et al.* [35] discuss advanced synthesis techniques used for the design and development of

MOF, with a focus on the research advancements in photocatalytic applications. The standard piezoelectric model does not account for dynamic non-covalent interactions, such as host–guest binding. This limitation hinders the rational design of environmentally benign piezoelectric supramolecular materials. Based on the research conducted by Chen *et al.* [36], it has been discovered that the electromechanical response of a flexible peptide metal–organic framework (MOF) can be enhanced through guest–host interactions, similar to the conformation-dependent mechanoresponse observed in Piezo channel proteins. This study introduces a novel adaptive piezoelectric supramolecular material utilizing the carnosine dipeptide as its foundation. The pursuit of novel materials with exceptional stability and capacity poses significant problems in the field of sustainable energy conversion and storage systems. Cao *et al.* [37] provide a comprehensive overview of the latest advancements in the synthesis techniques and electrochemical properties of MOF materials, along with their utilization in supercapacitor applications. The incorporation of nanostructured active materials, such as carbon-, metal-, redox-active-polymer, metal–organic, and covalent–organic framework-based electrode materials, along with electrolytes of both conventional aqueous and unconventional systems, has significantly enhanced their electrochemical stability and charge storage capabilities in devices. Metal–organic frameworks (MOFs) are gaining popularity due to their ability to be structurally modified before and after synthesis, their flexibility in ligand design and functional groups, and their capacity to incorporate various metal nodes. These characteristics enable the design of MOFs with specific qualities needed for different processes. Iniyan *et al.* [38] provide a comprehensive examination of electrocatalysts based on metal–organic frameworks. The focus is on the design and synthesis of these catalysts for electrochemical reactions including hydrogen evolution, oxygen evolution, and carbon dioxide reduction.

### 4.3.5 Electrolytes

To make supercapacitors more eco-friendly, you can use ionic liquids or aqueous electrolytes instead of the old-fashioned organic ones. Utilized tea leaves are employed to create carbon with a high surface area and electrochemical characteristics. Bhoyate *et al.* [39] propose that discarded tea leaves can be utilized to create eco-friendly, high-performance supercapacitor devices at a reduced expense. Kim *et al.* [40] describe the development of a new environmentally friendly polymer electrolyte called LiTFS-LiSMC. This electrolyte is created by combining lithium trifluoromethanesulfonate with chemically modified methyl cellulose. The electrolyte also demonstrates excellent compatibility with patterned carbon, allowing for straightforward production of micro-supercapacitors. Spinel $LiMn_2O_4$ is increasingly being recognized for its suitability in supercapacitor applications due to its abundant availability, low cost, eco-friendliness, high energy density, and excellent rate capability. Prasankumar *et al.* [41] focus on the rapid and environmentally friendly production of $LiMn_2O_4$ and its possible use in creating supercapacitors. The synthesis of nanostructured multimetal oxide $LiMn_2O_4$ was achieved utilizing a rapid and environmentally friendly technique involving glucose-mediated microwave irradiation. Supercapacitors offer significant environmentally friendly progress in energy conversion and storage, with immense potential to influence the future global economy. Recently, an overview of various modeling techniques, such as empirical modeling, dissipation transmission line models, continuum models, atomistic models, quantum models, and simplified analytical models, has been proposed for the theoretical investigation of supercapacitors [42].

## 4.4 Advancements in Electrode Materials

### 4.4.1 Carbon-Based Materials: Activated Carbon, Carbon Nanotubes, and Graphene

Lo *et al.* [43] present a study on the synthesis and analysis of composites consisting of graphene, carbon nanotubes (CNTs), and manganese oxide (MnO). The authors investigate how the chemical composition and phase transition of these composites affect the properties of the resulting electrode film. The MnO, which exhibits faradic activity, also served as a significant co-binder, resulting in a decreased requirement for polymeric binder. This led to improved conductivity and capacitance of the electrode. Graphite, graphene, carbon nanotubes, and activated carbon are among the frequently utilized carbon-based compounds. Verma *et al.* [44] present a concise yet thorough summary of the attributes of appropriate electrode materials for supercapacitor devices. Zang *et al.* [45] demonstrate the synthesis of graphene and carbon nanotubes directly within vertically aligned carbon nanotube (VACNT) forests using chemical vapor deposition (CVD). This process results in the formation of either a 3D CNT–graphene network or a 3D CNT–CNT network. The electrochemical capacitance is enhanced by the presence of large surface areas (such as graphene and

CNT), high electrochemical reaction sites (found in graphene), and the use of metal particles as pseudo-capacitor materials. Functional carbon-based materials, such as activated carbon, carbon nanotubes/fibers, and graphene, are commonly employed as electrodes in energy storage systems due to their low potential plateau, satisfactory capacity, and consistent cycling performance. However, these materials have notable drawbacks, including their dependence on fossil fuels and the energy-intensive conditions required for their synthesis, which result in environmental harm and high costs. Samantaray *et al.* [46], provide a complete analysis of the synthesis, characteristics, and performances of various carbonaceous materials suggested for use in DSSC-CE. The nano-carbon materials encompass carbon nanoparticles, activated carbon, carbon nanofibers, carbon nanotubes, two-dimensional graphene, and hybrid carbon material composites.

### 4.4.2 Conductive Polymers

Conductive polymers show great potential as pseudocapacitive materials for supercapacitors since they are inexpensive, are environmentally friendly, exhibit high pseudo-capacitance, and are easy to manufacture [33, 47]. Zhuo *et al.* [48] present a successful technique for producing porous composite electrodes using renewable cellulose, which is environmentally friendly, for supercapacitors. Gong *et al.* [49] have developed a set of conductive polymer hydrogels by grafting β-cyclodextrin onto cellulose hydrogels and embedding nanostructured polyaniline into three-dimensional networks. The resulting conductive hydrogels exhibit excellent flexibility and sufficient adhesive properties, enabling the convenient production of all-in-gel supercapacitors. These conductive hydrogels can be directly placed between carbon cloths to create comprehensive all-in-one supercapacitor devices, eliminating the need for additional packaging. Zhang *et al.* [50] highlight the advantages of utilizing tartaric acid, derived from natural sources, as a non-innocuous component in the MXene composite. A tartaric-acid-treated, water-stable MXene/PEDOT is shown to demonstrate its practical utility. A PSS conductive coating is created to function as electrodes for an ultrafast supercapacitor. Among all assemblies based on 2D materials, this supercapacitor has the highest known performance in an alternating-current filtering application. Energy production can achieve cleanliness and sustainability by relying on renewable resources and can be effectively utilized when stored efficiently. Lately, there has been a significant emphasis on the utilization of carbon and polymers obtained from biomass to create safer and more environmentally friendly electrodes for energy storage systems.

## 4.5 Challenges and Future Perspectives

### 4.5.1 Current Limitations in Sustainable Supercapacitor Technology

The present limits of sustainable supercapacitor technology include issues with materials and design.

(a) **Material Challenges:** A major area of concentration is the creation of novel electrode materials and electrolytes for high-performance supercapacitors. Novel materials with the potential to improve supercapacitive performance are being investigated, such as graphene oxide nanoribbons (GONRs) and $V_2CT_x$ MXene composites. These materials exhibit encouraging cycle life and specific capacitance values. To increase these materials' stability and efficiency in supercapacitors, more research and development are necessary [51].

(b) **Design and Application Challenges:** Research is being done on creating freestanding supercapacitors using novel materials such carbon fiber mats or 3D printed materials. These have prospective uses in structural and wearable supercapacitors, but their creation still faces considerable obstacles, particularly with regard to their physicochemical and mechanical properties.

(c) **Electrochemical Performance:** Studies are being done to extend the life of supercapacitors and increase their electrochemical capacitance. For instance, a straightforward electrochemical polymerization method has been used to create stable PEDOT derivative electrodes with the goal of outperforming the state of the art in electrochemical performance [51].

(d) **Energy Density and Efficiency:** Commercial energy storage devices and the design of inexpensive supercapacitors based on aqueous carbon are contrasted. This entails investigating how to raise the energy density of different aqueous electrolytes and redox additions. Finding a balance between high energy density and efficiency in supercapacitors is a difficult task, though, especially when compared to other energy storage technologies already in use, like batteries.

(e) **Cost and Environmental Considerations:** Sustainable supercapacitors must be both economical and ecologically benign. Low-carbon technologies and materials based on biopolymers are being investigated; however, in order to make these materials competitive with existing technologies, issues with their scalability and performance must be resolved [52].

The main focus of current research and development in the field of sustainable supercapacitors is represented by these restrictions. In order to advance supercapacitor technology and provide more effective and sustainable energy storage solutions, it is imperative that these issues be resolved.

### 4.5.2 Future Research Directions and Potential Breakthroughs

Future avenues for study and possible advances in sustainable supercapacitor technology are concentrated on a range of creative strategies and materials to get around existing constraints and improve efficiency. Important areas of attention consist of the following:

(a) **Development of New Materials:** Research into novel electrolytes and electrode materials is essential. Novel materials such as MXene composites and graphene oxide nanoribbons are becoming more and more the focus of research. Although these materials have the potential to increase specific capacitance and cycle life, more tuning is needed to ensure stability and efficiency in supercapacitor applications.

(b) **Innovative Design Approaches:** One area of interest is the development of freestanding supercapacitors made of 3D printed materials or carbon fiber mats. These designs show promise for use in structural and wearable supercapacitors. The secret to their development will be to solve the problems with their mechanical and physicochemical characteristics.

(c) **Energy Density and Efficiency:** Research is being done on designing supercapacitors based on aqueous carbon that are more reasonably priced than those made by commercial companies. This entails experimenting with different redox additives and aqueous electrolytes to balance efficiency and maximize energy density.

(d) **Cost and Environmental Considerations:** Sustainable supercapacitors must be both economical and ecologically benign. Low-carbon technologies and materials based on biopolymers are being investigated; however, in order for these materials to be practical substitutes for existing technologies, issues with their scalability and performance must be resolved.

In order to advance supercapacitor technology and provide more effective and sustainable energy storage solutions, it will be imperative to address these issues. To lay the foundation for the upcoming generation of supercapacitors, the emphasis is on novel materials, design strategies, and enhanced performance measures.

## 4.6 Conclusions

This chapter highlights the crucial importance of sustainable materials in the development of supercapacitors, which are becoming increasingly essential in the field of energy storage. The indispensability of supercapacitors in renewable energy applications is clear due to their emphasis on high power density and quick charging capabilities. Our comprehensive analysis of both traditional and environmentally sustainable materials utilized in the manufacturing of supercapacitors not only uncovers the current technological environment but also outlines a trajectory for future advancements. The deliberate investigation of sophisticated electrode materials, including carbon-based compounds, conductive polymers, and metal oxides, in conjunction with state-of-the-art electrolytes such as gel and ionic liquids, establishes the foundation for the future development of energy storage systems. The thorough assessment of supercapacitors, taking into account their energy density, power density, and cycle stability, demonstrates the significant potential of incorporating sustainable materials in their design and production. In essence, the research emphasizes the crucial significance of sustainable supercapacitors in advancing a more environmentally friendly future, specifically through their utilization in electric vehicles, portable gadgets, and renewable energy systems. This is in line with the overarching objective of shifting towards more sustainable energy storage options, therefore making a substantial contribution to the worldwide endeavor in environmental preservation and sustainability.

## References


1. Kumar, T., Verma, A., Vandana, V., *Sustainability of Green Nanocomposite-based Supercapacitors: Advances and Limitations*, vol. 2, ch. 17, pp. 462–476, Royal Society of Chemistry, 2023.
2. Rani, U. *et al.*, Exploring properties of organometallic double perovskite (CH3NH3) 2AgInCl6: A novel material for energy conversion devices. *Mod. Phys. Lett. B*, 38, 18, 2450144, 2023.
3. Kumar, T. *et al.*, Photolithographic Fabrication of Supercapacitor Electrodes, in: *Functionalized Nanomaterials Based Supercapacitor: Design, Performance and Industrial Applications*, pp. 161–175, Nature to Nanomaterials, Springer, Singapore, 2023.
4. Winter, M. and Brodd, R.J., What are batteries, fuel cells, and supercapacitors? *Chem. Rev.*, 104, 10, 4245–4270, 2004.
5. Walsh, F., Electrochemical technology for environmental treatment and clean energy conversion. *Pure Appl. Chem.*, 73, 12, 1819–1837, 2001.
6. Namisnyk, A. and Zhu, J., A survey of electrochemical super-capacitor technology, in: *Australian Universities Power Engineering Conference*, University of Canterbury, New Zealand, 2003.
7. Molina, M.G., Distributed energy storage systems for applications in future smart grids, in: *2012 Sixth IEEE/PES Transmission and Distribution: Latin America Conference and Exposition (T&D-LA)*, IEEE, 2012.
8. D'Souza, F., *Fullerenes, Nanotubes, and Carbon Nanostructures-213th ECS Meeting*, The Electrochemical Society, 2008.
9. Schneuwly, A. and Gallay, R., Properties and applications of supercapacitors from the state-of-the-art to future trends, in: *Proceeding PCIM*, Citeseer, 2000.
10. Sundberg, N., *Efficiency improvements with super capacitors in mechatronic systems*, Master of Science Thesis, Stockholm, Sweden 2007, diva-portal.org
11. Wang, Y., *Modeling of ultracapacitor short-term and long-term dynamic behavior*, OhioLINK Electronic Theses and Dissertations Center, University of Akron, 2008. http://rave.ohiolink.edu/etdc/view?acc_num=akron1217029983
12. Green, A. and Jehoulet, C., The non-battery battery-The potential role of supercapacitors in standby power applications, in: *The Battcon 2002 Proceedings*, 2002.
13. Dincer, I. and Acar, C., Innovation in hydrogen production. *Int. J. Hydrog. Energy*, 42, 22, 14843–14864, 2017.
14. Steinmann, W.-D., Thermo-mechanical concepts for bulk energy storage. *Renew. Sustain. Energy Rev.*, 75, 205–219, 2017.
15. Lei, C. *et al.*, Reduction of porous carbon/Al contact resistance for an electric double-layer capacitor (EDLC). *Electrochim. Acta*, 92, 183–187, 2013.
16. Libich, J. *et al.*, Supercapacitors: Properties and applications. *J. Energy Storage*, 17, 224–227, 2018.

17. Wang, R., Zhang, Y., Liao, Y., Performance of rolling piston type rotary compressor using fullerenes (C70) and NiFe 2 O 4 nanocomposites as lubricants additives. *Front. Energy*, 14, 644–648, 2020.
18. Qu, D. *et al.*, Synthesis of MnO nano-particle@ Flourine doped carbon and its application in hybrid supercapacitor. *Appl. Surf. Sci.*, 413, 344–350, 2017.
19. Makino, S. *et al.*, Room temperature performance of 4áV aqueous hybrid supercapacitor using multi-layered lithium-doped carbon negative electrode. *J. Power Sources*, 326, 711–716, 2016.
20. Zhang, S.S., Eliminating pre-lithiation step for making high energy density hybrid Li-ion capacitor. *J. Power Sources*, 343, 322–328, 2017.
21. Simon, P. and Gogotsi, Y., Materials for electrochemical capacitors. *Nat. Mater.*, 7, 11, 845–854, 2008.
22. Joshi, P. and Sutrave, D., Supercapacitor: basics and overview. *J. Inf. Comput. Sci.*, 9, 12, 2019.
23. Lu, P. *et al.*, Supercapacitor and nanoscale research towards electrochemical energy storage. *Int. J. Smart Nano Mater.*, 4, 1, 2–26, 2013.
24. Kuzmenko, V. *et al.*, Sustainable carbon nanofibers/nanotubes composites from cellulose as electrodes for supercapacitors. *Energy*, 90, 1490–1496, 2015.
25. Adam, A.A. *et al.*, State of the art and new directions on electrospun lignin/cellulose nanofibers for supercapacitor application: A systematic literature review. *Polymers*, 12, 12, 2884, 2020.
26. Thangavel, R. *et al.*, High-energy green supercapacitor driven by ionic liquid electrolytes as an ultra-high stable next-generation energy storage device. *J. Power Sources*, 383, 102–109, 2018.
27. Ramadoss, A. *et al.*, Fabrication and characterization of supercapacitors toward self-powered system, in: *Advancements in Energy Storage Technologies*, Books on Demand, Technology & Engineering 2, 202, IntechOpen, 2018.
28. Lin, L.-Y., Sustainable functional materials for next-generation supercapacitors, in: *Sustainable Materials for Next Generation Energy Devices*, pp. 3–26, Challenges and Opportunities, Elsevier, 2021.
29. Kumar, T. and Verma, A., Role of FNMs-Based Supercapacitors in the Food and Beverage Industry, in: *Functionalized Nanomaterials Based Supercapacitor: Design, Performance and Industrial Applications*, pp. 459–471, Springer Nature, Singapore, 2023.
30. Kausar, A., Green nanocomposites for energy storage. *J. Compos. Sci.*, 5, 8, 202, 2021.
31. Liu, X. *et al.*, Porous organic polymers for high-performance supercapacitors. *Chem. Soc Rev.*, 51, 8, 3181–3225, 2022.
32. Tang, J. *et al.*, Waste Adsorbent-Derived Interconnected Hierarchical Attapulgite@ Carbon/NiCo Layered Double Hydroxide Nanocomposites for Advanced Supercapacitors. *Langmuir*, 39, 7, 2739–2750, 2023.
33. Kumar, T. and Verma, A., Eco-Friendly Conducting Polymer-Based Functionalized Nanocomposites Dedicated for Electrochemical Devices, in:

*Functionalized Nanomaterials Based Supercapacitor: Design, Performance and Industrial Applications*, pp. 405–440, Nature to Nanomaterials, Springer, Singapore, 2023.
34. Wang, L. *et al.*, Flexible solid-state supercapacitor based on a metal–organic framework interwoven by electrochemically-deposited PANI. *J. Am. Chem. Soc.*, 137, 15, 4920–4923, 2015.
35. Kaur, H. *et al.*, Metal-organic frameworks for photocatalytic degradation of pollutants, in: *Nano-materials as photocatalysts for degradation of environmental pollutants*, pp. 91–126, Challenges and Possibilities, Elsevier, 2020.
36. Chen, Y. *et al.*, Guest molecule-mediated energy harvesting in a conformationally sensitive peptide–metal organic framework. *JACS*, 144, 8, 3468–3476, 2022.
37. Cao, Z. *et al.*, Metal–organic framework materials for electrochemical supercapacitors. *Nano Micro Lett.*, 14, 1, 181, 2022.
38. Iniyan, S. *et al.*, *An Overview of Metal-organic Framework Based Electrocatalysts: Design and Synthesis for Electrochemical Hydrogen Evolution, Oxygen Evolution, and Carbon Dioxide Reduction Reactions*, p. e202300317, The Chemical Record, Wiley, 2023.
39. Bhoyate, S. *et al.*, Eco-friendly and high performance supercapacitors for elevated temperature applications using recycled tea leaves. *Global Challenges*, 1, 8, 1700063, 2017.
40. Kim, S.K. *et al.*, Recyclable High-Performance Polymer Electrolyte Based on a Modified Methyl Cellulose–Lithium Trifluoromethanesulfonate Salt Composite for Sustainable Energy Systems. *ChemSusChem*, 13, 2, 376–384, 2020.
41. Prasankumar, T. *et al.*, Expeditious and eco-friendly synthesis of spinel LiMn2O4 and its potential for fabrication of supercapacitors. *J. Alloys Compd.*, 834, 155060, 2020.
42. Kumar, A. *et al.*, Theories and models of supercapacitors with recent advancements: impact and interpretations. *Nano Express*, 2, 2, 022004, 2021.
43. Lo, A.-Y. *et al.*, Effect of composition ratios on the performance of graphene/carbon nanotube/manganese oxide composites toward supercapacitor applications. *ACS Omega*, 5, 1, 578–587, 2019.
44. Verma, K.D. *et al.*, Characteristics of electrode materials for supercapacitors, in: *Handbook of Nanocomposite Supercapacitor Materials I: Characteristics*, pp. 269–285, Springer, 2020.
45. Zang, X. *et al.*, Chemical vapor deposition of 3D graphene/carbon nanotubes networks for hybrid supercapacitors. *Sens. Actuators, A*, 304, 111886, 2020.
46. Samantaray, M.R. *et al.*, Synergetic effects of hybrid carbon nanostructured counter electrodes for dye-sensitized solar cells: A review. *Materials*, 13, 12, 2779, 2020.
47. Kumar, T. *et al.*, Chemical Modifications for the Development of Conducting Polymer-Based Supercapacitors, in: *Functionalized Nanomaterials Based*

*Supercapacitor: Design, Performance and Industrial Applications*, pp. 353–370, Springer, Singapore, 2023.
48. Zhuo, H. *et al.*, Cellulose carbon aerogel/PPy composites for high-performance supercapacitor. *Carbohydr. Polym.*, 215, 322–329, 2019.
49. Gong, Q. *et al.*, A facile preparation of polyaniline/cellulose hydrogels for all-in-one flexible supercapacitor with remarkable enhanced performance. *Carbohydr. Polym.*, 245, 116611, 2020.
50. Zhang, M. *et al.*, Multitasking tartaric-acid-enabled, highly conductive, and stable MXene/conducting polymer composite for ultrafast supercapacitor. *Cell Rep. Phys. Sci.*, 2, 6, 12 pages, 100449, 2021.
51. Ahamed, M.B., Hussain, C.M., Deshmukh, K., *Low-carbon Supercapacitors: Towards Sustainability in Energy Storage Devices*, vol. 2, 524 pages, Royal Society of Chemistry, 2023.
52. Huang, S. *et al.*, Challenges and opportunities for supercapacitors. *APL Mater.*, 7, 10, 2019.

# 5

# Role of Material Selection and Fabrication Approach in the Performance of Sustainable Supercapacitors

**S. Sreehari[1,5], A. V. Mahadev[2], D. A. Nayana[3], Dinesh Raj R.[4], P. K. Manoj[3] and Arun Aravind[5*]**

[1]*Department of Surface and Plasma Science, Faculty of Mathematics and Physics, Charles University, Prague, Czech Republic*
[2]*Department of Astronomy, Aryabhatta Research Institute of Observational Sciences, Nainital, Uttarakhand, India*
[3]*Department of Physics, T K M College of Arts and Science, Kollam, Kerala, India*
[4]*Department of Botany and Biotechnology, Bishop Moore College, Mavelikara, Kerala, India*
[5]*Centre for Advanced Functional Materials (CAFM), Department of Physics, Bishop Moore College, Mavelikara, Kerala, India (Research Centre Affiliated to the University of Kerala)*

***Abstract***

The development of sustainable supercapacitors is of paramount importance in addressing the growing need for high-performance energy storage devices. This book chapter focuses on the role of material selection and fabrication approaches in achieving optimal performance and sustainability in supercapacitors. A detailed discussion on the significance of material selection, emphasizing the properties and characteristics required for sustainable electrode materials, is presented here. Various electrode materials commonly employed in supercapacitors, including conducting polymers, metal oxides and carbon-based nanomaterials, are explored in detail, highlighting their unique features and capabilities as supercapacitor electrode materials. Additionally, the chapter encompasses emerging materials such as black phosphorus, covalent organic frameworks, metal nitrides, MXenes, and metal–organic frameworks, which hold great promise for advancing supercapacitor technology. This chapter also delves into the crucial role of fabrication

**Corresponding author*: arun@bishopmoorecollege.org

Basheer Ahamed and Chaudhery Mustansar Hussain (eds.) Sustainable Supercapacitors: Next-Generation of Green Energy Storage Devices, (123–158) © 2025 Scrivener Publishing LLC

approaches in supercapacitor performance. We evaluate various physical and chemical deposition techniques concerning their impact on electrode morphology, porosity, and interfacial properties. Furthermore, the impact of these depositions on different substrates is analyzed, focusing on their perspectives in device fabrication. By providing an in-depth analysis of material selection and fabrication approaches, this chapter aims to enhance the understanding of researchers and engineers engaged in sustainable energy storage.

***Keywords*:** Supercapacitor, pseudocapacitor, carbon nanotube, activated carbon, electrode, electrolyte, negative Poisson's ratio

## 5.1 Introduction

Supercapacitors (SCs) have garnered significant interest owing to their distinctive attributes, including long cycle life, large power density ($P_d$), large specific capacitance ($C_s$), and environmentally favorable characteristics. SCs possess remarkable properties related to their rapid charge/discharge rates and extended life cycles [1]. Extensive research is currently underway to explore a range of materials and fabrication techniques aimed at enhancing the performance of SCs. Attaining significant advancements in this field through enhanced electrode materials and cost-effective, eco-friendly production hinges on a profound perception of the electron and ion transport pathways, techniques for charge storage, and electrochemically active sites [2].

SCs are commonly categorized under two groups: electrochemical double layer capacitors (EDLCs) and pseudocapacitors. Storage mechanism was the basis for this classification; Figures 5.1 and 5.2 represent the

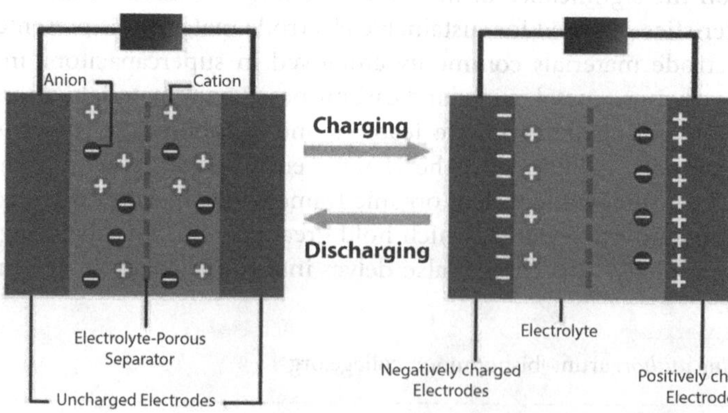


**Figure 5.1** Mechanism of ELDC.

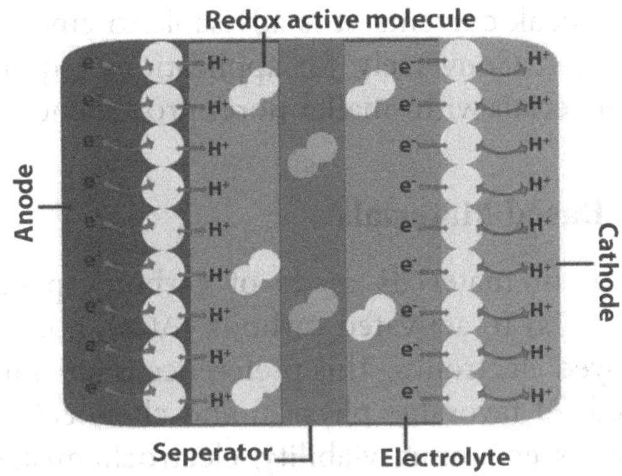


**Figure 5.2** Mechanism of a pseudocapacitor.

mechanism of ELDCs and pseudocapacitors, respectively. EDLCs achieve charge storage electrostatically through reversible ion adsorption at the electrolyte–electrode boundary [2]. The suitable materials for ELDCs require immense mechanical and chemical stability, substantial specific surface area (SSA), and acceptable electrical conductivity. Typically, EDLCs employ carbon-based electrode materials, such as carbon nanotubes (CNTs), graphene, and activated carbon. Pseudocapacitors employ chemical storage through redox reactions in proximity (within a few nanometers) to the surface. Materials demonstrating redox behavior are utilized in pseudocapacitors, including conducting polymers (CPs), transition metal sulfides, and transition-metal oxides (TMOs). Pseudocapacitors can achieve higher $C_s$ than ELDCs. The ability of electrodes to exhibit the pseudocapacitance phenomenon relies on the material's chemical attraction to the ions absorbed at the electrode surface, coupled with the configuration and size of the electrode pores [3]. The following sections comprehensively explore diverse electrode materials and their functioning.

## 5.2 Electrode Materials for Supercapacitors

SC electrodes must possess excellent conductivity, stability across varying temperatures, robust chemical inertness, a significant SSA, resistance to corrosion, environmentally friendly attributes, and cost-effectiveness [4]. The pore size is having a very important role in the working of SCs. Generally, as pore size decreases, $C_s$ and energy density ($E_d$) tend to increase. However, a reduction in pore size amplifies Equivalent Series Resistance (ESR), leading to a decline in $P_d$. Consequently, for applications demanding higher peak currents, it is advisable to employ SC electrodes featuring larger pores. Conversely, for applications requiring elevated $E_d$ levels, electrode materials with smaller pores prove beneficial [5, 6].

### 5.2.1 Carbon-Based Materials

A broad spectrum of materials made of carbon, spanning from carbon nanotubes (CNTs) to activated carbons (ACs), holds prominence as extensively employed electrodes. This preference is attributed to their frequently acknowledged favorable physical and chemical attributes. These attributes encompass economic viability, electrochemistry characterized by inertness, diverse structural forms (such as tubes, monoliths, sheets, composites, aerogels, and powders), convenient processability, manageable porosity, and the existence of active sites for electrocatalysis that enable a variety of redox processes [7].

#### *5.2.1.1 Activated Carbons (ACs)*

The primary choice for EDLC electrode material is the ACs. These porous carbon substances possess a significant SSA. Despite their lower electrical conductivity (ranging from 1250 to 2500 S/m), they remain sufficient for SC functionality. To achieve this extensive SSA, their porous architecture encompasses macropores (> 500 Å), mesopores (20–500 Å), and micropores (<20 Å) [4, 8]. ACs are commonly synthesized through the chemical or physical (thermal) activation of diverse carbon-rich sources like a nutshell, wood, and coal. The chemical activation occurs at temperatures between 400 and 700°C, with the use of activating agents like zinc chloride, sodium hydroxide, potassium hydroxide, and phosphoric acid [9–11]. On the other hand, the physical activation involves subjecting carbon precursors to elevated temperatures (ranging from 700 to 1200°C) along with oxidizing gases like air, $CO_2$, and steam. Research is underway to determine the optimal pore size in case of a specific ion size. Also, researchers are trying to develop enhanced techniques for adjusting the distribution of pore sizes during the manufacturing process. According to certain studies, organic electrolytes may benefit from pores with a size of 0.8 nm, while aqueous electrolytes may benefit from pores with a size of either 0.4 or 0.7 nm [11, 12]. Certain publications have demonstrated a connection between ion size and pore size by observing a peak $C_s$ value [13, 14]. Additionally, the surface of carbon materials containing functional groups may enhance faradaic redox processes, leading to a 5–10% rise in $C_s$ [15]. Electrodes with an SSA of around 1000 $m^2$ per gram yield a $C_s$ of approximately 10 μFcm$^{-2}$ (equivalent to 100 F $g^{-1}$).

A large percentage of micropores exist when the ACs are made using the coconut shell, compared to ACs made from charcoal. Activated carbon fibers made from activated carbon have micropores that are easily controlled and have a remarkably narrow size distribution (surface area of approximately 2500 $m^2\ g^{-1}$). These activated carbon fiber electrodes are particularly advantageous because of their low electrical resistance and efficient collector connection. Due to the existence of micropores, they primarily demonstrate double-layer capacitance with minor contributions from pseudocapacitance [16].

### *5.2.1.2 Carbide-Derived Carbons (CDCs)*

CDCs encompass a broad array of carbon-based materials generated from carbide precursors, which undergo conversion into elemental carbon through various methods, such as physical means (like thermal decomposition) or chemical techniques (such as halogenation). CDCs encompass various materials like graphene, graphite, CNTs, nanocrystalline diamond, and more. They display a significant SSA featuring customizable pore sizes, enhancing ion confinement and consequently elevating pseudocapacitance. When the pore distribution is appropriately tailored, CDC electrodes have the potential to offer approximately 75% more $E_d$ compared to ACs [17].

Graphene possesses a theoretical SSA of 2630 $m^2\ g^{-1}$, implying a potential theoretical $C_s$ of 550 F $g^{-1}$. El-Kady and colleagues [18] harnessed the potential of graphene sheets as electrodes in their research. They designed an SC that featured uniquely curved graphene sheets, preventing them from stacking and creating mesopores that could be saturated with ionic electrolytes at voltages as high as 4 V. Remarkably, this SC demonstrated an $E_d$ of 85.6 Wh/kg, equivalent to that of nickel metal hydride batteries, and boasted a significantly higher $P_d$ than conventional batteries [19]. The two-dimensional (2D) configuration of graphene amplifies the efficiency of the charge–discharge cycle, facilitating rapid entry and exit of charge carriers from the deep pores of the electrodes. This enhancement results in increased power output. These SCs could find valuable applications in 100/120 Hz filtering applications [20]. Materials derived from chemically modified graphene (CMG), consisting of carbon sheets only one atom thick, can be functionalized to match specific requirements. The CMG configuration, composed of individual sheets, achieves the expansive surface not by depending on the arrangement of pores within the solid substrate. Instead, each chemically altered sheet of graphene possesses the capability for adapting physically for accommodating various kinds of electrolytes, including both spatial distribution and size. This unique feature ensures that the electrolyte will always have continuous availability to the exceptionally large surface area offered by the CMG material [21, 22]. In an aqueous electrolyte, these CMG materials show the $C_s$ value to be 135 F $g^{-1}$, and for an organic electrolyte, this value reaches 99 F/g [23], as studied by Stoller *et al.* Figure 5.3 displays the characterization images and schematic that represent an ultracapacitor test cell having two electrodes used in their investigation.

In their study, Xu and colleagues [24] demonstrated that SCs, which are flexible, exhibited excellent capacitive performance by utilizing a graphene

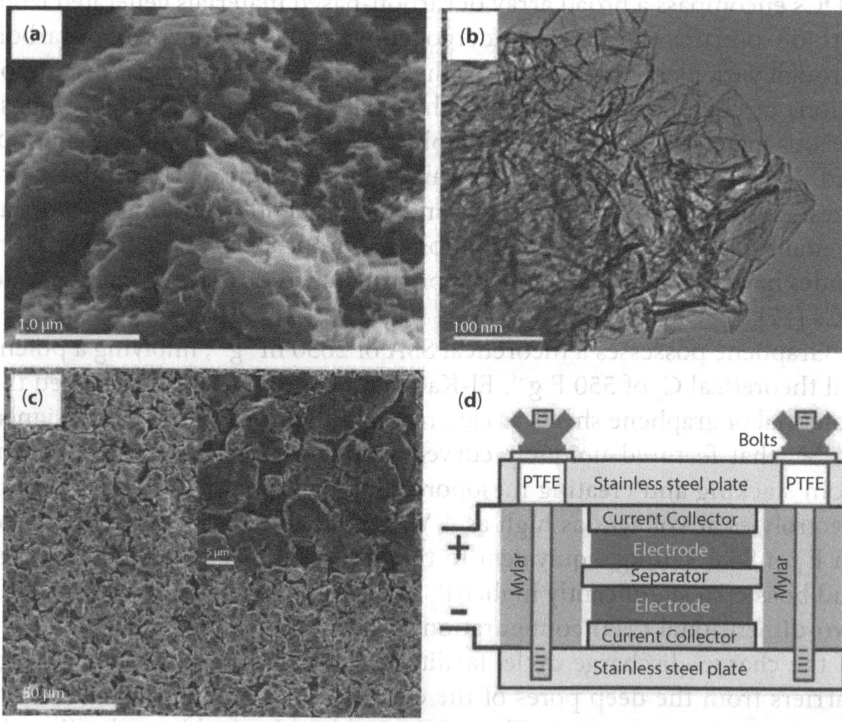


**Figure 5.3** (a) Scanning electron microscope (SEM) depiction of the CMG particle surface. (b) Transmission electron microscope (TEM) visualization that highlights, extending from CMG particle surface, how individual graphene sheets are depicted. (c) SEM image at low and high magnification (inset) illustrating electrode surface of CMG particles. (d) Schematic representation of the assembly of the test cell. Reprinted with permission from [23]. Copyright 2008 American Chemical Society.

film of different thicknesses. When using the 120-μm-thick film, they get a $C_s$ value of 186 F/g; when they use the 42-μm film, a $C_s$ value up to 196 F/g is achieved. These SCs also demonstrated exceptional mechanical flexibility, outstanding cycling stability, and very small leakage current (10.6 μA).

Carbon nanotubes (CNTs) are distinctive cylindrical nanostructures with diameters in the nanometer range and exceptionally high length-to-diameter ratios. These nanotubes can encompass a single shell or multiple concentric layers composed of carbon atoms, often numbering from just one to several tens or even hundreds of layers, with a separation of approximately 0.34 nm between adjacent layers. The arrangement of carbon atoms within these concentric shells closely mirrors the honeycomb pattern found in graphite sheets [25].

Figure 5.4A depicts the illustration of the fabrication process of the nanotube, where an endless strip of graphene is rolled up into a seamless cylinder. To define the nanotube, we use the chiral vector $C_h = na_1 + ma_2$, connecting lattice points O and A on the graphene sheet. The strip is cut from the sheet along this vector, perpendicular to it. The tube's properties are described by $T = t_1a_1 + t_2a_2$, which is the primitive translation vector [26]. Nanotube's unique characteristics can be explained using two sets of quantities. We can use the pair θ and $R = C_h/2\pi$, which are the chiral angle and the radius of the nanotube, respectively, or we can use n and m, which are a set of integer numbers. Here, θ signifies the angle made from the $C_h$ towards the closest zigzag arrangement of C–C bonds. The chiral angle ranges from zero to 30° for all distinct tubes. Electrodes composed of CNTs are engineered as an intricate network of intertwined nanotubes featuring an accessible array of interconnected mesopores. These mesopores within CNTs are intricately linked, forming a continuous distribution, thereby maximizing the utilization of their SSA [27].

Compared to ACs, CNT electrodes exhibit lower ESR because the electrolyte ions can readily permeate the interconnected mesoporous structure [28]. Li *et al.* [29] proposed that the accessibility of the electrode/electrolyte interface and ion size influences the effect of flexible single-walled carbon nanotube (SWCNT) SCs. SWCNTs possess a theoretical SSA of 1315 $m^2g^{-1}$, while compared with ACs, multi-walled carbon nanotubes (MWCNTs) have a smaller SSA. MWCNTs feature mesopores, which efficiently facilitate the flowing mechanism of ions at the electrode–electrolyte junction. When the size of the ions and pore size closely match, it results in a high packing density of ionic species. However, MWCNTs may experience reduced mechanical stability due to substantial volume changes resulting from repetitive ion intercalation and depletion [30]. Peng and colleagues [31] documented that for MWCNTs, the $C_s$ value is 50 F $g^{-1}$.

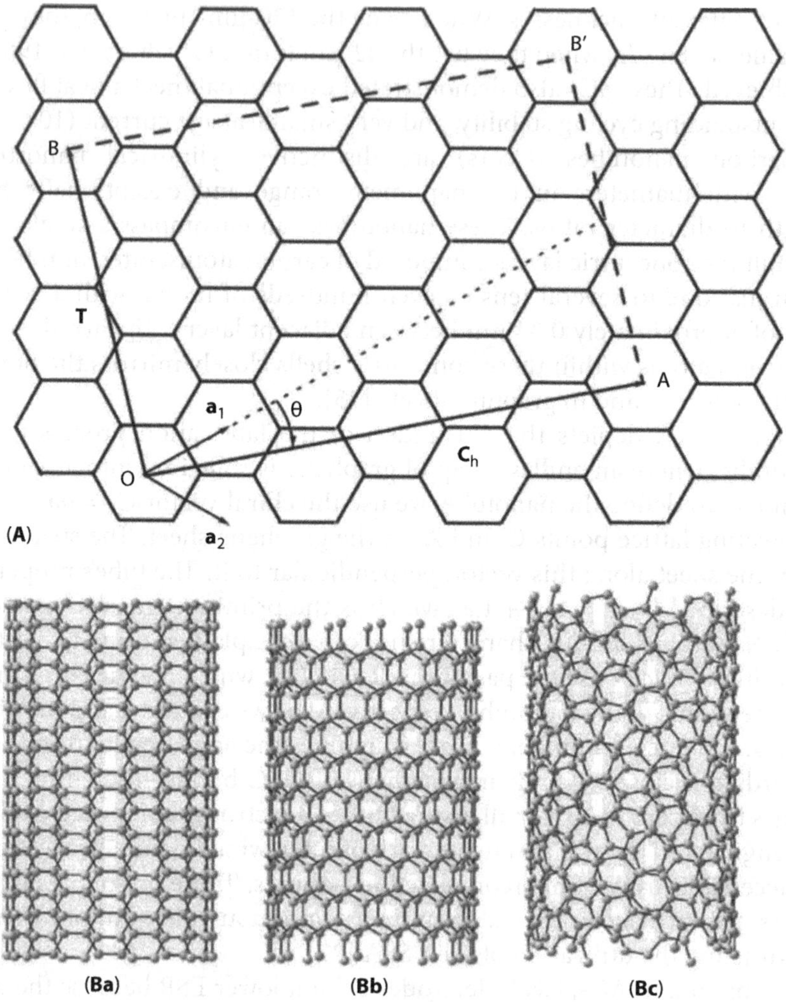


**Figure 5.4** (A) Illustration of the CNT construction. (B) Different types of tubes: (Ba) armchair tubes (n, n) (θ = 30°). (Bb) zigzag tubes (n, 0) (θ = 0°). (Bc) Chiral tubes (all other tubes are chiral). Reprinted with permission from [25]. Copyright 2004, Elsevier.

On the other hand, Xie and co-workers [6] successfully synthesized high-performance nanotubes derived from willow catkins via a straightforward carbonization process. These nanotubes, referred to as HPNCTs, at 1 A $g^{-1}$ show a capacitance value of 292 F $g^{-1}$ and an impressive SSA value of 1775.7 $m^2g^{-1}$. Furthermore, HPNCT-800, even at 10 A $g^{-1}$, displayed an excellent rate capability, maintaining 83.5% of its capacitance retention.

### 5.2.1.3 *Other Carbon Materials*

Various structures of carbon, including carbon onions, carbon aerogels (CAGs), and activated carbon fibers (ACFs), have garnered attention in research for their potential applications in SCs. When choosing materials for SC electrodes, the key criteria involve having a SSA that is easily accessible and having a strong conductive nature towards electricity. The ACFs generally exhibit substantial SSAs of about 3000 $m^2/g$. Also, it offers a relatively adjustable distribution of pore size [32]. Usually, an ACF is derived by carbonizing the fibrous carbon material, which is pre-existing, and then doing the activation procedure. Recently, there has been a report that in a KOH electrolyte solution, researchers got an impressive value of 371 F/g as the capacitance. Nonetheless, there might be questions about extended period stability due to the combination of surface functional groups and a large SSA [33]. Furthermore, it is worth noting that the production cost of ACFs is generally higher compared to standard ACs.

CAG, often referred to as "frozen smoke," is an exceptionally porous and extremely lightweight synthetic material. It consists of a seamless structure formed by interconnected carbon nanoparticles featuring a blend of mesopores. One of its remarkable properties is its ability to chemically bond with the current collector, eliminating the need for an additional binding agent. Consequently, this material boasts a low ESR, which contributes to its exceptional performance and high $P_d$ [34]. Electrodes made from aerogels, produced through the pyrolysis of resorcinol-formaldehyde aerogels, demonstrate superior conductivity compared to ACs. These aerogel electrodes offer the advantages of being both slim and robust, ensuring stability against mechanical stresses and vibrations, making them particularly suitable for use in high-vibration environments. In fact, CAG electrodes have achieved an impressive value of 104 $F/cm^3$ for $C_s$, resulting in $P_d$ and $E_d$ values being 20 W $kg^{-1}$ and 90 Wh $kg^{-1}$, respectively [35].

## 5.2.2 Metal Oxide-Based Materials

TMO materials possess numerous appealing qualities that make them highly attractive for utilization as energy storage device electrodes, including their abundant availability, environmentally friendly nature, easy accessibility, as well as a range of intriguing characteristics such as diverse compositions and structures, expansive surface areas, and impressive theoretical specific capacitance [36–38]. Enhancements in the physical and chemical attributes of metal oxide materials have been achieved through various avenues, including the strategic design of their compositions, the creation of innovative nanostructures, improvements in electroconductivity, and the deliberate introduction of oxygen vacancies. These advances have significantly elevated the overall performance of metal oxides. This includes chemical stability, availability of electro-active sites, specific surface area, and electrical conductivity [39]. B.E. Conway characterized TMOs like $Co_3O_4$, NiO, $MnO_2$, $Fe_3O_4$, $IrO_2$, $RuO_2$, and others, highlighting their notable pseudocapacitance properties [8]. Among these, $RuO_2$ stands out with its exceptionally low ESR compared to others. Compared to EDLCs, $RuO_2$ boasts more extensive power $P_d$ and $E_d$. However, $RuO_2$ is associated with a significant cost and exhibits subpar performance when subjected to high current densities [40].

TMOs have many advantages as materials for SC electrodes because of the variable valence states and inherent chemical stability. Cobalt oxide has an impressive theoretical capacitance (3560 F/g), superior electrochemical performance, and excellent reversibility [41]. Various cobalt oxide nanostructures have been developed, such as the ultra-layered $Co_3O_4$ structures made by Rao *et al.* [42], which show a $C_s$ of 548 F $g^{-1}$ at 8 A $g^{-1}$. Wang *et al.* [43] reported a 3D hollow $Co_3O_4$ structure with a Cs of 820 F $g^{-1}$ at 5 mV $s^{-1}$. Additionally, nanoporous $Co_3O_4$ prepared using the solvothermal method demonstrates a $P_d$ of 1.56 kW $kg^{-1}$, an $E_d$ of 42.3 Wh $kg^{-1}$, and a $C_s$ of 476 F/g.

Extensive research has been conducted on the utilization of TMOs in conjunction with carbon-based compounds to enhance their SC performance. Das *et al.* [44] presented noteworthy findings, achieving a remarkable $C_s$ of 1715 F $g^{-1}$, closely approaching the theoretically predicted value of 2000 F/g for $C_s$, which is the value of SCs constructed using $RuO_2$. To this configuration, $RuO_2$ was electrodeposited onto a SWCNT film electrode. Furthermore, we get a $P_d$ of 128.01 kW $kg^{-1}$, an $E_d$ of 39.28 Wh $kg^{-1}$, and a $C_s$ of 502.78 F/g, when $RuO_2$ was deposited onto a graphene foam, demonstrating remarkable stability over more than 8000 cycles [45]. Abundant research is being conducted on TMO electrodes in combination with various other materials.

### 5.2.3 Conducting Polymers (CPs)

CPs attain their electrical conductivity by forming a conjugated bond system along the backbone of the polymer. This conductive property is commonly obtained through the chemical oxidation or electrochemical oxidation of monomers. CPs have garnered significant interest for their ability to deliver substantial capacity (attributed to their redox properties), rapid charging and discharging capabilities, and cost-effectiveness in comparison with other carbon-based materials. Notably, the p/n type configuration of the polymer holds tremendous promise to achieve large values for $P_d$ as well as $E_d$. However, the advancement of pseudocapacitors based on CPs has been hindered by the limited cycling stability and the absence of efficient n-doped conductive materials [46]. Polyaniline (PANI) became famous for its low weight, excellent conductivity, mechanical flexibility, affordability, environmental friendliness, and impressive theoretical capacitance. However, a challenge with PANI is its tendency to undergo dimensional changes, shrinking, and swelling during the ion doping/dedoping processes during charge and discharge. To address this issue, researchers have developed a PANI/MOs/nanocarbon ternary hybrid by coating PANI onto MOs and carbon materials. This hybrid exhibits superior cyclic stability and impressive $C_s$. Furthermore, PANI displays a diverse spectrum of colors stemming from its various protonation and oxidation states. These electrochromic characteristics have practical applications in the creation of electrochromic SCs [47]. Polythiophene and polypyrrole (PPy) can be made directly in their doped form, offering excellent stability [46]. In comparison to other conducting polymers, PPy boasts higher flexibility and density. Not only do they show a remarkable electrical conductivity range of 10–500 S/cm, but they are also capable of rapid redox reactions, improving the charge storage efficiency [48].

Polyvinyl alcohol (PVA) stands out as an economical and environmentally friendly synthetic polymer. It is water-soluble, transparent, and lacks any noticeable odor. This versatile material exhibits impressive tensile strength and flexibility and excels in forming films, emulsifying substances, and bonding. These characteristics are influenced by humidity levels. Higher humidity can reduce the tensile strength; however, it enhances the tear resistance and elongation [49]. On a different note, among conducting polymers, poly-phenylene-vinylene (PPV) stands out as it can be transformed into extremely ordered film. It possesses a narrow band gap. Thus, it can be doped to create a polymer, which is electrically conductive in nature having a maximum conductivity of $10^{-3}$ S/cm. Additionally, by introducing functional side groups, we can alter their features [46].

### 5.2.4 Nanocomposites of Carbon Materials, CPs, and MOs

Composites comprise multiple materials, each with a unique mechanical, chemical, and physical properties. Nanocomposite electrodes blend carbon materials with either CPs or MOs, effectively combining faradaic (chemical) and non-faradaic (physical) charge storage mechanisms within a single electrode. The carbon materials contribute to a very large SSA, while pseudocapacitive materials further enhance the electrode's capacitance. CNTs serve as a structural foundation, ensuring uniform dispersion of MOs or CPs and resulting in increased pseudocapacitance and electric double-layer capacitance. These specialized electrodes achieve greater capacitance than their individual carbon, MOs, or CP counterparts [3, 50].

The combination of an AC cathode and a CPs anode offers superior $E_d$ and $P_d$ compared to EDLCs, while also delivering better durability during charge–discharge cycles than pseudocapacitors [50]. Additionally, a CNT hydrogel infused with PANI demonstrated an impressive $C_s$ of 680 mF cm$^{-2}$ at a current density of 1 mA cm$^{-2}$ [51]. Also, the amalgamation of TMOs and carbon proves highly beneficial for manufacturing high-performance SCs. A range of hybrid materials, including $Co_3O_4$/reduced graphene oxide (rGO) [52], $Co_3O_4$/CNTs [53], and $Co_3O_4$/CNFs [54], has been fabricated to achieve superior electrical conductivity and a significantly enlarged surface area.

The combination of MOs and CPs has been extensively explored to enhance electrochemical performance. Numerous research endeavors have been dedicated to the development of MOs integrated with CPs, aiming to significantly augment electrode conductivity. This enhancement translates into cyclic stability, rate capability, and improved $C_s$ [55, 56]. For instance, in a study conducted by Liu *et al.* [57], through *in situ* polymerization, they synthesized PPy-coated $MoO_3$. This material demonstrated an $E_d$ of 20 Wh kg$^{-1}$ at a $P_d$ of 75 W kg$^{-1}$ and a $C_s$ of 110 F g$^{-1}$ at 100 mA g$^{-1}$. Moreover, when employed as the positive electrode in an asymmetric SC (ASC), with PPy@ MoO3 as the positive electrode and AC as the negative electrode in a 0.5 M $K_2SO_4$ aqueous solution, the device exhibited an impressive $E_d$ of 12 Wh kg$^{-1}$ at a $P_d$ of 3 kW kg$^{-1}$. In a separate study led by Raj *et al.* [58], $Co_3O_4$-Pind was prepared, achieving remarkable $C_s$ values of 1805 F g$^{-1}$ at 2 A g$^{-1}$ and 1625 F g$^{-1}$ at 25 A g$^{-1}$, showcasing the potential of this composite for high-performance energy storage applications.

### 5.2.5 Modern-Day Materials

#### *5.2.5.1 MXenes*

In 2011, Yury Gogotsi introduced a groundbreaking class of materials known as MXenes, which are two-dimensional inorganic compounds. MXenes consist of ultra-thin layers of transition metal carbonitrides, carbides, or nitrides, typically just a few atoms deep. They possess a distinctive combination of properties, including high hydrophilicity and conductivity, thanks to their peculiar surface, which is hydroxyl-abundant, in addition to exceptional mechanical strength [59]. MXenes follow a general formula represented as $M_{n+1}AX_n$, with M representing a transition metal and A representing an element from groups 13 or 14. X signifies carbon and/or nitrogen, and n has a value of 1, 2, or 3. These materials are created through a selective process that involves removing A from their layered 3D MAX phase structure. To date, researchers have successfully synthesized approximately 20 different MXenes, all of which have demonstrated exceptional $C_s$ for the intercalation of various metal cations, including but not limited to $Na^+$, $Li^+$, $Al^{3+}$, $K^+$, $Mg^{2+}$, and more [60].

Fu *et al.* [61] developed a flexible paper electrode by employing 2D layered $Ti_3C_2T_x$, resulting in an impressive volumetric capacitance of 892 F $cm^{-3}$. Furthermore, their work demonstrated exceptional cycling durability, with no loss of capacitance observed after 10,000 cycles. Additionally, the researchers successfully crafted yarn SCs utilizing PEDOT-PSS and MXenes, showcasing remarkable device stability and efficiency, even under conditions of twisting and bending [62]. Rakhi and colleagues [63] utilized chemical synthesis to create nanocrystalline $\varepsilon$-$MnO_2$ on nanosheets of MXene, specifically $\varepsilon$-$MnO_2/Ti_3C_2T_x$ and $\varepsilon$-$MnO_2/Ti_2CT_x$. This approach resulted in the growth of nanocrystalline whiskers of $\varepsilon$-$MnO_2$, which significantly benefited the nanocomposite electrode by enhancing its SSA. As a result, the $C_s$ of the electrode improved by approximately threefold when compared to symmetric SCs based on pure MXene. Furthermore, SCs based on $\varepsilon$-$MnO_2$ and MXene exhibited robust cycling sturdiness, retaining approximately 88% of their capacitance after 10,000 cycles.

### *5.2.5.2 Metal Nitrides*

Metal nitrides have garnered attention as potential SC materials because of their strong chemical stability, favorable electrochemical characteristics, and well-established technological applicability. Specifically, transition metal nitrides (TMNs) stand out as promising candidates for SC electrode materials, thanks to their outstanding characteristics, including exceptional conductivity, intriguing electronic structure, expansive surface area, and impressive mechanical traits such as reasonable resistance to corrosion, ductile behavior, and high hardness [64].

The mechanism of energy storage in TMN relies on its pseudocapacitive behavior, which arises from surface redox reactions involving partially oxidized nitride species and its coaxial nanostructure. Presently, extensive research is underway to explore the supercapacitive properties of various TMN structures [65]. The compelling characteristics of high

SSA and electrochemical performance make molybdenum nitride a particularly attractive material for SC applications [64]. Researchers, such as Li *et al.* [66], have conducted electrochemical tests on ultrafine particles of $\gamma$-$Mo_2N$ using different electrolytes and varying molar concentrations. The results of the research demonstrate the feasibility of controlling the size of crystals in the desired product through the manipulation of experimental conditions. This leads to the achievement of a large specific capacitance. Specifically, these materials show a specific capacitance of 172 F/g in $H_2SO_4$ electrolyte over a wide potential range. Das and colleagues [67] reported the synthesis of nanoflakes composed of cobalt nitride (CoN), which exhibited impressive specific capacities of 990 mAh/g at a rate of 0.33 C and 690 mAh/g at 6.6 C following repeated charge–discharge cycles of 50 and 80, respectively. Remarkably, the material displayed a rise in discharge capacity as the cycle life extended, even when subjected to high current conditions. In another study [68], an asymmetric SC configuration was employed, combining metal nitrides, specifically chromium nitride (CrN) and cobalt nitride (CoN), with carbon-based materials, pitted against AC. Under non-aqueous electrolyte conditions, the specific capacitances of CrN/AC and CoN/AC were measured at 75 F/g and 37 F/g, respectively, when evaluated at a current density of 30 mA/g. Notably, despite its lower specific capacitance, CoN/AC exhibited a higher $E_d$ compared to CrN/AC due to its rapid faradaic charge storage mechanism.

#### 5.2.5.3 *Metal–Organic Frameworks (MOFs) and Covalent Organic Frameworks (COFs)*

The concept of reticular design has facilitated accurate forecasting of crystalline framework structures, the ability to adjust chemical composition, integration of diverse functionalities into the framework's structure, and, consequently, the refinement of properties in MOFs and COFs to a degree unmatched by any other material category. Exploiting the distinctive attributes of reticular materials has led to substantial progress, encompassing both fundamental insights and practical applications [69].

MOFs are used as templates for creating porous metal oxides and nanocomposites with porous metal/metal oxides and carbons [70]. This innovative method, first introduced by Yaghi *et al.* [71] in 1995, holds promise for enhancing the performance of materials based on metal oxides. While TMOs possess appealing properties like superior electrochemical response, good electrical conductivity, cost-effectiveness, and ease of processing, their practical use is hindered by a decrease in capacitive response during continuous faradaic reactions. This has spurred a growing demand for novel synthesis techniques capable of producing stable porous structures, offering precise control over the phase dimensions of metal oxides, and the use of MOFs comes into play. MOFs are created by chemically bonding inorganic and organic components. When polyvalent organic carboxylates combine with metal-containing units, they form 3D structures with precise pore size distributions and high SSA ranging from 1000 to 10,000 $m^2/g$. These MOFs are constructed using a variety of elements, including transition metals like Co, Zn, Fe, Ni, and Cu, alkaline earth elements such as Ba and Sr, p-block elements like In and Ga, and mixed metals [72].

MOFs' excellent surface area and versatility make them suitable for producing SC electrodes. Even so, they encounter several challenges, including reduced electrical conductivity during high charge–discharge rates, limited cycle life under such conditions, and hindered electrolyte diffusion within the highly crystalline MOFs [73]. Various approaches can be utilized to enhance the capacitive behavior of materials derived from MOFs: (i) Mixing MOF-derived materials with conductive carbon substances like rGO, graphene, and CNTs, alongside a secondary metal oxide to enhance electrical conductivity. (ii) Increasing the surface area of MOF-derived materials by initially heating MOF precursors under a nitrogen atmosphere before exposing them to air. This process prevents the rapid release of volatile gases, which could cause the collapse of the frameworks. (iii) Optimizing the pore size of MOF-derived materials to match the size of electrolyte ions by adjusting heating conditions. This allows ions to penetrate deeper into the MOF pores, a feat made possible by lowering the crystallinity of the material [73]. Wang and colleagues [70] introduced a MOF, specifically polyhedral ZIF-8, for SC applications. Their work demonstrated significantly improved $E_d$, $P_d$, and cycling life compared to contemporary hybrid capacitors. In another study, Kaur and team [74, 75] constructed a composite material by combining graphene with MOFs on a $TiO_2$/FTO substrate, as well as a nanocomposite involving quantum dot-MOFs (QD-MOFs). These 1-D and 2-D materials create pathways that promote an easy movement of ions during charge and discharge processes, ultimately enhancing the overall performance. Thus, the utilization of 3-D materials that offer numerous reaction sites within their 3-D networks may lead to even higher $E_d$ and $P_d$ in SC applications.

COFs represent a new class of microporous materials characterized by meticulously arranged structures created by linking organic building blocks through covalent bonding. These materials are becoming increasingly significant due to their precisely defined structures, exceptional porosity, flexibility in molecular design, and the ability to control the location and nature of redox-active groups with precision [76]. The first COF was introduced by Côté and his team back in 2005, employing boronate esters and boroxine linkages. This pioneering work yielded porous materials with robust crystalline structures [77].

Mulzer and colleagues [78] presented a more advanced technique using a PEDOT-modified COF (specifically, DAAQ-TFP COF) to achieve exceptionally fast charging rates and higher $E_d$. Xu and his team [79] have elucidated that the PEDOT-modified DAAQ-TFP COF films exhibit superior current response in cyclic voltammetry compared to the pristine DAAQ-TFP COF film, thanks to the electrical connectivity provided by the PEDOT chains. Furthermore, the transport of ions is enhanced through vertical pore channels, effectively reducing ion diffusion distances. Han and his collaborators [80] have created nanocoating of COFs on nickel nanowires, yielding a $C_s$ of 314 F/g at a high current density of 50 A/g, while retaining 74% of $C_s$ at 2 A/g. This high current density enables rapid charge and discharge processes. In another study, Kim and colleagues [76] produced N-doped carbon by carbonizing COFs through a 2-D network linked by azine bonds, referred to as ACOF1. Within ACOF1, micropores with diameters less than 1 nm have been formed, resulting in a substantial surface area (1596 $cm^2/g$). The $C_s$ of carbonized ACOF1 reach 234 F/g at a current density of 1.0 A/g, surpassing the $C_s$ of carbonized COF1, which stands at 191 F/g.

### 5.2.5.4 *Black Phosphorus (BP)*

BP, the latest addition to the family of 2D material, has recently become a subject of significant interest. This is primarily attributed to its remarkable theoretical $C_s$ of 2596 mAh/g. BP stands out due to its unique structural characteristics, featuring corrugated planes of phosphorus atoms connected by strong interlayer P-P bonds along with relatively weaker interlayer van der Waals forces. BP in bulk form can be transformed into thin sheets ranging from a single layer to a few layers, disrupting the feeble interlayer interactions [81, 82]. Hao and collaborators [83] developed an SC using liquid-exfoliated nanoflakes of BP, which is highly flexible and delivers remarkable performance metrics. This included a volumetric $C_s$ of 13.75 $F/cm^3$, a $P_d$ of 8.83 $W/cm^3$, an $E_d$ of 2.47 $mWh/cm^3$, and an impressive lifespan exceeding 30,000 cycles. Chen and colleagues [84] employed a sonochemical process to synthesize a BP–red phosphorus heterostructure. This BP/red phosphorus hybrid exhibited a $C_s$ of approximately 60.1 F/g and retained 83.3% of its actual capacitance after 2000 cycles.

### 5.2.6 The Electrolytes

Not only the choice of electrode materials, but also the selection of electrolytes has an impact on the performance of SCs. Achieving both high $E_d$ and $P_d$ in an SC requires a high cell operating voltage. However, this is constrained by the stability of the electrolyte when subjected to the applied potential [7]. A prevailing trend in SC advancement involves transitioning from aqueous electrolytes, limited by their low operating voltage (1 V) due to water decomposition, to non-aqueous mediums. Non-aqueous electrolytes, such as those based on aprotic solvents like acetonitrile or carbonate-based solvents (e.g., propylene carbonate), permit a much wider voltage window, around 2.5 V [85]. Ionic liquids (ILs), often referred to as room-temperature molten salts, have gained significant attention as potential next-generation electrolytes. They possess appealing properties for energy storage applications, including high thermal and electrochemical stability across a wide potential range (up to 4.3 V), non-toxicity, non-flammability, versatility in cation and anion combinations, and acceptable conductivity at elevated temperatures [86, 87].

Conductivity in an electrolyte system that hinges on two key factors: the salt's ability to dissociate into free charge carriers, such as anions and cations, and the mobility of these ions within the electrolyte. Therefore, future studies may focus on designing ILs with combinations of ions featuring extensive electrochemical stability window, low molecular weight, and high dissociation constants [7]. However, the current technology still falls short of meeting the practical application requirements for ILs; their ongoing development holds great promise for achieving energy storage devices with excellent performance, safety, and environmental friendliness.

## 5.3 Fabrication Techniques for Supercapacitors

The energy storage of a supercapacitor relies on three essential parts: electrodes, electrolytes, and separators. Each has a specific job, and they must work together for the device to store energy efficiently. Now, we are going to dwell on the fabrication techniques that are employed to produce supercapacitors.

### 5.3.1 Electrode Fabrication

#### *5.3.1.1 Laser Processing*

Laser processing stands out as an innovative and auspicious technology for producing supercapacitors. Recent investigations have concentrated on leveraging laser processing to create micro supercapacitors and nanostructured electrodes and elevate the overall performance of supercapacitors. Recent studies provided an overview of the accomplishments of laser-fabricated micro-supercapacitors, emphasizing the mechanism and strategy employed by lasers to enhance device performance [88]. Scientists are also working on laser-based nanostructured electrodes. This has been studied for the flexible in-plane micro-supercapacitors [89]. The Laser-Induced Hydrothermal Growth (LIHG) technique emerged as an up-and-coming method for the flexible in-plane micro-supercapacitors.

Additionally, laser processing has been instrumental in crafting supercapacitors on flexible substrates, such as silk and other materials [90]. A noteworthy outcome of this approach involved successfully creating a transparent conductive film (TCF) supercapacitor on a silk substrate, boasting a specific capacity of 35.8 mF/cm$^2$ and 59% transmittance. Furthermore, laser processing has been harnessed to fabricate graphene-based supercapacitors (G-SCs). They are getting attention for their remarkable flexibility and good electrochemical performance [91]. The study detailing the laser fabrication of G-SCs reported commendable outcomes that include excellent cycling stability and specific capacitance. In summary, laser processing emerges as a promising technology in the supercapacitor fabrication domain, motivating researchers to explore diverse techniques and materials to augment the performance and longevity of the devices.

#### *5.3.1.2 3D Printing*

We need to enhance the overall energy efficiency of stretchable supercapacitors. This must be done without affecting mechanical attributes. It involves the creation of 3D architectures featuring internal porous structures, thick electrodes, and active stretchability [92]. This underlying design strategy aims to augment both the interfacial area per unit footprint and the loading of electroactive materials. Notably, some key challenges persist for 3D stretchable structural SCs. A critical consideration pertains to the need for a meticulously designed 3D stretchable and flexible framework characterized by a significant surface area. The objective drives this design imperative to ensure both sufficient softness or toughness and large loading of electroactive materials, enabling the structure to endure recurrent deformations. Also, the requirement for firmly bonded hierarchical structured electroactive nanocomposites is essential for guaranteeing a substantial capacitance and energy output [93].

The pivotal aspect in manufacturing 3D-printed stretchable or flexible structural SCs is that the electrodes must have a three-dimensional structural design. Over recent years, considerable research efforts have been directed towards advancing electrode framework design. As investigations have deepened, structures with Negative Poisson's ratio (NPR) have garnered increasing recognition from researchers. Furthermore, they are recognized as auxetic structures. If the NPR structures are stretched, they exhibit lateral contraction under compression and lateral expansion. The occurrence of NPR is caused by the distinctive re-entrant unit cells within multicellular structures. Due to axial compressive loading, shrinkage deformation will happen to the concave section of the unit. This leads to the concentration of cells to a point and consequently induces negative Poisson's ratio effects [94, 95]. These effects contribute to the improvement in the robustness of the overall structure. At the same time, new topologies with varying Poisson's ratios can be developed so that the mechanical properties can be regulated. The 3D NPR multicellular structure acquires the inborn benefits of multicellular structures. This includes an impressive strength-to-weight ratio, a substantial surface area, efficient energy absorption, and commendable impact resistance. Also, it displays NPR effects, enhancing shape adaptability and producing large shear modulus [96, 97]. As an electrode framework, the 3D NPR multicellular structure is awarded an exemplary status for 3D-printed stretchable structural SCs due to its exceptional mechanical behaviors and physical properties.

#### *5.3.1.3 Sol-Gel Method*

Research on sol-gel processing of glass materials and inorganic ceramics has started with Ebelmen and Graham's studies on silica gels [98]. In this method, a 'sol' is formed that gradually evolves towards forming a gel-like diphasic system containing both liquid and solid phases. Colloidal particle dispersions in liquid media are referred to as sols. Solid particles with dimensions ranging from 1 to 100 nanometers are what define colloids [99].

The sol-gel process is started with a precursor preparation. Metal alkoxides are dissolved in a suitable solvent, typically alcohol or an alcohol mixture. Additional components, such as catalysts, acids, or bases, can be incorporated to regulate the reaction rate and final material properties. We use hydrolysis and condensation reactions for the sol formation, which can manifest as a viscous liquid or a gel, depending on the extent of reaction completion. Then, the sol undergoes a controlled aging period to facilitate further reaction and network formation, significantly impacting the final material properties. Solvent removal is achieved through drying processes, often involving heat and vacuum, to solidify the gel. Sometimes, we use the calcination process at high temperatures to improve crystallinity and enhance mechanical properties [100]. The sol-gel process offers several significant upper hands over other material synthesis techniques. This includes immense purity, exact control over the morphology of the material and uniform composition [101].

### *5.3.1.4 Chemical Vapor Deposition (CVD)*

Chemical vapor deposition (CVD) is a widely applied technique to synthesize materials like powders, thin films, or single crystals. The process involves depositing a solid material from vapor through chemical reactions taking place on or near the surface of a typically heated substrate. The final product will be a thin film, powder, or single crystal, depending on the procedure. We can manipulate various experimental parameters, including temperature, substrate material, pressure, reaction gas composition, and flow rates. These variations will result in diverse materials exhibiting modified physical and chemical properties. One of the pivotal features of CVD is that it is possible to produce uniform thickness coatings and selective deposition on substrates [102].

### *5.3.1.5 Electrochemical Deposition*

Electrochemical deposition, or electroplating, is a technique for depositing conducting or semiconducting materials, typically a conductor, onto a substrate. This process utilizes an electric field and a redox reaction (reduction–oxidation) to convert dissolved ions in a solution into a solid film on the surface of an electrode. Three key electrodes are crucial in this process: the working, counter, and reference electrodes. The working electrode is where the desired material gets deposited. The counter electrode provides the electrons necessary for the reduction reaction at the working electrode. We use the reference electrode to measure the potential at the working electrode, which works as a stable reference point [103].

In electrochemical deposition, we first prepare the electrolyte. The electrolyte is a solution of a salt of the material to be coated, and it contains the ions that will be reduced to form the coating. Electricity ignites the process, flowing through the electrolyte via dedicated electrodes. The part to be coated is taken as the cathode. The anode, often made of the deposition material or an inert conductor, assists in the flow of electricity. As electricity pulses through the solution, the dissolved ions are attracted to the cathode. Drawn by the electric field, they abandon their liquid existence and transform into a solid film, adhering to the surface of the cathode.

Electrochemical deposition boosts fiber electrode capacitance by crafting nanostructures and conductive polymers. This approach simplifies the fabrication of high-performance carbon nanotube-based fiber electrodes.

### 5.3.2 Electrolyte Fabrication

The electrolyte is a crucial element of a supercapacitor, as it facilitates the movement of ions between the electrodes. There are various techniques for fabricating electrolytes for supercapacitors.

#### *5.3.2.1 Gel Polymer Electrolyte (GPE)*

A gel polymer electrolyte (GPE) is a highly flexible and efficient electrolyte commonly employed in supercapacitors. It typically comprises a polymeric material that acts as a matrix and an electrolyte salt that provides mobile ions. The fabrication process of GPE involves several steps, which are outlined below.

***Polymer Selection*:** The choice of polymeric material for the Gel Polymer Electrolyte (GPE) necessitates chemical stability, mechanical robustness, and compatibility with the electrolyte salt. Conventional materials encompass polyvinyl alcohol (PVA), polyacrylate, and polyether sulfone [104, 105].

***Electrolyte Salt*:** The selection of electrolyte salt is contingent on the targeted ionic conductivity and compatibility with the chosen polymeric material. Frequently employed salt options encompass hydrogen sulfate ($H_2SO_4$), potassium hydroxide (KOH), and lithium chloride (LiCl) [105].

***Gelation Process*:** The gelation process commences by amalgamating the polymer and electrolyte salt and then subjecting the mixture to heat or exposure to UV light. This process initiates the formation of a three-dimensional network of polymer chains, effectively entrapping the electrolyte salt within [104].

***Mechanical Strengthening*:** Cross-linking agents or physical forces are employed to enhance the mechanical robustness of the GPE. This can be accomplished by introducing a second polymer or utilizing techniques such as vacuum filtration, electrospinning, or blade coating [106].

***Drying*:** The GPE undergoes a drying phase, eliminating excess solvent and establishing a stable gel. The drying process can be conducted under reduced pressure or at room temperature, tailored to the specific requirements of the supercapacitor [104].

***Electrolyte Adjustment*:** If deemed necessary, the ionic conductivity of the GPE may be fine-tuned by incorporating additives or modifying the electrolyte salt [105].

***Supercapacitor Assembly***: The GPE is interposed between the electrodes, and the assembly is finalized by affixing current collectors to both electrodes [106].

### 5.3.2.2 *Aqueous Electrolytes*

Aqueous electrolytes have become prevalent in modern research due to their ease of handling in laboratory settings compared to ionic liquids and organic electrolytes, which necessitate a purification process. Aqueous electrolytes [107, 108] usually exhibit a very large conductivity value of 0.8 $Scm^{-2}$ for 1 M $H_2SO_4$, surpassing the conductivity of ionic liquids and organic electrolytes by at least one order of magnitude. This heightened conductivity proves advantageous in enhancing power density as well as reducing the ESR. Commonly used aqueous electrolytes, including $H_2SO_4$, potassium chloride (KCl), and potassium hydroxide (KOH), offer ease of handling in open environments and low ionic resistance [16, 109]. Despite these advantages, aqueous electrolytes present certain drawbacks. The main issues are poor stability window and corrosion. These setbacks can heavily affect the stability and system performance. The pH levels within a system, whether acidic or basic, can lead to corrosion of packaging materials and collectors. This initiates corrosive reactions that compromise the performance of the system and reduce cycle lifetimes. Due to the water breakdown at 1.23 V, aqueous electrolytes often confine their cell voltage to 1 V, consequently imposing a limitation on the voltage stability of these electrolytes. This restriction significantly constrains an energy storage system's attainable energy and power density [110]. For aqueous electrolytes, Tomiyasu *et al.* demonstrated the broadest voltage window of 3.2 V through a saturated aqueous solution of sodium perchlorate (SSPAS) [111]. It suggests that SSPAS has the potential to serve as a substitute for non-aqueous electrolytes, thereby achieving superior electrochemical properties.

The choice of electrolyte fabrication technique varies with the specific requirements of the supercapacitor, such as its intended use, performance, and cost.

### 5.3.2.3 *Redox Additive Electrolyte (RAE)*

The performance of supercapacitors experiences a significant boost when the electrode is constructed with a redox additive electrolyte (RAE). Especially when compared to the aqueous electrolytes, this attribute will enhance the increased redox active sites on both the electrolyte and the electrode surface. The utilization of RAE contributes to substantially expanding the voltage window, elevating the energy density, and rendering it applicable in various domains. This improvement is evident in the work of Sundriyal *et al.* [112], who detailed the synthesis of a ZIF-67/rGO composite in the presence of RAE. Their study demonstrated notable enhancements in surface area and conductivity even after undergoing 1000 charge–discharge cycles.

### 5.3.3 Separator Fabrication

The separator acts as a barrier, preventing electrical shorts while facilitating ion transport between the electrodes. Its porosity, thickness, and chemical compatibility are important parts of maintaining the structural integrity and safety of the supercapacitor. Balancing these factors ensures efficient ion movement while minimizing parasitic reactions. Despite its critical role in supercapacitor safety and performance, the optimization of separator membranes (chemical composition and morphology) has received limited research attention [113, 114].

Separators are crafted from various materials such as rubber, aqua gel, and polyolefins. To prevent short circuits between electrodes, separators must possess electrical insulation and should be thin [115]. Also, they must exhibit electrochemical stability and serve as suitable dielectric materials.

Additionally, they should facilitate large ionic mobility between the electrode surface and the electrolyte. The type of separator employed varies based on our needs; separators used in structural members must bear mechanical loads, while flexible separators are essential for wearable electronics. Separators must withstand high temperatures for applications like oil drilling [116]. Different materials used for separator membranes are listed below.

#### *5.3.3.1 Polymer-Based Membranes*

They exhibit high chemical and mechanical resistance, and are cost-effective, abundant, and quickly processed. Among the most widely utilized commercial separators in SCs are polypropylene (Celgard) separators. There are other polymer-based separators such as polyethene (PE), PVDF, PVDF-HFP, polyethylene oxide (PEO), polyethene glycol (PEG), and polyvinyl chloride (PVC) [117–119]. Electrospinning is one of the methods used to fabricate them, and it has the advantage of controlling the pore size and porosity of the separator membrane [120, 121].

#### 5.3.3.2 *Ceramic-Based Separators*

Conventional separators from cellulose or polymers are unsuitable for high-temperature applications. This is because they shrink and deform at very high temperatures, leading to unreliable performance. Certain SCs with solid organic or ionic liquid electrolytes are employed for these specific applications [122, 123]. They are used together with either ceramic materials or natural clay [124, 125].

Nickel oxide (NiO) and zirconium dioxide ($ZrO_2$) are the main components of the ceramic separators. NiO has a low resistance at very high temperatures due to the negative temperature coefficient of resistance. Also, as a pore-forming agent, graphite powder was added. This will increase the pore size and the operating temperature, leading to quick ion diffusion and increased capacitance [103, 116].

#### 5.3.3.3 *Bio-Based Separator Membranes*

Driven by their abundance, affordability, biodegradability, and ease of access, researchers have explored the potential of tree leaves as separators for supercapacitors. Yao *et al.* investigated the use of *Bradyrhizobium japonicum* (BJ) biomass, a 3D structure augmented with nitrogen content, as an electrode material. They employed zinc chloride ($ZnCl_2$) activation (BJPC-1.5) to carbonize the BJ biomass in a single step, demonstrating the feasibility of this bio-derived material for supercapacitor applications [126]. The primary reason why tree leaves cannot be used as separators is because of their poor porosity. With the use of alkaline solutions, chemical treatments are carried out to increase the porosity of tree leaves. Jin *et al.* used this method, where they took four distinct types of tree leaves: *Platanus orientalis* (PO), *Osmanthus fragrans* (OF), *Magnolia grandiflora* (MG), and *Cinnamomum camphora* (CC), which were subjected to a NaOH activation procedure on their surface [127].

Avian eggshell membranes (ESMs), boasting non-toxicity, low cost, abundance, biodegradability, and highly porous structures, have become viable options for the production of separators for SCs. Exhibiting stability in both alcoholic and aqueous media, ESMs offer a compelling alternative to conventional separators [128]. Notably, the outer shell membrane, which can be readily separated from eggshells, is commonly utilized. Researchers have successfully fabricated separators from the eggshells of ducks (DESM), chickens (CESM), and geese (GESM), subsequently employing them in coin-type supercapacitors [129]. Also, there is use of cellulose-based membranes [130] and graphene oxide films [131].

A substance demonstrating significant thermal stability, commendable mechanical characteristics, and an elevated SSA is considered as an optimal separator material. Additionally, it functions as an electrolyte reservoir, maintains electrochemical inertness over an extended potential range, and exhibits robust cycling stability with substantial capacity retention. These characteristics collectively contribute to the material's desirability. Moreover, its compatibility with straightforward and cost-effective fabrication processes enhances its appeal for various applications.

## 5.4 Conclusion

The role of material selection and fabrication approaches in developing sustainable supercapacitors is crucial for advancing energy storage technology. This book chapter has provided a comprehensive overview of the significance of material selection, emphasizing the properties required for sustainable electrode materials in supercapacitors. Various electrode materials, including conducting polymers, metal oxides, and carbon-based nanomaterials, have been discussed, highlighting their unique features and capabilities. Moreover, emerging materials such as black phosphorus, covalent organic frameworks, metal nitrides, MXenes, and metal–organic frameworks hold great promise for further enhancing supercapacitor technology.

We have also explored diverse methodologies employed in fabricating supercapacitors. This encompassed an examination of various techniques utilized in the creation of electrodes, electrolytes, and separators, all crucial components contributing to the overall performance of these energy storage devices. Various physical and chemical deposition techniques have been evaluated in this context, emphasizing their potential in device fabrication. The selection of fabrication methods is underpinned by scientific rationale, with researchers conscientiously opting for approaches best suited to their specific experimental objectives and material requirements. Understanding the impact of fabrication approaches on electrode morphology, porosity, and interfacial properties is essential for optimizing supercapacitor performance. This conscientious approach ensures that the resulting devices are tailored to meet the performance demands and environmental conditions they are intended to encounter, thereby maximizing their efficacy and utility in real-world contexts.

This chapter aims to provide valuable insights for researchers and engineers engaged in sustainable energy storage by synthesizing the knowledge on material selection and fabrication approaches. Continued exploration and advancement in this field will undoubtedly contribute to developing high-performance and environmentally friendly supercapacitors, meeting the growing demand for efficient energy storage solutions in various applications.

## References

1. dos Reis, G.S. *et al.*, Flexible supercapacitors of biomass-based activated carbon-polypyrrole on eggshell membranes. *J. Environ. Chem. Eng.*, 9, 5, 106155, 2021, doi: 10.1016/j.jece.2021.106155.
2. Yu, Z., Tetard, L., Zhai, L., Thomas, J., Supercapacitor electrode materials: Nanostructures from 0 to 3 dimensions. *Energy Environ. Sci.*, 8, 3, 702–730, 2015, doi: 10.1039/c4ee03229b.
3. Poonam, Sharma, K., Arora, A., Tripathi, S.K., Review of supercapacitors: Materials and devices. *J. Energy Storage*, 21, 801–825, Feb. 2019, doi: 10.1016/j.est.2019.01.010.
4. Frackowiak, E. and Béguin, F., Carbon materials for the electrochemical storage of energy in capacitors. *Carbon*, 39, 6, 937–950, 2001, doi: 10.1016/S0008-6223(00)00183-4.
5. Zhi, J., Wang, Y., Deng, S., Hu, A., Study on the relation between pore size and supercapacitance in mesoporous carbon electrodes with silica-supported carbon nanomembranes. *RSC Adv.*, 4, 76, 40296–40300, 2014, doi: 10.1039/c4ra06260d.
6. Xie, L. *et al.*, Hierarchical porous carbon microtubes derived from willow catkins for supercapacitor applications. *J. Mater. Chem. A Mater.*, 4, 5, 1637–1646, 2016, doi: 10.1039/c5ta09043a.
7. Zhang, L.L. and Zhao, X.S., Carbon-based materials as supercapacitor electrodes. *Chem. Soc. Rev.*, 38, 9, 2520, 2520–2531, 2009, doi: 10.1039/b813846j.
8. Conway, B.E., Transition from 'supercapacitor' to 'battery' behavior in electrochemical energy storage, in: *Proceedings of the International Power Sources Symposium*, p. 1539, 1991.
9. Qu, D. and Shi, H., Studies of activated carbons used in double-layer capacitors. *J. Power Sources*, 74, 1, 99–107, 1998, doi: 10.1016/S0378-7753(98)00038-X.
10. Endo, M. *et al.*, Capacitance and Pore-Size Distribution in Aqueous and Nonaqueous Electrolytes Using Various Activated Carbon Electrodes. *J. Electrochem. Soc.*, 148, 8, A910, 2001, doi: 10.1149/1.1382589.
11. Raymundo-Piñero, E., Kierzek, K., Machnikowski, J., Béguin, F., Relationship between the nanoporous texture of activated carbons and their capacitance properties in different electrolytes. *Carbon N Y*, 44, 12, 2498–2507, 2006, doi: 10.1016/j.carbon.2006.05.022.

12. Salitra, G., Soffer, A., Eliad, L., Cohen, Y., Aurbach, D., Carbon Electrodes for Double-Layer Capacitors I. Relations Between Ion and Pore Dimensions. *J. Electrochem. Soc.*, 147, 7, 2486, 2000, doi: 10.1149/1.1393557.
13. Zhong, C., Deng, Y., Hu, W., Qiao, J., Zhang, L., Zhang, J., A review of electrolyte materials and compositions for electrochemical supercapacitors. *Chem. Soc. Rev.*, 44, 21, 7484–7539, 2015, doi: 10.1039/c5cs00303b.
14. Ania, C.O., Khomenko, V., Raymundo-Piñero, E., Parra, J.B., Béguin, F., The large electrochemical capacitance of microporous doped carbon obtained by using a zeolite template. *Adv. Funct. Mater.*, 17, 11, 1828–1836, 2007, doi: 10.1002/adfm.200600961.
15. Kötz, R. and Carlen, M., Principles and applications of electrochemical capacitors. *Electrochim. Acta*, 45, 15–16, 2483–2498, 2000, doi: 10.1016/S0013-4686(00)00354-6.
16. Pandolfo, A.G. and Hollenkamp, A.F., Carbon properties and their role in supercapacitors. *J. Power Sources*, 157, 1, 11–27, 2006, doi: 10.1016/j.jpowsour.2006.02.065.
17. Presser, V., Heon, M., Gogotsi, Y., Carbide-derived carbons - from porous networks to nanotubes and graphene. *Adv. Funct. Mater.*, 21, 5, 810–833, 2011, doi: 10.1002/adfm.201002094.
18. El-Kady, M.F., Strong, V., Dubin, S., Kaner, R.B., Laser scribing of high-performance and flexible graphene-based electrochemical capacitors. *Sci. (1979)*, 335, 6074, 1326–1330, 2012, doi: 10.1126/science.1216744.
19. Liu, C., Yu, Z., Neff, D., Zhamu, A., Jang, B.Z., Graphene-based supercapacitor with an ultrahigh energy density. *Nano Lett.*, 10, 12, 4863–4868, 2010, doi: 10.1021/nl102661q.
20. Miller, J.R., Outlaw, R.A., Holloway, B.C., Graphene double-layer capacitor with ac line-filtering performance. *Sci. (1979)*, 329, 5999, 1637–1639, 2010, doi: 10.1126/science.1194372.
21. Stankovich, S. *et al.*, Graphene-based composite materials. *Nature*, 442, 7100, 282–286, 2006, doi: 10.1038/nature04969.
22. Stankovich, S. *et al.*, Synthesis of graphene-based nanosheets via chemical reduction of exfoliated graphite oxide. *Carbon N Y*, 45, 7, 1558–1565, 2007, doi: 10.1016/j.carbon.2007.02.034.
23. Stoller, M.D., Park, S., Yanwu, Z., An, J., Ruoff, R.S., Graphene-Based ultracapacitors. *Nano Lett.*, 8, 10, 3498–3502, 2008, doi: 10.1021/nl802558y.
24. Xu, Y., Lin, Z., Huang, X., Liu, Y., Huang, Y., Duan, X., Flexible solid-state supercapacitors based on three-dimensional graphene hydrogel films. *ACS Nano*, 7, 5, 4042–4049, 2013, doi: 10.1021/nn4000836.
25. Popov, V.N., Carbon nanotubes: Properties and application. *Mater. Sci. Eng., R*, 43, 3, 61–102, 2004, doi: 10.1016/j.mser.2003.10.001.
26. Dresselhaus, M.S., Dresselhaus, G., Saito, R., Carbon fibers based on C60 and their symmetry. *Phys. Rev. B*, 45, 11, 6234, 1992, doi: 10.1103/PhysRevB.45.6234.

27. Frackowiak, E., Metenier, K., Bertagna, V., Beguin, F., Supercapacitor electrodes from multiwalled carbon nanotubes. *Appl. Phys. Lett.*, 77, 15, 2421–2423, 2000, doi: 10.1063/1.1290146.
28. An, K.H. *et al.*, Electrochemical properties of high-power supercapacitors using single-walled carbon nanotube electrodes. *Adv. Funct. Mater.*, 11, 5, 387–932, 2001, doi: 10.1002/1616-3028(200110)11:5<387::AID-ADFM387>3.0.CO;2-G.
29. Li, X., Rong, J., Wei, B., Electrochemical behavior of single-walled carbon nanotube supercapacitors under compressive stress. *ACS Nano*, 4, 10, 6039–6049, 2010, doi: 10.1021/nn101595y.
30. Dillon, A.C., Carbon nanotubes for photoconversion and electrical energy storage. *Chem. Rev.*, 110, 11, 6856–6872, 2010, doi: 10.1021/cr9003314.
31. Peng, C., Jin, J., Chen, G.Z., A comparative study on electrochemical codeposition and capacitance of composite films of conducting polymers and carbon nanotubes. *Electrochim. Acta*, 53, 2, 525–537, 2007, doi: 10.1016/j.electacta.2007.07.004.
32. Xu, B. *et al.*, Highly mesoporous and high surface area carbon: A high capacitance electrode material for EDLCs with various electrolytes. *Electrochem. Commun.*, 10, 5, 795–797, 2008, doi: 10.1016/j.elecom.2008.02.033.
33. Simon, P. and Gogotsi, Y., Materials for electrochemical capacitors, in: *Materials for Sustainable Energy: A Collection of Peer-Reviewed Research and Review Articles from Nature Publishing Group*, pp. 845–854, 2010, doi: 10.1142/9789814317665_0021.
34. Aricò, A.S., Bruce, P., Scrosati, B., Tarascon, J.M., Van Schalkwijk, W., Nanostructured materials for advanced energy conversion and storage devices. *Nat. Mater.*, 4, 5, 366–377, 2005, doi: 10.1038/nmat1368.
35. Chien, H.C., Cheng, W.Y., Wang, Y.H., Lu, S.Y., Ultrahigh specific capacitances for supercapacitors achieved by nickel cobaltite/carbon aerogel composites. *Adv. Funct. Mater.*, 22, 23, 5038–5043, 2012, doi: 10.1002/adfm.201201176.
36. Wang, G., Zhang, L., Zhang, J., A review of electrode materials for electrochemical supercapacitors. *Chem. Soc. Rev.*, 41, 2, 797–828, 2012, doi: 10.1039/c1cs15060j.
37. Yu, G. *et al.*, Enhancing the supercapacitor performance of graphene/MnO2 nanostructured electrodes by conductive wrapping. *Nano Lett.*, 11, 10, 4438–4442, 2011, doi: 10.1021/nl2026635.
38. Lu, Q., Chen, J.G., Xiao, J.Q., Nanostructured electrodes for high-performance pseudocapacitors. *Angew. Chem. Int. Ed.*, 52, 7, 1882–1889, 2013, doi: 10.1002/anie.201203201.
39. An, C., Zhang, Y., Guo, H., Wang, Y., Metal oxide-based supercapacitors: progress and prospectives. *Nanoscale Adv.*, 1, 12, 4644–4658, 2019, doi: 10.1039/c9na00543a.
40. Kim, I.H. and Kim, K.B., Ruthenium oxide thin film electrodes for supercapacitors. *Electrochem. Solid-State Lett.*, 4, 5, A62, 2001, doi: 10.1149/1.1359956.

41. Yan, X., Wang, Y., Ma, Z., Synthesis, characterization and electrochemical performance of cobalt oxides for supercapacitor. *Int. J. Electrochem. Sci.*, 13.1, 1074–1083, 2018.
42. Meher, S.K. and Rao, G.R., Ultralayered Co3O4 for high-performance supercapacitor applications. *J. Phys. Chem. C*, 115, 31, 15646–15654, 2011, doi: 10.1021/jp201200e.
43. Wang, Y., Lei, Y., Li, J., Gu, L., Yuan, H., Xiao, D., Synthesis of 3D-nanonet hollow structured Co3O4 for high capacity supercapacitor, in: *ACS Appl. Mater. Interfaces*, pp. 6739–6747, 2014.
44. Das, R.K., Liu, B., Reynolds, J.R., Rinzler, A.G., Engineered macroporosity in single-wall carbon nanotube films. *Nano Lett.*, 9, 2, 6739–6747, 2009, doi: 10.1021/n1803168s.
45. Wang, W. *et al.*, Hydrous ruthenium oxide nanoparticles anchored to graphene and carbon nanotube hybrid foam for supercapacitors. *Sci. Rep.*, 4, 4452, 2014, doi: 10.1038/srep04452.
46. Snook, G.A., Kao, P., Best, A.S., Conducting-polymer-based supercapacitor devices and electrodes. *J. Power Sources*, 196, 1, 1–12, 2011, doi: 10.1016/j.jpowsour.2010.06.084.
47. Han, G. *et al.*, MnO2 nanorods intercalating graphene oxide/polyaniline ternary composites for robust high-performance supercapacitors. *Sci. Rep.*, 4, 4824, 2014, doi: 10.1038/srep04824.
48. Zhou, C., Zhang, Y., Li, Y., Liu, J., Construction of high-capacitance 3D CoO@Polypyrrole nanowire array electrode for aqueous asymmetric supercapacitor. *Nano Lett.*, 13, 5, 2078–2085, 2013, doi: 10.1021/nl400378j.
49. Gaaz, T.S. *et al.*, Properties and applications of polyvinyl alcohol, halloysite nanotubes and their nanocomposites. *Molecules*, 20, 12, 22833–22847, 2015, doi: 10.3390/molecules201219884.
50. Banerjee, S. and Kar, K.K., Conducting polymers as electrode materials for supercapacitors, in: *Springer Series in Materials Science*, vol. 302, pp. 333–352, 2020.
51. Zeng, S., Chen, H., Cai, F., Kang, Y., Chen, M., Li, Q., Electrochemical fabrication of carbon nanotube/polyaniline hydrogel film for all-solid-state flexible supercapacitor with high areal capacitance. *J. Mater. Chem. A Mater.*, 3, 47, 23864–23870, 2015, doi: 10.1039/c5ta05937b.
52. Xiang, C., Li, M., Zhi, M., Manivannan, A., Wu, N., A reduced graphene oxide/Co3O4 composite for supercapacitor electrode. *J. Power Sources*, 226, 65–70, 2013, doi: 10.1016/j.jpowsour.2012.10.064.
53. Shan, Y. and Gao, L., Formation and characterization of multi-walled carbon nanotubes/Co3O4 nanocomposites for supercapacitors. *Mater. Chem. Phys.*, 103, 2–3, 206–210, 2007, doi: 10.1016/j.matchemphys.2007.02.038.
54. Abouali, S. *et al.*, Electrospun Carbon Nanofibers with *in Situ* Encapsulated Co3O4 Nanoparticles as Electrodes for High-Performance Supercapacitors. *ACS Appl. Mater. Interfaces*, 7, 24, 13503–13511, 2015, doi: 10.1021/acsami.5b02787.

55. Chen, H., Cong, T.N., Yang, W., Tan, C., Li, Y., Ding, Y., Progress in electrical energy storage system: A critical review. *Prog. Nat. Sci.*, 19, 3, 219–312, 2009, doi: 10.1016/j.pnsc.2008.07.014.
56. Xie, K. and Wei, B., Materials and structures for stretchable energy storage and conversion devices. *Adv. Mater.*, 26, 22, 3592–3617, 2014, doi: 10.1002/adma.201305919.
57. Liu, Y., Zhang, B., Yang, Y., Chang, Z., Wen, Z., Wu, Y., Polypyrrole-coated α-MoO3 nanobelts with good electrochemical performance as anode materials for aqueous supercapacitors. *J. Mater. Chem. A Mater.*, 1, 43, 13582–13587, 2013, doi: 10.1039/c3ta12902k.
58. Raj, R.P., Ragupathy, P., Mohan, S., Remarkable capacitive behavior of a Co3O4-polyindole composite as electrode material for supercapacitor applications. *J. Mater. Chem. A Mater.*, 3, 48, 24338–24348, 2015, doi: 10.1039/c5ta07046e.
59. Naguib, M. *et al.*, Two-dimensional nanocrystals produced by exfoliation of Ti3ALC2, in: *MXenes: From Discovery to Applications of Two-Dimensional Metal Carbides and Nitrides*, pp. 15–29, 2023.
60. Chaudhari, N.K., Jin, H., Kim, B., San Baek, D., Joo, S.H., Lee, K., MXene: An emerging two-dimensional material for future energy conversion and storage applications. *J. Mater. Chem. A*, 5, 47, 24564–24579, 2017, doi: 10.1039/c7ta09094c.
61. Fu, Q. *et al.*, Free-standing Ti3C2Tx electrode with ultrahigh volumetric capacitance. *RSC Adv.*, 7, 20, 11998–12005, 2017, doi: 10.1039/c7ra00126f.
62. Zhang, J., Seyedin, S., Gu, Z., Yang, W., Wang, X., Razal, J.M., MXene: A potential candidate for yarn supercapacitors. *Nanoscale*, 9, 47, 18604–18608, 2017, doi: 10.1039/c7nr06619h.
63. Rakhi, R.B., Ahmed, B., Anjum, D., Alshareef, H.N., Direct Chemical Synthesis of MnO2 Nanowhiskers on Transition-Metal Carbide Surfaces for Supercapacitor Applications. *ACS Appl. Mater. Interfaces*, 8, 29, 18806–18814, 2016, doi: 10.1021/acsami.6b04481.
64. Sreehari, S., George, N.S., Jose, L.M., Nandakumar, S., Subramaniam, R.T., Aravind, A., A Review on 2D Transition Metal Nitrides: Structural and Morphological Impacts on Energy Storage and Photocatalytic Applications. *J. Alloys Compd.*, 950, 169888, Mar. 2023, doi: 10.1016/j.jallcom.2023.169888.
65. Mansoor, N.S., Fattah-alhosseini, A., Elmkhah, H., Shishehian, A., Electrochemical behavior of TiN, CrN and TiN/CrN nanostructured coatings on the nickel-chromium alloy used in dental fixed prosthesis. *J. Asian Ceram. Soc.*, 8, 3, 694–710, Jul. 2020, doi: 10.1080/21870764.2020.1776915.
66. Li, X., Xing, Y., Wang, H., Wang, H., Wang, W., Chen, X., Synthesis and characterization of uniform nanoparticles of γ-Mo2N for supercapacitors. *Trans. Nonferrous Met. Soc. China*, 19, 3, 620–625, Jun. 2009, doi: 10.1016/S1003-6326(08)60323-4.

67. Das, B., Reddy, M.V., Malar, P., Osipowicz, T., Subba Rao, G.V., Chowdari, B.V.R., Nanoflake CoN as a high capacity anode for Li-ion batteries. *Solid State Ion*, 180, 17–19, 1061–1068, Jul. 2009, doi: 10.1016/j.ssi.2009.05.007.
68. Das, B., Behm, M., Lindbergh, G., Reddy, M.V., Chowdari, B.V.R., High performance metal nitrides, MN (M = Cr, Co) nanoparticles for non-aqueous hybrid supercapacitors. *Adv. Powder Technol.*, 26, 3, 783–788, May 2015, doi: 10.1016/j.apt.2015.02.001.
69. Freund, R. *et al.*, The Current Status of MOF and COF Applications. *Angew. Chem. Int. Ed.*, 60, 45, 23975–24001, 2021, doi: 10.1002/anie.202106259.
70. Wang, R. *et al.*, Engineering metal organic framework derived 3D nanostructures for high performance hybrid supercapacitors. *J. Mater. Chem. A Mater.*, 5, 1, 292–302, 2017, doi: 10.1039/c6ta09143a.
71. Yaghi, O.M. and Li, H., Hydrothermal Synthesis of a Metal-Organic Framework Containing Large Rectangular Channels. *J. Am. Chem. Soc.*, 117, 41, 10401–10402, 1995, doi: 10.1021/ja00146a033.
72. Qiu, T., Liang, Z., Guo, W., Tabassum, H., Gao, S., Zou, R., Metal–Organic Framework-Based Materials for Energy Conversion and Storage. *ACS Energy Lett.*, 5, 2, 520–532, 2020, doi: 10.1021/acsenergylett.9b02625.
73. Salunkhe, R.R., Kaneti, Y.V., Yamauchi, Y., Metal-Organic Framework-Derived Nanoporous Metal Oxides toward Supercapacitor Applications: Progress and Prospects. *ACS Nano*, 11, 6, 5293–5308, 2017, doi: 10.1021/acsnano.7b02796.
74. Kaur, R., Kim, K.H., Deep, A., A convenient electrolytic assembly of graphene-MOF composite thin film and its photoanodic application. *Appl. Surf. Sci.*, 396, 1303–1309, 2017, doi: 10.1016/j.apsusc.2016.11.150.
75. Kaur, R., Rana, A., Singh, R.K., Chhabra, V.A., Kim, K.H., Deep, A., Efficient photocatalytic and photovoltaic applications with nanocomposites between CdTe QDs and an NTU-9 MOF. *RSC Adv.*, 7, 46, 29015–29024, 2017, doi: 10.1039/c7ra04125j.
76. Kim, G., Yang, J., Nakashima, N., Shiraki, T., Highly Microporous Nitrogen-doped Carbon Synthesized from Azine-linked Covalent Organic Framework and its Supercapacitor Function. *Chem. Eur. J.*, 23, 69, 17504–17510, 2017, doi: 10.1002/chem.201702805.
77. Alahakoon, S.B., Thompson, C.M., Occhialini, G., Smaldone, R.A., Design Principles for Covalent Organic Frameworks in Energy Storage Applications. *ChemSusChem*, 10, 10, 2116–2129, 2017, doi: 10.1002/cssc.201700120.
78. Mulzer, C.R. *et al.*, Superior charge storage and power density of a conducting polymer-modified covalent organic framework. *ACS Cent. Sci.*, 2, 9, 667–673, 2016, doi: 10.1021/acscentsci.6b00220.
79. Xu, Q., Dalapati, S., Jiang, D., Charge up in wired covalent organic frameworks. *ACS Cent. Sci.*, 2, 9, 586–587, 2016, doi: 10.1021/acscentsci.6b00264.
80. Han, Y. *et al.*, Nanocoating covalent organic frameworks on nickel nanowires for greatly enhanced-performance supercapacitors. *Nanotechnology*, 28, 33, 33LT01, 2017, doi: 10.1088/1361-6528/aa7bb6.

81. Lin, S., Chui, Y., Li, Y., Lau, S.P., Liquid-phase exfoliation of black phosphorus and its applications. *FlatChem*, 2, 15–37, 2017, doi: 10.1016/j.flatc.2017.03.001.
82. Ren, X. *et al.*, Properties, preparation and application of black phosphorus/phosphorene for energy storage: a review. *J. Mater. Sci.*, 52, 17, 10364–10386, 2017, doi: 10.1007/s10853-017-1194-3.
83. Hao, C. *et al.*, Flexible All-Solid-State Supercapacitors based on Liquid-Exfoliated Black-Phosphorus Nanoflakes. *Adv. Mater.*, 28, 16, 3194–3201, 2016, doi: 10.1002/adma.201505730.
84. Chen, X. *et al.*, A black/red phosphorus hybrid as an electrode material for high-performance Li-ion batteries and supercapacitors. *J. Mater. Chem. A Mater.*, 5, 14, 6581–6588, 2017, doi: 10.1039/c7ta00455a.
85. Miller, J.R. and Simon, P., Materials science: Electrochemical capacitors for energy management. *Science*, 321, 5889, 651–652, 2008, doi: 10.1126/science.1158736.
86. Chen, R., Wu, F., Li, L., Xu, B., Qiu, X., Chen, S., Novel binary room-temperature complex system based on LiTFSI and 2-oxazolidinone and its characterization as electrolyte. *J. Phys. Chem. C*, 111, 13, 5184–5194, 2007, doi: 10.1021/jp066429f.
87. Balducci, A. *et al.*, High temperature carbon-carbon supercapacitor using ionic liquid as electrolyte. *J. Power Sources*, 165, 2, 922–927, 2007, doi: 10.1016/j.jpowsour.2006.12.048.
88. Wang, X. and Zhang, Q., Recent progress on laser fabrication of on-chip microsupercapacitors. *J. Energy Storage*, 34, 101994, 2021, doi: 10.1016/j.est.2020.101994.
89. Liu, H., Sun, Z., Chen, Y., Zhang, W., Chen, X., Wong, C.P., Laser Processing of Flexible In-Plane Micro-supercapacitors: Progresses in Advanced Manufacturing of Nanostructured Electrodes. *ACS Nano*, 16, 7, 10088–10129, 2022, doi: 10.1021/acsnano.2c02812.
90. Yu, Y., Wang, S., Ma, D., Joshi, P., Hu, A., Recent Progress on Laser Manufacturing of Microsize Energy Devices on Flexible Substrates. *JOM*, 70, 9, 1816–1822, 2018, doi: 10.1007/s11837-018-2986-x.
91. Fu, X.-Y., Chen, Z.-D., Han, D.-D., Zhang, Y.-L., Xia, H., Sun, H.-B., Laser fabrication of graphene-based supercapacitors. *Photonics Res.*, 8, 4, 20–23, 2020, doi: 10.1364/prj.382401.
92. Zhao, C., Wang, C., Gorkin, R., Beirne, S., Shu, K., Wallace, G.G., Three dimensional (3D) printed electrodes for interdigitated supercapacitors. *Electrochem. Commun.*, 41, 20–23, 2014, doi: 10.1016/j.elecom.2014.01.013.
93. Chang, P., Mei, H., Tan, Y., Zhao, Y., Huang, W., Cheng, L., A 3D-printed stretchable structural supercapacitor with active stretchability/flexibility and remarkable volumetric capacitance. *J. Mater. Chem. A Mater.*, 8, 27, 13646–13658, 2020, doi: 10.1039/d0ta04460a.

94. Jiang, J.W., Chang, T., Guo, X., Tunable negative Poisson's ratio in hydrogenated graphene. *Nanoscale*, 8, 35, 15948–15953, 2016, doi: 10.1039/c6nr04976a.
95. Yu, L., Yan, Q., Ruzsinszky, A., Negative Poisson's ratio in 1T-type crystalline two-dimensional transition metal dichalcogenides. *Nat. Commun.*, 8, 1–8, 2017, doi: 10.1038/ncomms15224.
96. Huang, C. and Chen, L., Negative Poisson's Ratio in Modern Functional Materials. *Adv. Mater.*, 28, 37, 8079–8096, 2016, doi: 10.1002/adma.201601363.
97. Taylor, M., Francesconi, L., Gerendás, M., Shanian, A., Bertoldi, K., Low porosity metallic periodic structures with negative poisson's ratio. *Adv. Mater.*, 26, 15, 2365–2370, 2014, doi: 10.1002/adma.201304464.
98. Graham, T., On the properties of silicic acid and other analogous colloidal substances. *J. Chem. Soc.*, 17, 318–327, 1864, doi: 10.1039/JS8641700318.
99. Hutchinson, E., Chemistry: Interfacial Phenomena. J. T. Davies and E. K. Rideal. Academic Press, New York, ed. 2, 1963. 494 pp. $15, in: *Science (1979)*, vol. 142, 1963.
100. Hench, L.L., The sol-gel process - Chemical Reviews. *Chem. Rev.*, 90, 33–72, ACS Publications), 1990.
101. Ozer, N., Sol-Gel Optics: Processing and applications. *Sol. Energy Mater. Sol. Cells*, 43, 3, 223–320, 1996, doi: 10.1016/0927-0248(95)00162-x.
102. Martin, P.M., *Handbook of Deposition Technologies for Films and Coatings Science, Applications and Technology*, 3rd Edition, Chapter 7, pp. 314–361, Elsevier, 2010.
103. Prakash, S. and Yeom, J., Advanced Fabrication Methods and Techniques, in: *Nanofluidics and Microfluidics*, pp. 87–170, 2014.
104. Yan, C., Jin, M., Pan, X., Ma, L., Ma, X., A flexible polyelectrolyte-based gel polymer electrolyte for high-performance all-solid-state supercapacitor application. *RSC Adv.*, 10, 16, 9299–9308, 2020, doi: 10.1039/c9ra10701k.
105. Peng, X. *et al.*, A zwitterionic gel electrolyte for efficient solid-state supercapacitors. *Nat. Commun.*, 7, 1–8, 2016, doi: 10.1038/ncomms11782.
106. Dai, H. *et al.*, Polymer gel electrolytes for flexible supercapacitors: Recent progress, challenges, and perspectives. *Energy Storage Mater.*, 34, 320–355, 2021, doi: 10.1016/j.ensm.2020.09.018.
107. Kang, Y.J., Chung, H., Kim, W., 1.8-V flexible supercapacitors with asymmetric configuration based on manganese oxide, carbon nanotubes, and a gel electrolyte. *Synth. Met.*, 166, 1, 40–44, 2013, doi: 10.1016/j.synthmet.2013.01.013.
108. Lufrano, F. and Staiti, P., Conductivity and capacitance properties of a supercapacitor based on Nafion electrolyte in a nonaqueous system. *Electrochem. Solid-State Lett.*, 7, 11, A447, 2004, doi: 10.1149/1.1808095.
109. Naoi, K. and Morita, M., Advanced polymers as active materials and electrolytes for electrochemical capacitors and hybrid capacitor systems. *Electrochem. Soc. Interface*, 17, 1, 44, 2008, doi: 10.1149/2.f06081if.

110. Rathod, D. *et al.*, Design of an 'all solid-state' supercapacitor based on phosphoric acid doped polybenzimidazole (PBI) electrolyte. *J. Appl. Electrochem.*, 39, 7, 1097–1103, 2009, doi: 10.1007/s10800-008-9764-3.
111. Tomiyasu, H., Shikata, H., Takao, K., Asanuma, N., Taruta, S., Park, Y.Y., An aqueous electrolyte of the widest potential window and its superior capability for capacitors. *Sci. Rep.*, 7, 45048, 2017, doi: 10.1038/srep45048.
112. Sundriyal, S., Shrivastav, V., Kaur, H., Deep, A., High-Performance Symmetrical Supercapacitor with a Combination of a ZIF-67/rGO Composite Electrode and a Redox Additive Electrolyte. *ACS Omega*, 3, 12, 17348–17358, 2018, doi: 10.1021/acsomega.8b02065.
113. Wang, H. *et al.*, Suppressing the Self-Discharge of Supercapacitors by Modifying Separators with an Ionic Polyelectrolyte. *Adv. Mater. Interfaces*, 5, 10, 170154, 2018, doi: 10.1002/admi.201701547.
114. Peng, H., Xiao, L., Sun, K., Ma, G., Wei, G., Lei, Z., Preparation of a cheap and environmentally friendly separator by coaxial electrospinning toward suppressing self-discharge of supercapacitors. *J. Power Sources*, 435, 226800, 2019, doi: 10.1016/j.jpowsour.2019.226800.
115. Zhang, X., He, B., Zhao, Y., Tang, Q., A porous ceramic membrane tailored high-temperature supercapacitor. *J. Power Sources*, 379, 60–67, 2018, doi: 10.1016/j.jpowsour.2018.01.033.
116. Ahankari, S., Lasrado, D., Subramaniam, R., Advances in materials and fabrication of separators in supercapacitors. *Mater. Adv.*, 3, 3, 1472–1496, 2022, doi: 10.1039/d1ma00599e.
117. Solarajan, A.K., Murugadoss, V., Angaiah, S., High performance electrospun PVdF-HFP/SiO2 nanocomposite membrane electrolyte for Li-ion capacitors. *J. Appl. Polym. Sci.*, 134, 32, 45177, 2017, doi: 10.1002/app.45177.
118. Solarajan, A.K., Murugadoss, V., Angaiah, S., Montmorillonite embedded electrospun PVdF–HFP nanocomposite membrane electrolyte for Li-ion capacitors. *Appl. Mater. Today*, 5, 33–40, 2016, doi: 10.1016/j.apmt.2016.09.002.
119. Murugadoss, V., Arunachalam, S., Elayappan, V., Angaiah, S., Development of electrospun PAN/CoS nanocomposite membrane electrolyte for high-performance DSSC. *Ionics (Kiel)*, 24, 12, 4071–4080, 2018, doi: 10.1007/s11581-018-2540-4.
120. Laforgue, A. and Robitaille, L., A Comparative Study of the Performance of Ultra-Porous Membranes as Separators in Supercapacitor Devices. *ECS Meet. Abstr.*, MA2011-01, 29, 13, 2011, doi: 10.1149/ma2011-01/29/1634.
121. Laforgue, A. and Robitaille, L., Electrochemical Testing of Ultraporous Membranes as Separators in Mild Aqueous Supercapacitors. *J. Electrochem. Soc.*, 159, 7, A929, 2012, doi: 10.1149/2.020207jes.
122. Sivaraman, P. *et al.*, A supercapacitor based on longitudinal unzipping of multi-walled carbon nanotubes for high temperature application. *RSC Adv.*, 5, 102, 83546–83557, 2015, doi: 10.1039/c5ra13136g.

123. Chang, C.H. *et al.*, High-temperature all solid-state microsupercapacitors based on SiC nanowire electrode and YSZ electrolyte. *ACS Appl. Mater. Interfaces*, 7, 48, 26658–26665, 2015, doi: 10.1021/acsami.5b08423.
124. Na, R. *et al.*, A flexible solid-state supercapacitor based on a poly(aryl ether ketone)-poly(ethylene glycol) copolymer solid polymer electrolyte for high temperature applications. *RSC Adv.*, 6, 69, 65186–65195, 2016, doi: 10.1039/c6ra11202a.
125. Mo, S. *et al.*, High-temperature performance of all-solid-state battery assembled with 95(0.7Li2S-0.3P2S5)-5Li3PO4 glass electrolyte. *Solid State Ion*, 296, 37–41, 2016, doi: 10.1016/j.ssi.2016.09.002.
126. Yao, Q., Wang, H., Wang, C., Jin, C., Sun, Q., One Step Construction of Nitrogen-Carbon Derived from Bradyrhizobium japonicum for Supercapacitor Applications with a Soybean Leaf as a Separator. *ACS Sustain. Chem. Eng.*, 6, 4, 4695–4704, 2018, doi: 10.1021/acssuschemeng.7b03777.
127. Jin, L. *et al.*, Tree leaves-derived three-dimensional porous networks as separators for graphene-based supercapacitors. *Mater. Today Energy*, 14, 100348, 2019, doi: 10.1016/j.mtener.2019.100348.
128. Yang, D., Qi, L., Ma, J., Eggshell membrane templating of hierarchically ordered macroporous networks composed of TiO2 tubes. *Adv. Mater.*, 14, 21, 1543–1546, 2002, doi: 10.1002/1521-4095(20021104)14:21 <1543::AID-ADMA1543>3.0.CO;2-B.
129. Taer, E. *et al.*, Eggs shell membrane as natural separator for supercapacitor applications. *Adv. Mater. Res.*, 896, 66–69, 2014, doi: 10.4028/www.scientific.net/AMR.896.66.
130. Torvinen, K. *et al.*, Pigment-cellulose nanofibril composite and its application as a separator-substrate in printed supercapacitors. *Electron. Mater. Lett.*, 11, 6, 1040–1047, 2015, doi: 10.1007/s13391-015-5195-6.
131. Jung, I., Dikin, D.A., Piner, R.D., Ruoff, R.S., Tunable electrical conductivity of individual graphene oxide sheets reduced at 'Low' temperatures. *Nano Lett.*, 8, 12, 4283–4287, 2008, doi: 10.1021/nl8019938.

# 6

# Electronics and Communication Applications

**Umesh V. Shembade[1], Mayuri G. Magadum[1], Sandeep B. Wategaonkar[2], Gopinath S. Khansole[3] and Annasaheb V. Moholkar[1]***

*[1]Thin Film Nanomaterials Laboratory, Department of Physics, Shivaji University, Kolhapur, India*
*[2]Department of Chemistry, Kisanveer Mahavidyalaya, Wai, India*
*[3]Department of Chemistry, D. A. B. N. Arts and Science College Chikhali, Sangli, India*

***Abstract***
One of the most reliable power sources for wearable electronics is supercapacitors (SCs), which have a high capacitance and power density, remarkable flexibility and deformability, and modest weight and size. They can be incorporated into various functional materials or woven into breathable textiles to meet curved surfaces for daily use. Considerable work has gone into creating high-performance flexible/stretchable energy storage devices to power integrated active devices as a result of the miniaturization of personal wearable electronics. SCs can perform this function, because of their straightforward construction, high power density, and cyclic stability. Furthermore, SCs can attain high electrochemical performance, and their applications can be increased by adding more unique functionalities. The components, principles of operation, and structure of SCs are described in detail to provide a more thorough analysis. To create a techno-economic analysis of SCs, the pros and cons, market profile, and new technologies with manufacturing firms are examined. This chapter examines the use of SCs in power systems, electric vehicles, hybrid energy storage systems that integrate renewable energy sources, and other applications. Furthermore, fundamentals of SC modeling design with charge and discharge experiments are investigated. As a summary, a few conclusions have been compiled from the examined papers, assessing the areas in which more study is required to enable the integration of energy storage technologies.

**Corresponding author*: avmoholkar@gmail.com; avm_phy@unishivaji.ac.in

Basheer Ahamed and Chaudhery Mustansar Hussain (eds.) Sustainable Supercapacitors: Next-Generation of Green Energy Storage Devices, (159–184) © 2025 Scrivener Publishing LLC

***Keywords:*** SCs, specifications, structures, electronic and communication sector, modeling of supercapacitors

## 6.1 Introduction

The idea of self-lacing electromechanical shoes and a self-powered, self-drying jacket attracted a lot of attention from the public. However, until a lot of work was put into creating wearables and electrical fabrics, it remained a fantasy from an adaptation. Currently, one of the emerging technologies with a significant potential impact on wearable, flexible, and conformable electronic systems is the intersection of electronics and telecommunications. This technique develops electrical and optoelectronic devices, through the use of flexible substrates like metal oxides, carbon nanotubes (CNTs), polymers, etc. As a result of the ongoing advancements in technology, consumers can now easily carry and utilize highly functional personal electronic devices. Simultaneously, the need for electronic gadgets to be small and light in addition to having good functionality has grown significantly. Similar to other electronic equipment, wearable electronics have developed towards miniaturization, multifunctionality, and user comfort. Wearable electronics can be incorporated into everyday clothing or applied directly to the skin to carry out a variety of tasks. These tasks include monitoring environmental pollutants, measuring pulses, analyzing bodily fluids like sweat, administering medication, and alerting the wearer of such activities. Many of the functions that were previously impossible to accomplish with a mobile phone alone were made available by smartphones nowadays [1–3].

Shortly, a lot of the tasks that smartphones currently complete will likely be divided into two categories and carried out by wearable technology that is either worn on the body or implanted. Apart from their current capabilities, these novel wearable devices will execute numerous jobs that are currently unimaginable. One of the main obstacles to completing this task is the rigidity and bulkiness of conventional energy storage devices utilized in electronic products. This highlights the essential need to create fresh approaches to energy harvesting and storage that meet the needs of wearable technology both now and in the future. In recent years, energy storage systems have emerged as a crucial tool for streamlining industrial operations and have become one of the most widely utilized alternatives. By enhancing the energy efficiency, dependability, and flexibility of electrical systems, these devices improve the performance of industrial systems [4–6].

As a result of the significant integration of renewable energy expected in electrical grids, this problem is highly relevant today. Energy storage device systems (ESDs) come in a variety of forms, which are categorized in Table 6.1 based on their nature or length of working cycle. By taking into account technologies with short cycles to applications with quick responses, like frequency stability or regenerative braking, the market sectors for each ESD may be identified from those tables. Meanwhile, those devices with long cycles could be used in long-term applications, such as massive energy storage or backup systems for critical loads [7].

Furthermore, the effectiveness of modern adaptive protection algorithms depends critically on the communication infrastructure's resilience and availability. Consequently, communication failure in these applications is an expensive risk with dire repercussions. Accordingly, the objective of a robust adaptive protection algorithm is to withstand communication failures without causing major harm to hardware assets, expenses, or productivity losses. In this work, an adaptive protection mechanism for microgrid SCs is suggested that takes into account the aforementioned concerns. This scheme can withstand communication outages by utilizing energy storage systems.

Here, we look into a variety of features that can be added straight to SCs to improve their applicability and usefulness in flexible wearable electronics (Figure 6.1). The materials and architectures utilized to create flexible/stretchable SCs are first described, followed by a discussion of the

**Table 6.1** Classification of ESDs based on their nature and life cycles.

| Sr. no. | Electrochemical energy storage device systems | Nature and life cycle |
|---|---|---|
| 1 | Flywheels (FESS)<br>Compressed air (CAES)<br>Hydro pumped energy storage (HPES) | Electrochemical<br>Short (1 s to 15 min)<br>Long days |
| 2 | Superconducting magnetic energy storage (SMES) | Electromagnetic<br>Short (1 s to 15 min) |
| 3 | Batteries | Electrochemical<br>(Medium 5 min to 24 h) |
| 4 | Fuel cells | Chemical<br>(Very short) |
| 5 | Supercapacitors | Electrochemical and electrostatics<br>Short (1 s to 15 min) |

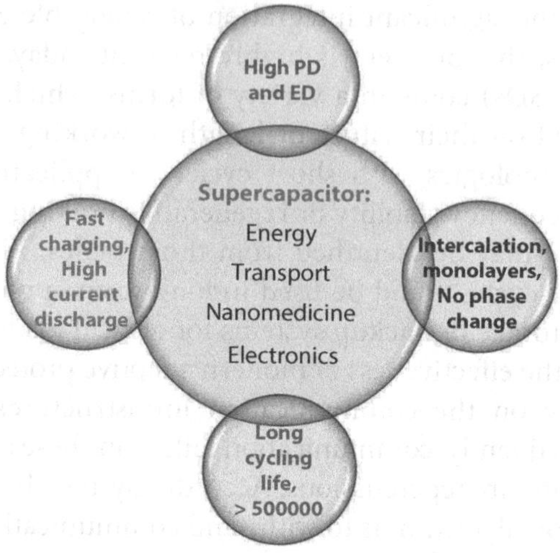


**Figure 6.1** The schematic of the SCs with their applicability and usefulness in the electronics and communication sector.

capabilities that can be exploited, complete with examples and comprehensive procedures. To create better future stretchy wearable electronics, we conclude by offering perspectives on high-performance device systems that can be combined with multifunctional SCs. To attain great energy harvesting and storage capacities, SCs should be carefully designed and constructed. A few conceptual concepts have been put out to generate fully integrated energy harvesting and storage systems.

## 6.2 Fundamentals of SCs

SCs are energy-storage devices that use the electrostatic separation of charges to store and release energy. The following are some essential features and basic aspects of SCs: SCs have a higher capacitance when compared to conventional capacitors. This is essential to their capacity to hold vast quantities of electrical charge. The amount of energy that an SC can store per unit of weight or volume is referred to as its energy density. In general, SCs have a lower energy density than batteries, even when they have a high PD (delivery energy quickly). SCs may charge and discharge quickly due to their high PD. They are ideal for uses that need sudden power spikes, including electronic gadgets that release energy quickly or regenerative braking in electric cars. Generally speaking, SCs have lower voltage ratings than batteries. The voltage rating is a crucial factor to take into account when developing systems with SCs [8].

A crucial factor in the functioning of SCs is the selection of electrode material. The negative electrode is typically made of conducting polymer or metal oxide, while the positive electrode is typically made of activated carbon. To help ions move between the electrodes in SCs, an electrolyte is used. The functioning voltage range, safety, and performance of the SCs are influenced by the kind of electrolyte—organic or aqueous.

To keep the positive and negative electrodes from coming into direct contact while yet enabling ion flow, a separator is employed. In addition to improving the SCs' overall performance and safety, it is essential to prevent short circuits. SCs may withstand numerous cycles of charging and discharging without experiencing appreciable degradation because they typically have a long cycle life. They are therefore appropriate for applications that need frequent cycling. The effect of temperature on SCs performance is possible and it is crucial for their practical uses and they operate steadily throughout a wide temperature range. The price of raw materials and production methods have an impact on the total cost of SCs. The cost-effectiveness of SCs can be increased by improvements in materials and production methods [9]. However, the overall fundamental aspects are shown in Figure 6.2.

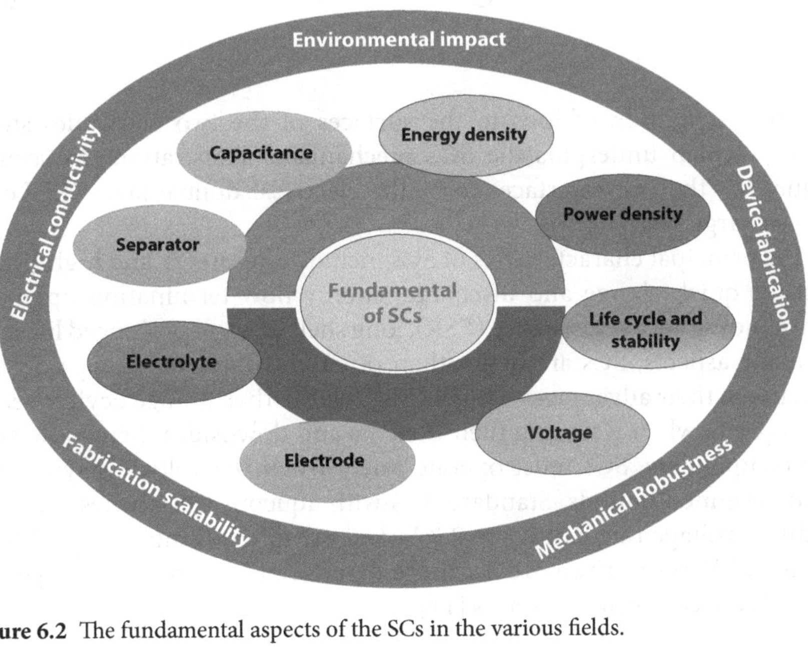


**Figure 6.2** The fundamental aspects of the SCs in the various fields.

## 6.3 Environmental Impact

In the development of energy storage technologies, particularly SCs, consideration of the environmental impact, including the recyclability and toxicity of materials utilized, is becoming more and more relevant. It is essential to comprehend these foundational elements to develop SCs that satisfy the unique demands of diverse applications, ranging from transportation systems to electronics. To enhance the functionality and properties of SCs, scientists and engineers are constantly investigating novel materials and technological advancements.

### 6.3.1 Structure and Specifications

This section provides an overview of the composition, mechanisms, attributes, categories, and components of SCs. Equation 6.1 presents the basic idea of SCs, which are based on electrostatic capacitors. The permittivity of air ($\varepsilon_0$), the surface area (*A*), the distance between two electrodes (*d*), and the relative permittivity of the dielectric material ($\varepsilon r$) are all shown in this equation. Using Equation 6.1, the capacitance can be modified by varying the thickness and surface area of the dielectric material [10].

$$C = \frac{\varepsilon_o * \varepsilon_r * A}{d} \tag{6.1}$$

The distribution of ions to the surfaces of the two electrodes stores energy, which underpins the SC's mechanism of operation. As seen in Figure 6.3 the two interfaces form the electrical double layer (EDLc), a space charge zone.

The principal characteristics of SCs include low energy and high-power density, quick charge and discharge, energy flow termination upon full charge, low internal resistance (ESR), long shelf life, and prolonged lifespan. In many aspects, SCs are better than other conventional storage devices because of their advantages. Utilizing SCs with other storage devices seems appropriate when weighing their benefits and downsides. Being low-voltage components, SCs must operate safely when the voltage stays within predetermined bounds. Standard SCs with aqueous electrolytes are rated within a voltage range of 2.1 to 2.3 V, and organic solvents are rated from 2.5 to 2.7 V, respectively. To meet the demands for increased voltage, the SC cells are connected in series [11].

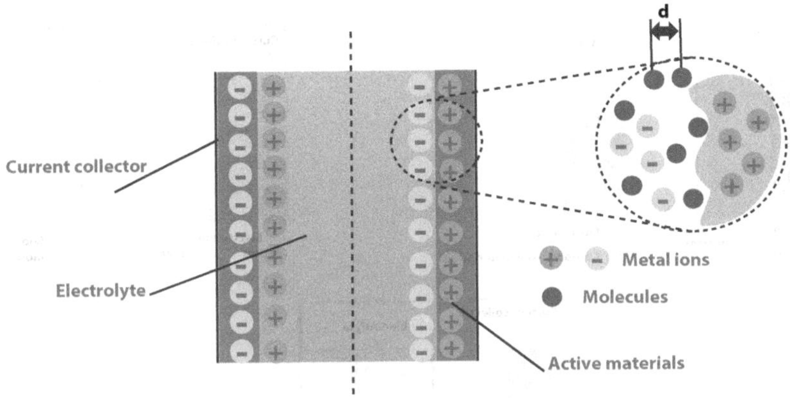


**Figure 6.3** Structure of SCs.

### 6.3.2 Classifications

According to methods of energy storage, SCs are divided into three groups: redox capacitors, which use both Faradaic and non-Faradaic processes to store charge, pseudo-capacitors (PSCs), which use faradic reactions, and electric-double layer capacitors (EDLC), which use the non-Faradaic process. For charge storage, the EDLC materials depend on surface area, porosity, and the availability of various carbon-based resources like activated carbon (AC), carbon nanotubes (CNT), carbon aerogels (CA), graphene (GO), and related derivatives. On the other hand, PSCs are made of substances like conducting polymers, metal oxides, and transition metal oxides (TMOs), respectively [12, 13]. By promoting the transfer of electrons from an electrode to an electrolyte contact via redox reactions, these materials produce capacitance. According to the energy storage mechanism, the SCs are classified into three groups, which are shown in Figure 6.4.

### 6.3.3 Electrode Materials and Their Role in SCs

SCs are energy storage devices that store electrical energy through the separation of charges at an electrochemical interface. The materials used in the construction of SCs play a crucial role in determining their performance.

**Carbon-based materials**

In the realm of SCs, which are energy storage devices that store electrical energy by separating charge in an electrochemical system, carbon-based materials are essential. Because of their superior electrical conductivity,

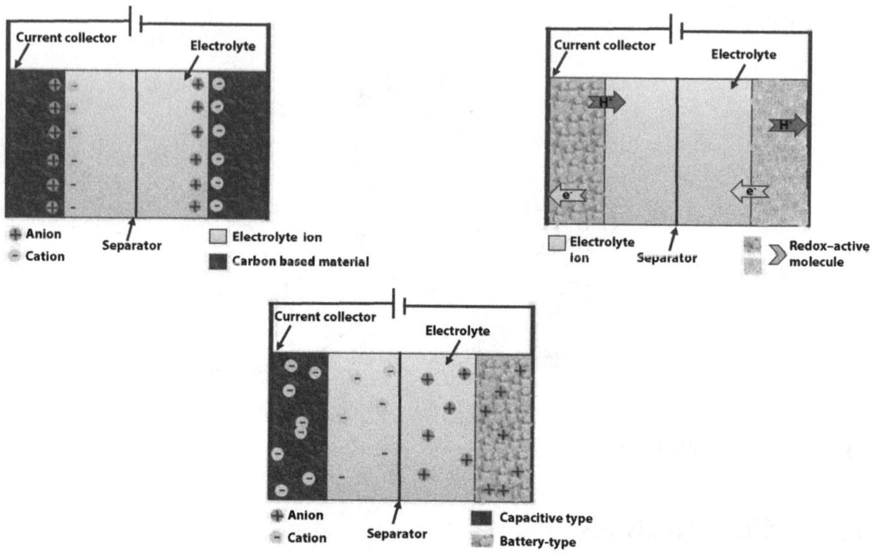


**Figure 6.4** Schematic for the classification of the SCs.

large surface area, and chemical stability, carbon-based materials play a major role in supercapacitors. The capacitive performance of the SCs is improved by this huge surface area, which offers more sites for the adsorption of ions. Generally speaking, carbon compounds are good electrical conductors. For the charge carriers (ions and electrons) inside the materials that make up the SC electrode to travel quickly, this characteristic is crucial. The long-term functionality and cycle stability of SCs depend on this stability. The durability of carbon materials is ensured by the severe electrochemical processes involved in the charge/discharge cycles. It is possible to adjust or modify the properties of carbon-based materials to maximize particular features for SCs applications. For instance, the structure of graphene can be designed to increase its conductivity, while the pore size and distribution in activated carbon can be adjusted to optimize ion adsorption. Because of this diversity, scientists can select materials according to the particular needs of the SCs application. In general, carbon-based materials support the electrodes in SCs and improve their cycle life, power density, and energy storage capacity. The efficiency and general performance of SCs are being further enhanced by ongoing research into new carbon-based materials and design techniques [14–16].

**Metal oxides**

SC devices renowned for their high PD and quick charge/discharge capabilities are one area in which metal oxides are heavily involved. Its distinct electrochemical characteristics explain the function of metal oxides in SCs. A significant quantity of electrical energy can be stored by metal oxides due to their high specific capacitance. During the cycles of charging and discharging, several metal oxides experience Faradaic reactions. These processes involve the reversible transfer of electrons between the metal oxide and electrolyte, contributing to the overall capacitance of the device. When compared to capacitive devices, this Faradaic approach can result in higher energy densities. Additionally, pseudocapacitance is a kind of capacitance that is produced by quick and reversible redox reactions, and it can be seen in some metal oxides [17, 18].

As a result, there is more charge storage available than with conventional EDLCs. In SCs, this characteristic is essential to obtaining low internal resistance and high PD output. However, the nanostructured metal oxides, such as metal oxide nanoparticles or nanowires, give a huge surface area for electrochemical processes. The capacitance is increased because there are more active sites for charge storage due to the larger surface area. However, some metal oxides are dependable for energy storage applications because they show strong stability throughout a variety of charge/discharge cycles. SCs can be made with a large range of metal oxides, which gives the flexibility to modify the characteristics of the electrode materials to meet the needs of applications. In conclusion, metal oxides provide high capacitance, pseudocapacitance, improved conductivity, and stability, all of which improve the electrochemical performance of SCs. Furthermore, the researchers continue to study and develop new metal oxide-based materials to improve the ED, PD, and overall performance of SCs for various applications. To increase performance, ED, and overall efficiency, new materials and combinations are constantly being investigated in the field of SCs [19].

**Metal tungstate**

Metal tungstates have gained attention in the field of SCs due to their unique electrochemical properties. The role of metal tungstates in SCs can be attributed to their high specific capacitance, good electrical conductivity, and ability to undergo reversible redox reactions. However, tungsten oxides and tungsten often exhibit high specific capacitance, which is a measure of the amount of electrical charge a material can store per unit mass or surface area. Additionally, the metal tungstates can undergo pseudocapacitive behavior, meaning they can store charge through fast and reversible redox reactions at the electrode–electrolyte interface. This leads to higher capacitance compared to typical electrical double-layer capacitors (EDLCs), resulting in improved energy storage performance [20].

Metal tungstate-based oxides generally possess good electrical conductivity, facilitating rapid electron transport and enhancing the overall efficiency of the device. Metal tungstates can undergo reversible redox reactions during the charging and discharging cycles, which is essential for the long-term stability and cycling life of SCs. This reversibility contributes to the reliability and durability of the energy storage system. However, tungsten oxides and tungstate-based oxides can be synthesized in various forms, such as nanoparticles, nanowires, or thin films, allowing for the optimization of their properties for specific SCs applications. Furthermore, the researchers continue to explore and develop new metal tungstate-based materials to enhance the performance of SC applications and address challenges such as ED, cycle life, and cost respectively. The unique combination of properties in metal tungstates makes them promising candidates for advancing the field of SC technology [21].

**Metal phosphate**

Metal phosphates can play a significant role in the field of SCs, which are energy storage devices that store electrical energy via the separation of charge in an electric double layer. These materials can provide high capacitance due to their ability to store charge at the interface between the electrode and the electrolyte. The choice of metal phosphate can influence the specific capacitance and ED of the SCs. Further, metal phosphates may offer high ED compared to traditional carbon-based electrodes. This is crucial for SCs as it determines how much energy can be stored per unit of mass or volume. The development of metal phosphate-based electrodes can contribute to improving the overall energy density of SCs. Some metal phosphates can be modified or combined with other materials to enhance their electrical conductivity [7].

Metal phosphates may exhibit good cycling stability, ensuring that the SCs can undergo numerous charge and discharge cycles without significant degradation in performance. This is a crucial factor for the long-term reliability and durability of SCs. Some metal phosphates are composed of elements that are abundant and environmentally friendly. This makes them attractive for the development of sustainable energy storage solutions, aligning with the growing focus on eco-friendly technologies. Researchers can modify the structure and composition of metal phosphates to tailor their properties for specific applications. This ability to tune the properties allows for the optimization of SCs' performance based on the desired characteristics, such as capacitance, voltage range, and cycling stability. Research in the field of SCs is ongoing, and scientists are continually exploring new materials, including metal phosphates, to improve the efficiency and capabilities of these energy storage devices [22].

**Metal–organic framework (MOF)**

Metal–organic frameworks (MOFs) have gained attention in the field of SCs due to their unique properties and potential applications in energy storage. MOFs typically possess a high surface area, which is advantageous for SC applications. The large surface area allows for more active sites where electrochemical reactions can occur, enhancing the overall capacitance of the device. MOFs are characterized by their porous structure, which can provide additional sites for ion adsorption and desorption. This property contributes to increased ion accessibility and diffusion within the electrode material, leading to improved charge and discharge rates. The tunable nature of MOFs allows for the optimization of pore size and surface chemistry. This tunability can be tailored to specific ion sizes and electrochemical properties, optimizing the performance of the SCs in terms of ED and PD [23].

MOFs can be integrated with other conductive materials, such as carbon-based materials (carbon nanotubes, graphene, etc.), to create composite structures. These composites often exhibit synergistic effects that combine the high surface area and tunable properties of MOFs with the excellent conductivity of carbon materials, resulting in improved overall performance of the SCs. Some metal oxides within MOFs exhibit redox activity, allowing for reversible electrochemical reactions during charge and discharge cycles. This redox behavior can contribute to the capacitance of the device and may lead to improved energy storage capabilities. MOFs can be designed using environmentally friendly and abundant elements, making them more sustainable compared to some other materials used in supercapacitors. Despite these advantages, it is important to note that challenges such as stability and scalability need to be addressed for the practical implementation of MOFs in SCs. Ongoing research is focused on optimizing the synthesis methods, improving the cycling stability, and enhancing the overall performance of MOF-based SC electrodes [24]. Furthermore, the schematic for the electrode materials and their role in SCs is highlighted in Figure 6.5.

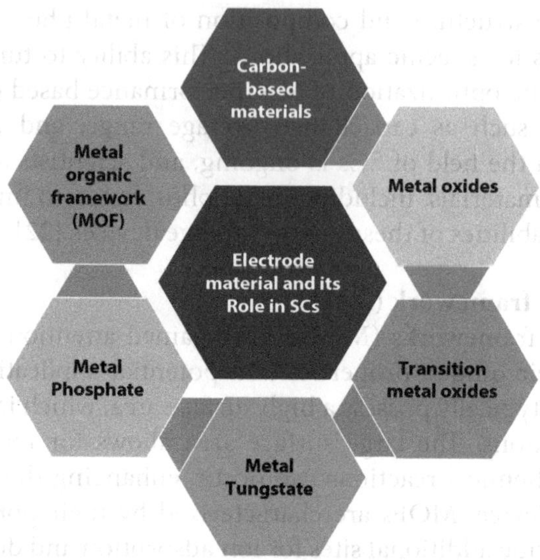


**Figure 6.5** Schematic for the electrode materials and their role in SCs.

## 6.4 Technological Aspects for SCs

Unlike traditional capacitors, SCs use a different mechanism to store energy, relying on the electrochemical double layer at the electrode–electrolyte interface and/or pseudocapacitance. The construction of an SC involves several key components:

**Electrodes**

SCs have two electrodes, typically made of porous materials with a high surface area. Common materials include activated carbon, carbon nanotubes, or graphene. An increased amount of charge storage is possible due to the high surface area. To form a double layer with the greatest possible quantity of electrolyte ions on the electrodes of SCs, electrochemical inert materials with the highest specific surface area are used, as capacitance is linked to the surface area. Graphite, metal oxides, and carbon are the most intriguing materials with significant surface activity. The main difficulty is to find cheap materials, which are chemically and electrically compatible with the electrolyte [25].

High surface area-activated carbon electrodes with the right surface and pore geometry are needed for high-energy capacitor applications. Activated carbon fibers, carbon black, active carbon, carbon fibers, carbon gel, skeletal carbon, mesocarbon, and microbeads are examples of carbonaceous materials. Surface areas of the best carbon electrodes reach up to 3,000 $m^2/g$ of material. Usually, high surface area carbon fibers or powders are used to manufacture them. For example, the powders are added to the metallic current collector as a paste. However, a configuration like this results in a significant amount of contact resistance between the grains and the support. Pressure must be used to solve these issues, or the carbon powder must be combined with metal fibers or powders to improve conductivity [26].

### Electrolyte

The electrodes are separated by an electrolyte, which is a conductive solution that allows the flow of ions between the electrodes. The choice of electrolyte depends on the type of SCs, as there are different types based on the storage mechanism (electrostatic double layer or pseudocapacitance). The electrolyte is another crucial element of an SC. A conducting salt and a solvent that separates the conducting salt and moves the charges between the electrodes make up the electrolyte. The ionic conductivity, ion mobility, diffusion coefficient, ion size, solvation, electrical and thermal stability, and dispersion interaction are all ultimately determined by the electrolyte's performance, which is determined by the properties of the conducting salt, ion concentration, type of solvent, and interaction between electrolyte and electrode [27].

Based on the solvent, the electrolytes employed in supercapacitors can be broadly divided into two groups: aqueous and non-aqueous electrolytes. Ionic liquid-based and organic electrolytes are examples of non-aqueous electrolytes. Wide operating potential window and temperature ranges, high ionic conductivity, strong chemical and mechanical stability, low flammability, and environmental friendliness are predicted characteristics of the perfect electrolyte for SCs. Considerable work has also been put into improving the performance of SCs with the addition of redox additives. To create electrolytes that are appropriate for stretchy SCs, it is also required to change the electrolytes' physical state from liquid to a stretchable solid to prevent leakage during mechanical deformation [28].

### Separator

To physically permit ion flow while separating the electrodes, a separator is employed. By limiting direct contact between the positive and negative electrodes, it stops short circuits. Polymer membranes are a common material for separators. The primary purpose of many of the separators that are sold commercially is for battery use. Therefore, to attain the remarkable performance of EDLCs, a precise assessment of the separator is necessary.

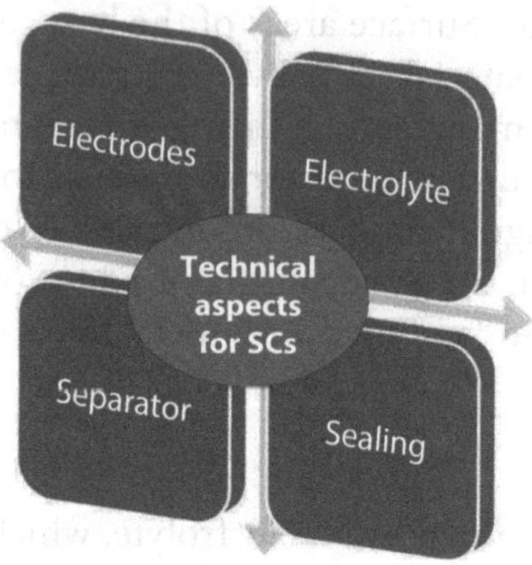


**Figure 6.6** Schematic of the technical aspects of the SCs.

Paper separators or polymers (usually PP) are needed if organic electrolytes are being used. Glass fiber separators and ceramic separators can both be used with aqueous electrolytes [29]. The separator prevents electrical contact between the electrodes but permits the transfer of charged ions. Large electrode surface, low separator, electrode thickness, high ionic electrolyte conductance, high ionic separator conductance, high electronic separator resistance, and high electrode electronic conductance are the fundamental requirements for achieving a competitive EDLC. Assembling these parts in a way that maximizes the SCs' performance and efficiency is the building process [8].

**Sealing**

To stop leaks and preserve the integrity of the interior components, the completed components are sealed. For the SCs to remain stable and reliable throughout time, sealing is essential. The resulting structure makes it easier to store and release electrical energy by enabling the efficient passage of ions through the electrolyte between the electrodes. The particular construction specifications can change based on the materials and kind of SCs. Additionally, the technical aspects of the SCs are shown in Figure 6.6.

## 6.5 Role of SCs in the Electronics Sector

SCs or electrochemical capacitors play a significant role in the electronics sector due to their unique energy storage capabilities. Unlike traditional batteries, SCs store energy via both electrostatic and electrochemical fields through chemical reactions, allowing for faster charge and discharge cycles. SCs excel in applications that require rapid bursts of energy because their high-power density allows them to deliver and absorb energy quickly and makes them suitable for applications such as regenerative braking systems in electric vehicles, where energy needs to be rapidly captured during braking and then released during acceleration. However, SCs are used in conjunction with batteries in various portable electronic devices that can provide quick bursts of energy to meet the high-power demands of these devices, enhancing overall performance and extending battery life [30].

Furthermore, SCs are employed in electronic systems as backup power sources to provide short-term power during outages, allowing critical functions to continue until a more stable power source (such as a battery or the main power supply) is restored. Additionally, SC devices play a crucial role in various fields that need rapid charge/discharge capabilities where frequent and high-energy pulses are required, namely, in pulse lasers, defibrillators, and certain medical devices. SCs are used in combination with batteries in power systems to handle peak loads, and they support the system during high-demand periods and then recharge during low-demand periods [31].

However, SCs can be used in renewable energy systems, especially for smoothing out fluctuations in power output from sources like solar and wind to store and release energy to compensate for intermittent power generation, improving overall system stability. In photography, SCs are employed in electronic flash units to provide the necessary burst of energy for the flash, ensuring rapid and reliable performance. Nowadays, most of the SC devices are used in hybrid and electric vehicles for various functions, such as capturing and releasing energy during braking, providing additional power during acceleration, and improving the overall efficiency of the energy management system for electric vehicles [32].

### 6.5.1 Starter

In specific automotive applications, they can function in tandem with conventional lead-acid batteries, even though they are not usually utilized as the main energy supply for car starters. For fast and effective engine cranking in the context of vehicle starting systems, SCs can be added to batteries as a source of energy. Usually, the SCs are linked in parallel with the conventional lead-acid battery in an automobile starting system. The supercapacitor fills in during the high-power demands of starting the engine, while the battery continues to supply the consistent, lower-power energy required for regular vehicle operations. Particularly in cold weather, this arrangement guarantees a prompt and dependable start while reducing the strain on the battery [33]. Therefore, they are usually utilized as supplemental parts in particular applications, such as starter systems, rather than serving as the main energy storage option for automobiles.

### 6.5.2 Hybrid Vehicle

Hybrid vehicles typically use a mix of an internal combustion engine and an electric motor, and the integration of SCs helps boost the overall performance and efficiency of hybrid vehicles. Additionally, the SCs play a crucial role in the following aspects, which are given below. In a hybrid car, the electric motor can function as a generator to transform kinetic energy into electrical energy during braking or deceleration. Because SCs can quickly store and release significant amounts of energy, they are perfect for storing the energy produced during regenerative braking. The electrical systems or the acceleration of electric vehicles can be powered by this stored energy. In brief spurts, SCs can produce high power. For high-demand tasks like accelerating or climbing hills, this can help supply more power. The vehicle's temporary power needs can be partially satisfied by SCs working with a battery. In a hybrid car, SCs can help the primary battery last longer. They are more capable than batteries of handling high rates of charge and discharge. SCs can lessen the strain on the battery and extend its lifespan by managing peak power demands and freeing it up to undertake more consistent, lower-power operations [34].

Fuel efficiency is enhanced by SCs through the effective management of energy flows within the vehicle. They maximize the usage of electric power in conjunction with internal combustion engines because of their rapid energy storage and release capabilities. In hybrid cars, the usage of SCs can help cut emissions. In situations where electric power is more efficient, such as stop-and-go traffic, the vehicle can rely more on electric power and less on its internal combustion engine by permitting more efficient energy capture and release [35].

Compared to conventional batteries, SCs are often lighter. Because of their low weight, they are a desirable option for hybrid cars, helping to reduce total weight and enhancing the economy and performance of the vehicle. Even though SCs have certain benefits, batteries are frequently paired with them to form a balanced energy storage system that meets the high-power and high-energy needs of hybrid cars. An improved and more adaptable energy management system is made possible in these vehicles by the combination of SCs and batteries [36].

### 6.5.3 Uninterruptable Power Supplies (UPS)

Through its rapid and effective energy storage solution, SCs contribute to Uninterruptible Power Supplies (UPS). In certain situations, SCs offer some advantages over batteries, which are the major energy storage components used in traditional UPS systems. Compared to conventional batteries, SCs can charge and discharge far more quickly. This quality comes in handy when quick action is needed to handle brief power outages or fluctuations. When compared to conventional batteries, SCs often have a longer cycle life. They are an enduring choice for UPS applications since they can withstand a high number of charge and discharge cycles without experiencing any noticeable degradation [37].

Compared to other batteries, SCs offer a higher power density. This indicates that they can deliver a large amount of power quickly, which is advantageous during unexpected power outages or surges. Unlike batteries, which are more prone to wear and tear, SCs usually operate without the requirement for routine replacement or maintenance cycles. Compared to batteries, which provide equal energy storage, SCs are typically lighter and more compact. This may be useful in situations where there is a shortage of space. It is crucial to remember that supercapacitors have drawbacks as well, chiefly about their energy density—that is, the quantity of energy they can store per unit mass or volume—when compared to batteries. They might not be appropriate for uses like prolonged power outages that call for long-term backup power. Both SCs and batteries may be used in tandem in certain UPS systems. Battery energy is required for extended backup periods, while SCs handle sudden, high-power needs. The goal of this hybrid method is to combine the best features of both technologies to produce a UPS system that is dependable, effective, and customized to meet the needs of a particular application [10].

### 6.5.4 Mobile Handsets

Even though SCs are less popular than conventional lithium-ion batteries in mobile phones, they can nevertheless have a specific impact on improving the functionality and general performance of these devices. SCs can be incorporated into mobile devices in the following ways: SCs can be used to provide rapid bursts of energy for quick charging. They can store energy quickly and release it rapidly when needed. This can be particularly useful for implementing fast charging features in mobile handsets, allowing users to charge their devices in a shorter amount of time. During periods of high-power demand or in the event of unexpected power outages, SCs can serve as a backup power supply. To ensure continuous operation during transient power events, this can help maintain a constant power supply to important components [38].

To increase battery life, SCs can be used in conjunction with conventional batteries. SCs can lessen the burden on the primary battery and extend its overall life by managing brief spikes in power requirements. Power consumption on mobile devices might fluctuate when engaging in different activities, such as gaming or using resource-intensive apps. By swiftly releasing energy during periods of high demand and storing it during periods of low demand, SCs can help even out these power peaks and offer a steadier power supply. To collect and store energy from ambient sources like solar or kinetic energy, SCs can be combined with energy harvesting devices. The main battery can then be recharged or some operations can be powered by the energy that has been captured, increasing overall energy efficiency. To strike a balance between energy density and power delivery capabilities, they are therefore frequently used in conjunction with lithium-ion batteries in mobile phones. The objectives and priorities of the device manufacturer's design will determine how precisely SCs are integrated into mobile phones. However, the schematic for the role of SCs in the electronics sector is shown in Figure 6.7.

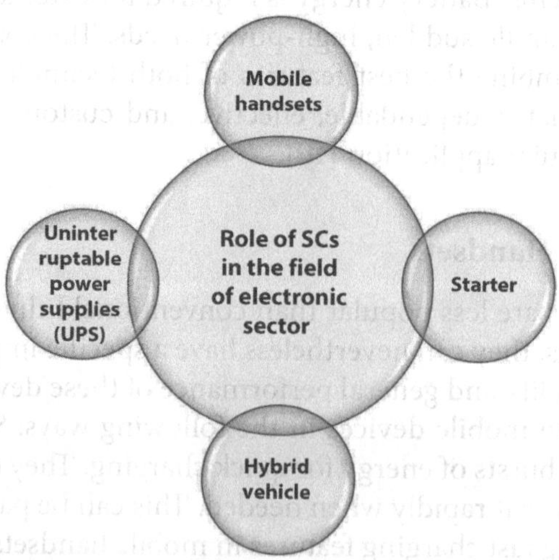


**Figure 6.7** Schematic for the role of SCs in the electronics sector.

## 6.6 Future Prospects of SCs in the Electronics Sector

In certain scenarios, SCs could completely replace conventional batteries. They are well suited for applications requiring rapid bursts of energy, like consumer electronics, renewable energy systems, and electric cars, because of their high PD and fast charge/discharge cycles. To improve the performance of electric vehicles, SCs can be utilized in tandem with batteries. By effectively managing the quick charge and discharge linked to regenerative braking, SCs help EVs operate more energy-efficiently overall. SCs are appropriate for wearable electronic devices since they are flexible and lightweight. They can handle different wearable devices' sporadic power demands and offer a rapid energy boost. SCs have a benefit in IoT devices because of their quick power burst capability. Through their assistance, the power needs of sensors and actuators in IoT networks can be effectively managed.

To improve the performance of laptops, smartphones, and other portable electronic devices, SCs can be utilized in conjunction with batteries. They can give sudden energy spikes, cutting down on charging times and raising the general efficiency of the gadget. By regulating voltage and supplying temporary power during variations, SCs can be used to enhance power quality. This is especially crucial for applications that depend on a steady power source. SCs have potential uses in implantable medical devices and defibrillators, among other devices where rapid energy supply is necessary. It is noteworthy that whereas SCs present certain benefits, they also have certain drawbacks, such as a lower energy density in comparison to batteries. To get over these restrictions and enhance the functionality and affordability of supercapacitors in a range of electronic applications, more research and development work will be necessary in the future.

## 6.7 Role of SCs in the Communications Sector

SCs are important in the communications industry and the field of telecommunications due to their numerous benefits for a range of applications. In the event of power outages or fluctuations, SCs are utilized to supply temporary backup power, which delivers a quick burst of energy, guaranteeing unbroken communication at crucial times. Additionally, telecommunication systems often experience fluctuations in power demand, especially during peak usage times. In addition to having the ability to charge and discharge cycles quickly, SCs may be used to control and smooth out these peak power demands, guaranteeing a steady and dependable power supply. This makes them appropriate for applications that call for brief energy bursts. Devices that need to send or receive data in quick bursts will find this functionality useful [39].

Nonetheless, the SCs can be used with conventional batteries to improve their efficiency. SCs can prolong battery life by lessening the strain on batteries by managing high power demands during peak hours. To stabilize voltage and current levels, avoid voltage sags and surges, and guarantee a steady and dependable power supply to delicate electronic components, SCs can function as energy buffers and filters in communication systems. SCs can be used for cold starts in off-grid or remote telecommunication stations to supply the first energy burst needed to power up the communication equipment, even in harsh weather circumstances. In situations where renewable energy sources are used, SCs can store excess energy created during periods of high availability (e.g., sunny or windy days) and release it when needed. As a result, system efficiency is increased overall and energy utilization is optimized in telecommunication systems by offering a dependable and prompt source of battery backup power. This is especially crucial for maintaining continuous communication services during an emergency [40]. To summarize, SCs enhance the dependability, effectiveness, and efficiency of communication and communications systems by resolving issues with power fluctuations, energy storage, and backup power needs. Because of their distinct qualities, they are useful parts of many applications in this industry.

## 6.8 Future Prospects of SCs in the Communications Sector

SCs are a vital component in the communications industry that can help solve specific problems and improve efficiency. The communications industry may see some of the following opportunities for SCs: In wireless communication devices like smartphones and wearables, SCs could be employed to store and release energy quickly. Faster charging times and enhanced device performance may result from this. SCs can function as dependable backup power sources for base stations and cell towers, two essential components of communication infrastructure. In the event of a power outage, they can deliver brief energy surges to maintain communication services. SCs are well-suited for applications requiring high-power pulses, such as radar systems or satellite communication. They are useful for sustaining constant communication in difficult conditions because of their capacity to provide brief energy spurts. The performance of communication devices can be maximized by combining SCs and batteries in hybrid energy storage systems. Batteries can store energy for a longer period, whereas SCs can sustain quick charge and discharge cycles. With the growing reliance of the communications industry on electric vehicles for deployment and maintenance, SCs can be used in these vehicles to offer fast energy storage and release, improving operational efficiency. Green energy sources like solar and wind can be used in conjunction with SCs to power communication infrastructure. When production peaks and demand is high or there are power oscillations, they can store extra energy and discharge it. Being more environmentally friendly than many conventional battery technologies, SCs may be selected to reduce the environmental impact of communication equipment. It is noteworthy that several factors, including but not limited to cost, technological advancement, and the specific requirements of diverse applications, will influence the acceptability of SCs in the communication business. The ongoing study and advancement in the constantly changing field of energy storage may lead to new applications for SCs in the communications sector. Consistently consider consulting the most recent sources for the latest information on technical developments.

## 6.9 Summary and Conclusion

This book chapter provided an overview of the evolution and history of SCs. Further, essential components of SCs such as their structure, principles, requirements, categories, and materials were mentioned in this book chapter. The benefits, markets, new technologies, manufacturers, and other aspects of the techno-economic assessments of SCs were all examined. To enlighten the readers, the use of SCs in several applications including communication systems, hybrid power systems, and electric vehicles was explored. This extensive book chapter on SCs synthesized all the studies in a similar format to guide and motivate the field's research efforts. Finally, this study will close the gap and add to the body of knowledge because it addresses the most recent advancements in both the market and the literature. The most significant obstacles to the widespread use of SCs appear to be the advancements in material technology discussed in this article, as well as addressing application-specific issues and improving efficiency and price-related market attractiveness. The maximal instantaneous output power and battery lifetime are both increased when SCs are used in conjunction with batteries. It requires a series connection to raise the voltage across an SC device. A voltage repartition device that is active has been defined to guarantee optimal efficiency and prevent over-voltage over any SCs.

## References


1. Wang, S., Ma, J., Shi, X., Zhu, Y., Wu, Z.S., Recent status and future perspectives of ultracompact and customizable micro-supercapacitors. *Nano Res. Energy*, 1, 500–526, 2022, https://doi.org/10.26599/NRE.2022.9120018.
2. Afif, A., Rahman, S.M., Tasfiah Azad, A., Zaini, J., Islan, M.A., Azad, A.K., Advanced materials and technologies for hybrid supercapacitors for energy storage – A review. *J. Energy Storage*, 25, 100852, 2019, https://doi.org/10.1016/j.est.2019.100852.
3. Manjunatha, C., Shwetha, K.P., Athreya, Y., Kumar, S.G., Kamath, M.K.S., Perspective—Supercapacitor-Powered Flexible Wearable Strain Sensors. *ECS Sensors Plus*, 2, 017002, 2023, https://doi.org/10.1149/2754-2726/acb27a.
4. Shembade, U.V., Gurav, S.R., Gaikwad, M.A., Wategaonkar, S.B., Ghatage, S.R., Sonkawade, R.G., Kim, J.H., Moholkar, A.V., Hydrothermal synthesis of graphene oxide interspersed in non-uniform tungsten oxide nanorod and its performance towards highly efficient hybrid supercapacitor. *Ceram. Int.*, 50, 340–350, 2023, https://doi.org/10.1016/j.ceramint.2023.10.107.
5. Shembade, U.V., Dhas, S.D., Gurav, S.R., Sonkawade, R.G., Wategaonkar, S.B., Ghatage, S.R., Gaikwad, M.A., Kim, J.H., Parale, V.G., Park, H.H., Moholkar, A.V., Acid substitutions for WO3 nanostructures synthesis by the hydrothermal route and its effect on physio-chemical and electrochemical properties for supercapacitors. *J. Energy Storage*, 72, 108432, 2023, https://doi.org/10.1016/j.est.2023.108432.
6. Dubal, D.P., Kim, J.G., Kim, Y., Holze, R., Lokhande, C.D., Kim, W.B., Supercapacitors Based on Flexible Substrates: An Overview. *Energy Technol.*, 2, 325–341, 2014, https://doi.org/10.1002/ente.201300144.
7. Patil, S.S. and Patil, P.S., Status review of nickel phosphides for hybrid supercapacitors. *Nanoscale*, 14, 16731–16748, 2022, https://doi.org/10.1039/d2nr05139g.
8. Şahin, M.E., Blaabjerg, F., Sangwongwanich, A., A Comprehensive Review on Supercapacitor Applications and Developments. *Energies*, 15, 1–26, 2022, https://doi.org/10.3390/en15030674.
9. Redekar, R.S., Avatare, A.T., Chouhan, J.L., Patil, K.V., Pawar, O.Y., Patil, S.L., Bhoite, A.A., Patil, V.L., Patil, P.S., Tarwal, N.L., Review on recent advancements in chemically synthesized manganese cobalt oxide (MnCo2O4) and its composites for energy storage application. *Chem. Eng. J.*, 450, 137425, 2022, https://doi.org/10.1016/j.cej.2022.137425.

10. Keum, K., Kim, J.W., Hong, S.Y., Son, J.G., Lee, S.S., Ha, J.S., Flexible/Stretchable Supercapacitors with Novel Functionality for Wearable Electronics. *Adv. Mater.*, 32, 1–34, 2020, https://doi.org/10.1002/adma.202002180.
11. Conte, M., Supercapacitors technical requirements for new applications. *Fuel Cells, 10*, 5, 806–818, 2010, https://doi.org/10.1002/fuce.201000087.
12. Shembade, U.V., Wategaonkar, S.B., Moholkar, A.V., Exploring the synergistic effect of temperature on hydrothermally synthesized tungsten oxide (WO3) nanostructures and its role in asymmetric liquid-state hybrid supercapacitors. *Colloids Surf., A*, 682, 132916, 2024, https://doi.org/10.1016/j.colsurfa.2023.132916.
13. Shembade, U.V., Dhas, S.D., Gurav, S.R., Wategaonkar, S.B., Ghatage, S.R., Gaikwad, M.A., Parale, V.G., Sonkawade, R.G., Kim, J.H., Park, H.H., Moholkar, A.V., Chemically synthesized graphene oxide nanosheet (GONs) is an efficient electrode material for supercapacitors: Effects of current collectors. *Diam. Relat. Mater.*, 141, 110602, 2024, https://doi.org/10.1016/j.diamond.2023.110602.
14. Kong, H.X., Hybrids of carbon nanotubes and graphene/graphene oxide. *Curr. Opin. Solid State Mater. Sci.*, 17, 31–37, 2013, https://doi.org/10.1016/j.cossms.2012.12.002.
15. Veerapandian, M., Lee, M.H., Krishnamoorthy, K., Yun, K., Synthesis, characterization and electrochemical properties of functionalized graphene oxide. *Carbon N. Y.*, 50, 4228–4238, 2012, https://doi.org/10.1016/j.carbon.2012.05.004.
16. Johra, F.T. and Jung, W.G., Hydrothermally reduced graphene oxide as a supercapacitor. *Appl. Surf. Sci.*, 357, 1911–1914, 2015, https://doi.org/10.1016/j.apsusc.2015.09.128.
17. An, C., Zhang, Y., Guo, H., Wang, Y., Metal oxide-based supercapacitors: progress and prospectives. *Nanoscale Adv.*, 1, 4644–4658, 2019, https://doi.org/10.1039/c9na00543a.
18. Dhas, S.D., Maldar, P.S., Patil, M.D., Hubali, K.M., Shembade, U.V., Abitkar, S.B., Waikar, M.R., Sonkawade, R.G., Agawane, G.L., Moholkar, A.V., Hydrothermal synthesis of mesoporous NiMnO3 nanostructures for supercapacitor application: Effect of electrolyte. *J. Energy Storage*, 35, 102277, 2021, https://doi.org/10.1016/j.est.2021.102277.
19. Gurav, S.R., Chodankar, G.R., Sawant, S.A., Shembade, U.V., Moholkar, A.V., Sonkawade, R.G., Exploring the potential of simultaneous nanoarchitectonics and utilization of Co-MOFs electrode as well as powder for aqueous supercapacitors. *J. Energy Storage*, 73, 109254, 2023, https://doi.org/10.1016/j.est.2023.109254.
20. Rajpurohit, A.S., Punde, N.S., Rawool, C.R., Srivastava, A.K., Cobalt-Nickel bimetallic tungstate nanoparticles decorated Phosphorus. *Chem. Eng. J.*, 371, 679–692, 2019, https://doi.org/10.1016/j.cej.2019.04.100.

21. Jha, S., Mehta, S., Chen, E., Sankar, S.S., Kundu, S., Liang, H., Bimetallic tungstate nanoparticle-decorated-lignin electrodes for flexible supercapacitors. *Mater. Adv.*, 1, 2124–2135, 2020, https://doi.org/10.1039/d0ma00494d.
22. Chodankar, N.R., Pham, H.D., Nanjundan, A.K., Fernando, J.F.S., Jayaramulu, K., Golberg, D., Han, Y.K., Dubal, D.P., True Meaning of Pseudocapacitors and Their Performance Metrics: Asymmetric versus Hybrid Supercapacitors. *Small*, 16, 1–35, 2020, https://doi.org/10.1002/smll.202002806.
23. Salunkhe, R.R., Kaneti, Y.V., Yamauchi, Y., Metal-Organic Framework-Derived Nanoporous Metal Oxides toward Supercapacitor Applications: Progress and Prospects. *ACS Nano*, 11, 5293–5308, 2017, https://doi.org/10.1021/acsnano.7b02796.
24. Gurav, S.R., Sawant, S.A., Chodankar, G.R., Shembade, U.V., Moholkar, A.V., Sonkawade, R.G., Exploration of aqueous electrolyte on the interconnected petal-like structure of Co-MOFs for high-performance paper-soaked supercapacitors. *Electrochim. Acta*, 467, 143027, 2023, https://doi.org/10.1016/j.electacta.2023.143027.
25. Brousse, T., Bélanger, D., Chiba, K., Egashira, M., Favier, F., Long, J., Miller, J.R., Morita, M., Naoi, K., Simon, P., Sugimoto, W., Materials for electrochemical capacitors, in: *Springer Handbooks*, pp. 495–561, 2017, https://doi.org/10.1007/978-3-662-46657-5_16.
26. Liang, C., Wang, S., Sha, S., Lv, S., Wang, G., Wang, B., Li, Q., Yu, J., Xu, X., Zhang, L., Novel semiconductor materials for advanced supercapacitors. *J. Mater. Chem. C.*, 11, 4288–4317, 2023, https://doi.org/10.1039/d2tc04816g.
27. Zhang, L., Yang, S., Chang, J., Zhao, D., Wang, J., Yang, C., Cao, B., A Review of Redox Electrolytes for Supercapacitors. *Front. Chem.*, 8, 1–7, 2020, https://doi.org/10.3389/fchem.2020.00413.
28. Iro, Z.S., Subramani, C., Dash, S.S., A brief review on electrode materials for supercapacitor. *Int. J. Electrochem. Sci.*, 11, 10628–10643, 2016, https://doi.org/10.20964/2016.12.50.
29. Sharma, S. and Chand, P., Supercapacitor and electrochemical techniques: A brief review. *Results Chem.*, 5, 100885, 2023, https://doi.org/10.1016/j.rechem.2023.100885.
30. Navarro, G., Torres, J., Blanco, M., Nájera, J., Santos-Herran, M., Lafoz, M., Present and future of supercapacitor technology applied to powertrains, renewable generation and grid connection applications. *Energies*, *14*, 3060, 2021, https://doi.org/10.3390/en14113060.
31. Habib, H.F., Mohamed, A.A.S., El Hariri, M., Mohammed, O.A., Utilizing supercapacitors for resiliency enhancements and adaptive microgrid protection against communication failures. *Electr. Power Syst. Res.*, 145, 223–233, 2017, https://doi.org/10.1016/j.epsr.2016.12.027.
32. Chen, D., Jiang, K., Huang, T., Shen, G., Recent Advances in Fiber Supercapacitors: Materials, Device Configurations, and Applications. *Adv. Mater.*, 32, 1–19, 2020, https://doi.org/10.1002/adma.201901806.

33. Lv, Y., Le Zhang, L., Wei, X., Qiu, B., Zhang, W., Qin, Q., Jia, D., He, X., Liu, Z., Wei, F., The emerging of zinc-ion hybrid supercapacitors: Advances, challenges, and future perspectives, Sustain. *Mater. Technol.*, 35, e00536, 2023. https://doi.org/10.1016/j.susmat.2022.e00536.
34. Chen, H., Du, X., Liu, X., Wu, R., Li, Y., Xu, C., Facile growth of nickel foam-supported MnCo2O4 porous nanowires as binder-free electrodes for high-performance hybrid supercapacitors. *J. Energy Storage*, 50, 104297, 2022, https://doi.org/10.1016/j.est.2022.104297.
35. Fleischmann, S., Mitchell, J.B., Wang, R., Zhan, C., Jiang, D.E., Presser, V., Augustyn, V., Pseudocapacitance: From Fundamental Understanding to High Power Energy Storage Materials. *Chem. Rev.*, 120, 6738–6782, 2020, https://doi.org/10.1021/acs.chemrev.0c00170.
36. Troschke, E., Oschatz, M., Ilic, I.K., Schiff-bases for sustainable battery and supercapacitor electrodes. *Exploration*, 1, 20210128, 2021, https://doi.org/10.1002/exp.20210128.
37. Varma, S.J., Sambath Kumar, K., Seal, S., Rajaraman, S., Thomas, J., Cells, Fiber-Type Solar, Nanogenerators, Batteries, and Supercapacitors for Wearable Applications. *Adv. Sci.*, 5, 1800340, 2018. https://doi.org/10.1002/advs.201800340.
38. Chen, P., Shen, G., Shi, Y., Chen, H., Zhou, C., Preparation and Characterization of Flexible Asymmetric Supercapacitors. *ACS Nano*, 4, 4403–4411, 2010.
39. Biswas, S. and Chowdhury, A., Organic Supercapacitors as the Next Generation Energy Storage Device: Emergence, Opportunity, and Challenges. *ChemPhysChem*, 24, e202200567, 2023, https://doi.org/10.1002/cphc.202200567.
40. Ramachandran, R., Saranya, M., Grace, A.N., Wang, F., MnS nanocomposites based on doped graphene: simple synthesis by a wet chemical route and improved electrochemical properties as an electrode material for supercapacitors. *RSC Adv.*, 7, 2249–2257, 2017, https://doi.org/10.1039/c6ra25457h.

# 7

# Energy Storage Breakthroughs: Supercapacitors in Healthcare Applications

**Jyoti Prakash Das[1], Sang Jae Kim[1]* and Ananthakumar Ramadoss[2†]**

*[1]Nanomaterials & System Lab, Major of Mechatronics Engineering, Faculty of Applied Energy System, Jeju National University, Jeju, South Korea*
*[2]School for Advanced Research in Polymers (SARP), Advanced Research School for Technology & Product Simulation (ARSTPS), Central Institute of Petrochemicals Engineering & Technology (CIPET), Chennai, India*

## *Abstract*
Modern instruments and analytical techniques trigger revolutionization in the healthcare sector. The rapid demand for wearable medical devices, implants, and point-of-care diagnosis instruments is emerging worldwide to provide state-of-the-art medical facilities for early detection, cure, or precaution of the disease. However, the ever-evolving healthcare sector needs a reliable power supply to provide uninterrupted service for various instruments, systems, and remote location patient monitoring devices. Supercapacitors (SCs) lead the board among energy storage devices due to their high power density, fast charging ability, and excellent operational life. This not only makes SCs a reliable source of energy to run the healthcare instruments but also puts an end to the hazardous battery wastes. The advancement in SCs provides multiple design configurations, tunable specific capacitance, and many material choices for healthcare applications. SCs can power implantable, like pacemakers and neurostimulators, for longer life without frequent operations. Wearable health monitoring devices provide regular health updates utilizing the SCs. Portable medical instruments in remote locations without any external sources can be charged easily with external sources like solar, wind, or ambient mechanical forces. Further, this chapter discusses the evolution of SCs, material selection, innovations, and design aspects for healthcare applications. The significant role of SCs in building a self-sustainable, energy-efficient healthcare system is depicted thoroughly.

**Corresponding author*: kimsangj@jejunu.ac.kr
†*Corresponding author*: ananth@cipet.gov.in

Basheer Ahamed and Chaudhery Mustansar Hussain (eds.) Sustainable Supercapacitors: Next-Generation of Green Energy Storage Devices, (185–226) © 2025 Scrivener Publishing LLC

***Keywords*:** Energy storage, implantable devices, health monitoring, design, supercapacitors

## 7.1 Introduction

Advancements in the healthcare sector are highly reliant on power sources to provide a quick, accurate diagnosis, detection of disease, and timely remedies [1–5]. The advanced medical equipment of the Internet of Things (IoT) comprises miniature electronics like sensors and actuators that need continuous power to function for a prolonged period [6–10]. Apart from diagnosis devices, implantable bioelectronics is an emerging trend in the healthcare sector that is popular for critical diagnosis and therapies [4, 11]. The emerging technology of bioelectronics is not limited to external diagnosis [12, 13]. The implantable bioelectronics needs a power source to function as a diagnostic and therapeutic tool [14]. Recent developments in biomaterial and instrumentation and the implementation of robotics have induced considerable potential in the power source for medical equipment and implants. Batteries and supercapacitors are the significant leaders among various power sources. Though batteries capture a huge market, they still lack biocompatibility due to the use of relatively toxic electrode materials and electrolytes [15–17]. The larger volume-limited cycle life, leakage, and surgery for painful replacement urge the development of alternative energy storage devices. The transient electronics act as a passive implant or targeted drug delivery utilizing metal or polymer-based materials. Compared to metal foils, the water-activated primary battery has a lower voltage and shorter work life and restricts *in vivo* application [18, 19]. There is a threat of explosion inside the body due to the degradation of the transition metals. A few energy storage devices and capacitors are developed for implantable power sources that can be charged wirelessly. However, dielectric-based capacitors have lower charge capacity and consist of metal and ceramics, which leads to the depletion of heavy particles inside the body [16]. Benefiting from the shortcomings of batteries, supercapacitors emerge as a potential power source in various industries like the automobile, manufacturing, and healthcare sectors. It provides higher specific capacitance, fast charging ability, no pollution, and the ability to charge wirelessly and from ambient mechanical vibrations or body pulses [20, 21]. Miniaturization is the beauty of future energy storage devices where supercapacitor fabrication techniques help fabricate micro-sized devices with a wide range of materials and tailor-made designs. According to medical needs, biodegradable, implantable medical electronics (BIMEs) devices' supercapacitors can supply power for a certain period. The miniaturizing of implantable devices has recently driven the need for suitable power sources. The challenge for such a power source is biocompatibility and stable performance for a prolonged period. The supercapacitors can be designed and fabricated without complications for the externally utilized power source and can be replaced easily or recharged as required [20].

On the other hand, the implantable supercapacitors need to be biocompatible, decomposable, or rechargeable wirelessly to power the implant effectively [22, 23]. Proper selection of material and its utilization for external and internal applications is challenging [24–26]. Integrating supercapacitors with energy harvesters opens up vast opportunities for application in the healthcare sector. For instance, transition metal-based thin film (W, Fe, or Mo) and agarose gel electrolyte-based supercapacitors were proposed as biodegradable supercapacitors. The electrochemical performance of such energy storage devices is enhanced due to the metal oxide coating generated from electrochemical corrosion that takes place at the electrode–electrolyte interface. However, the limited life cycle restricted the *in vivo* application. The development of CNT-based supercapacitors is emerging as an implantable power source for next-generation healthcare. The transition metal-based implantable power sources have been developed for implantable supercapacitors, but their impact on tissue and internal body parts is under debate. PEDOT:PSS is another potential candidate for the implantable power source, but the real-time application is unclear [27, 28]. The slight toxicity and non-biodegradability restrict the *in vivo* application of these materials. Researchers have attempted to develop a conductive hydrogel-based implantable power source that can be mechanically robust, biocompatible, and highly absorbable [29]. Sericin, another silkworm protein, and conductive polymers are potential candidates for supercapacitor application. The prime application of this protein is in wound healing and cardiac tissue engineering. The lower conductivity of this protein restricts its real-time application in energy storage implantable applications. Along with the conductive polymer, it can act as a power source, but the biodegradability of the polymer restricts its real-time application [30].

Contrary to implantable energy sources, the demand for external health monitoring devices and their power sources is emerging very quickly. The *ex vivo* applications like body motion sensors, pulse monitors, smart bands, glucose sensors, etc, are becoming a part of day-to-day life, and powering such miniature devices needs microlevel devices designed and fabricated in a body-compatible configuration [31]. The miniaturized devices need a very low power rating to run, but they can be powered easily with external power sources like solar or mechanical vibrations. Wearable devices integrated with textiles that are directly exposed to human skin are another trend in recent times. The micro-supercapacitors achieve fast charging and a higher work life [32].

Further, the advancement of bio-degradable supercapacitors, fiber type, and IDT electrodes with photochromic sensors is emerging and attracting researchers to explore various advanced fabrication techniques [33]. Combined with an energy storage application, the photochromic display can rely on real-time health monitoring. Further, the research on the wide-temperature-range supercapacitors and power from body sweat facilitate the wearables for healthcare and real-time medical monitoring applications. Figure 7.1 depicts the basic idea of utilizing the supercapacitor in various healthcare sections, including disease diagnosis, implantable, and curing equipment as a power source [9, 34, 35].

This chapter highlights the evolution of energy storage devices in healthcare applications and focuses on the power sources in healthcare applications. Various strategies for material selection for the supercapacitors for the healthcare sector are discussed. *In vivo* and *ex vivo* SCs for health monitoring are highlighted, pointing out the challenges and milestones. The future scope of SCs in the healthcare sector is summarized to think beyond and achieve the targets in the energy storage sector for future healthcare applications.

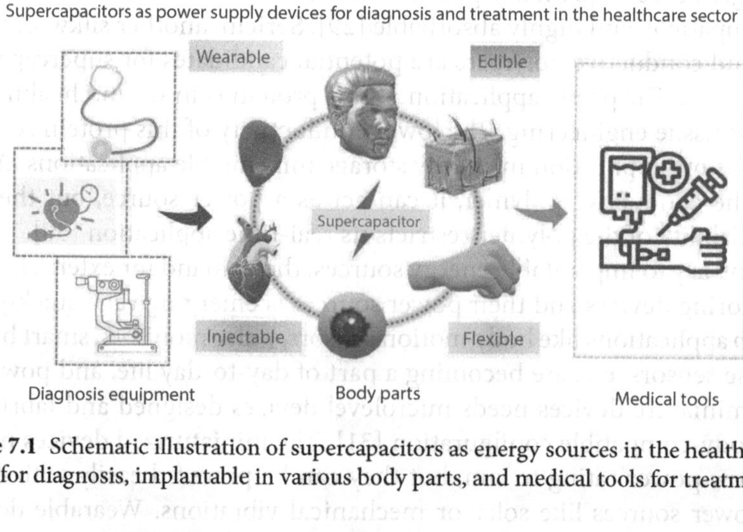


**Figure 7.1** Schematic illustration of supercapacitors as energy sources in the healthcare sector for diagnosis, implantable in various body parts, and medical tools for treatments.

## 7.2 Supercapacitors

Supercapacitors are better for electrochemical energy storage devices because of their reasonable energy density and high-power density [36]. Supercapacitors' superior power density allows for quick charge–discharge cycles [37–39]. Supercapacitors emerge as a viable substitute for batteries. The three categories of SCs are determined by their charge storage mechanism: (i) EDLC, (ii) pseudocapacitors, and (iii) hybrid-SCs. EDLC-type SCs are capacitors that function as charge storage devices by forming ion layers at the electrode interface [40–43]. The electrode material's porosity is an essential factor in the SCs' performance [44–46]. The surface redox reactions that occur concurrently in the electrodes enable the energy storage mechanism in pseudocapacitors. The hybrid supercapacitor type, which operates on an asymmetric configuration of two dissimilar type electrodes, is the third category of SCs [46–51]. SC primarily operates via surface absorption desorption at the cathodes (EDLC) and redox reaction at the anode (phase-changing materials). To put it simply, the EDLC has the least capacitance when compared to the other two SCs, but it can scan a million times on a single cycle and has a 90% rate capability [52–54]. Compared to EDLCs, the pseudocapacitive has a lower cyclic life but a higher energy storage capacity. Out of all the SCs, the hybrid ones with the greatest capacity and shortest cycle life are made of phase-changing materials and an EDLC electrode. In addition to these, microseconds are becoming more and more popular as a means of storing energy for use in microelectronics and Internet of Things devices [33, 55–59]. Mainly, material properties determine the types of SCs. Carbon-based materials and their allotropes, such as graphene and carbon nanotubes, make up the majority of the EDLCs. Transition metal oxides like $RuO_2$, $MnO_2$, FeOOH, etc. make up the majority of pseudocapacitors [60–64]. Because of their high surface-to-volume ratio, which offers a great active site for energy storage applications, nanomaterials are sought after as the ideal candidate [65, 66]. A variety of synthesis strategies, such as top-down and bottom-up are followed for the electrode material synthesis for energy storage applications [67–71]. Apart from the conventional processes, the 3D printing of SCs is carried out for complicated tailor-made designs utilizing carbon-based materials, metal oxides, transition metals, etc [72]. However, the unique characteristics required for biocompatible supercapacitors, especially in the healthcare sector, are discussed in detail later in this chapter.

## 7.3 Material Selection for Bio-Compatible Supercapacitor

The healthcare sector is highly material-sensitive. Biological organisms comprise soft tissues, sensitive skin, and flexible scaffolds [73]. These parts get exposed during any diagnosis or detection of disease and further undergo the curing process. Thus, the primary objective for such an application is to treat the target without affecting or damaging its functional properties. Figure 7.2 depicts the desirable properties of biocompatible supercapacitors for effective application in the healthcare sector [74]. The SCs undergo adverse body conditions and mechanical stretching and bending. The desired SCs must be mechanically robust, flexible, and stretchable without compromising the electrochemical properties. The biodegradability can make the storage device easily decomposable after application or implantation. The photo-rechargeable and photochromic smart SCs highly rely on the inherent properties of the selected materials. Wide temperature tolerance is a fundamental criterion for the material for

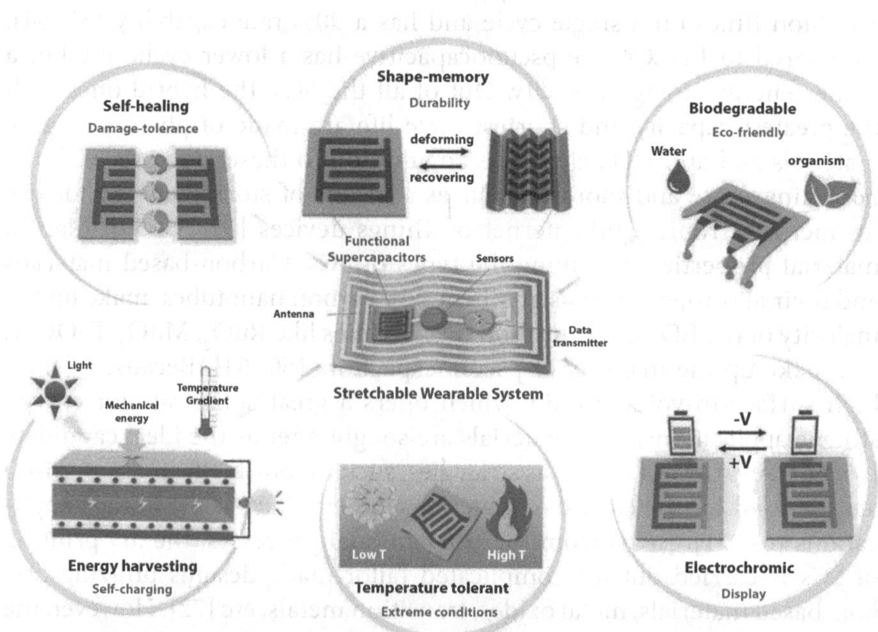


**Figure 7.2** Biocompatible supercapacitor, mechanical property, energy harvesting strategy from various sources, and the smart electrodes for energy storage. Reproduced with permission [74].

application in varying temperatures and for uniformity in a wide range of temperatures to provide a sustainable power source. The supercapacitor in healthcare differs from conventional energy storage devices as it has stringent necessary criteria. The material properties decide the primary application of the SCs in specific sectors. The material selection is prioritized based on the material's inherent properties. Based on the required parameters for the healthcare sector, the electrode materials need to be biocompatible, biodegradable, and chemically inert. Based on the biocompatibility of the material, the SC is further employed for either *in situ* or *ex situ* energy sources. The detailed strategy of the biocompatible supercapacitor is described based on the following features.

### 7.3.1 Biocompatibility

Biocompatibility is a basic need for a material for medical applications. In the Dorland medical dictionary, it is defined as "a quality that has no toxic or injurious effect on a biological system" [22]. When it comes to energy storage applications, based on the type of materials, the active materials are selected. To enhance the performance of the supercapacitor, various additives are hybridized or doped with the conventional material resulting in toxicity or reactive to the human body. The criteria for biomaterial are mainly to select the inert material while coming in contact with the body and portray unmatched performance without any toxic doping to the electrode or electrolyte. Figure 7.3a outlines the basic requirements for biocompatible or implantable energy storage devices [75]. The materials should not elicit inflammatory responses at the active sites. Thus, the biocompatibility of the device must be encompassed based on the cytotoxicity, anticoagulant efficiency, antibacterial, and *in vivo* response. Most of the polymers are biocompatible and inert to the human body but a few of the polymers acquire conductivity and are suitable for energy storage. Biomolecular modification can make some of the polymers suitable for energy storage applications. Figure 7.3b shows the PLA substrate's implementation to fabricate a bioresorbable supercapacitor [76]. The supercapacitor comprises nanoporous ZnO and PVA/PBS hydrogel as electrodes and electrolytes, respectively. The bio-based materials exhibit excellent electrochemical properties and biocompatibility while implanted into a rat. It works for 50 days with 70% capacity retention over 3000 cycles. The biocompatible materials provide a pathway toward reliable implantable devices for future bio-implants. The Zn-ion microsupercapacitor is fabricated over a gelatin film using activated carbon and a micro-Zn anode.

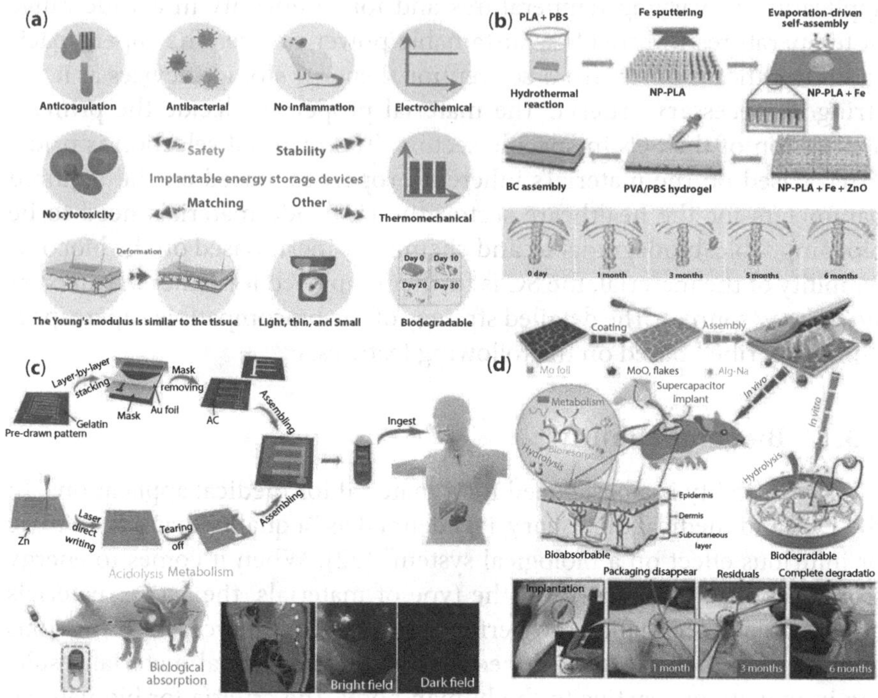


**Figure 7.3** (a) Basic required properties for biocompatible and implantable supercapacitors. Reproduced with permission [75]. (b) PLA-based bio-absorbable supercapacitor fabrication for healthcare application. Reproduced with permission [76]. (c) Implantable supercapacitor fabrication technique and implantable application [77]. (d) MoO-based supercapacitor for implantable energy storage application [78].

The micro-capacitor exhibits 605 mF $cm^{-2}$ with a higher energy density of 215.1 uWh $cm^{-2}$ with a potential window of 1.8 V, enabling it to power microelectronics effectively.

### 7.3.2 Stable Performance

In the long run, stable performance is crucial for energy storage devices. Thus, as medical applications, energy storage devices need stable performance without compromising the energy storage ability. Biocompatibility and stable cyclic performance are necessary for external applications to monitor health or disease detection. The implantable energy storage device undergoes various adverse biological conditions inside the body, which may influence its electrochemical properties. The implants and patches need to work for prolonged periods, which, once fixed, cannot be easily substituted. Hence, the energy storage device needs to exhibit stable performance under variable body temperatures and pH of body fluids without interrupting the energy supply. *In vivo* power supply is carried out successfully inside a pig stomach and displays potential application in future implantable electronics for diagnosis or prevention, as shown in Figure 7.3c [77]. Further researchers have grown molybdenum oxide (MoO) flakes over Mo foil utilizing a green electrochemical strategy. The materials exhibit excellent biocompatibility and stable performance over a prolonged period. The MSC exhibits a specific capacitance of 112.5 mF $cm^{-2}$. The energy density and power density of the MSC are 15.6 Wh $cm^{-2}$ and 2.53 mW $cm^{-2}$, respectively. The materials work in aqueous electrolytes, which involve faradic redox reactions, including surface oxidation and reduction. The capacitance retention of the MSC of 92.7% over 500 cycles provides insight into the development of stable MSC for implantable applications. Along with the stable performance, the device has shown excellent *in vivo* degradation and decomposed inside the body, leading to a sustainable power source for implants, as shown in Figure 7.3d [78].

### 7.3.3 Mechanical Endurance

The mechanical strength of the energy storage devices matters as they undergo varying mechanical loads like stretching and bending body parts internally and externally. Mostly, energy storage devices need to be flexible and stretchable to meet the desired Young's modulus of our tissue and scaffolds to avoid any structural failure, which can lead to performance decay. Adapting the energy storage device to the mechanical properties of the body parts and tissue is essential for optimal operation. The adverse effect of the implant can be mitigated by matching the mechanical properties. Elastomer-based SCs are highly used in mechanically robust energy storage devices. They can be fabricated by various printing or transferring methods. A highly flexible supercapacitor was developed by researchers using laser engraving techniques as shown in Figure 7.4a [79]. The device is highly flexible and can be stretched up to 50%, exhibiting excellent mechanical capabilities. Yarn-type supercapacitors attract the attention of researchers due to their unique structure, which can be stretchable and bendable. This type of SC is widely used in the fabrication of flexible/wearable applications with gel-type electrolytes as shown in Figure 7.4b. Acrylate rubber-based materials exhibit excellent mechanical properties, which were utilized as a flexible precursor for the stretchable electrode. The ACM is mixed with MWCNT to form the conductive electrode (Figure 7.4c). The device shows an excellent mechanical stability of

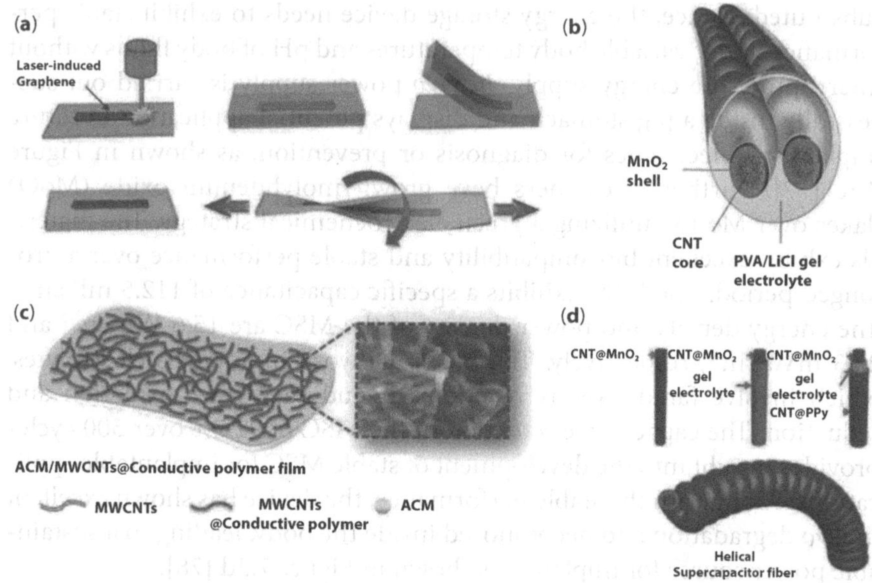


**Figure 7.4** (a) Laser-induced graphene-based flexible supercapacitor fabrication technique [79]. (b) Wire-type supercapacitor with core CNT-$MnO_2$ shell [81]. (c) MWCNT-based conductive polymer film fabrication strategy [82]. (d) Helical fiber-type supercapacitor with LiCl gel electrolyte [83].

50% without changing the electrochemical impedance. The energy storage device exhibits stable performance under 50% strain for 300 cycles. Such flexible devices acquire higher mechanical strength and are electrochemically stable. 10% MWCNT with PDMS has a conductivity of 4.19 S $cm^{-1}$. All-gel state stretchable fiber-based SC was fabricated with rGO/PANI hybrid hydrogel electrodes that exhibit stable electrochemical performance over repeated stretching cycles. The intermolecular stacking and hydrogen bonding provide excellent stability and electrochemical performance. The supercapacitor is 200% stretchable without compromising the electrochemical performance [80]. An SC without polymer substrate was fabricated as shown in Figure 7.4d utilizing helically twisted fibers consisting of PPy and $MnO_2$ coated CNT. The fiber-based SC retains 84% capacitance with a strain of 37.5%. The mechanical endurance for the design and electrochemical performance provides insight into new energy storage device fabrication, which can be used for implantable and health monitoring devices. The excellent mechanical strength enables the devices to be used in *in vivo* medical applications.

### 7.3.4 Design Flexibility

The design of the supercapacitor is another key factor for successful application in the healthcare sector. The device needs to be compact, thin, and flexible, and it can be fit to a tiny space for an implant or any external patch-type application. The performance of the energy storage device also relies highly on the electrode design, such as mass and surface area. The supercapacitor design can be optimized based on the required power, and the tailor-made electrodes can be fabricated for the designated power rating for specific applications. The implantable energy storage devices can be designed for a specific life span. To avoid any internal inflammation, the device can be fabricated with flexible electrodes to adjust the structure to fit the growing tissues and muscles inside the body. The device structure also plays a significant role in the biomedical application. Due to the technological advancements in the manufacturing sector, the supercapacitor's design can be framed in various dimensions like 1D, 2D, or 3D. The wire or fiber-type electrodes can be fabricated in a 1D configuration. In contrast, sheet or thin film-type devices can be framed for 2D configuration. 3D printed or assembled devices can be fabricated in 3D configuration with higher mass loading, which can provide better specific capacitance. Owing to the excellent mechanical and electrochemical properties, different types of biocompatible SCs are fabricated. Optical images of omnidirectional stretchable buckled CNT electrode over PDMS film are shown in Figure 7.5a [84]. The supercapacitor is capable of running under 200% strain. The buckled CNT over PDMS works as an excellent flexible SC. The conductive array of CNT with polyurethane allows for application during tension over the nitrogen-doped core-sheath (N-CNTs) array and performs well under expansion, which increases gradually from 0% to 500%, which provides insight into real-time powering applications under mechanical impact, as shown in Figure 7.5b [85]. Another design strategy to enhance the flexibility of the MSC array is the island structure. SU-8 substrate coated with Au acts as the platform for the island array. The strain applied was around 0.06 to 0.08% as the design can stretch 134%–185% along with the island structure. The strategy lacks performance enhancement as the applied strain does not provide any capacitance enhancement, and larger island fabrication leads to a reduction in the flexibility of the device, as shown in Figure 7.5c. Apart from island-type structures, various stacking designs or kirigami structures are popular designs for fabricating the SCs. CNT-based films retain 98.3% of their original capacitance for 3000 cycles. The ultra-flexible device can be stretched up to 2000%, showing

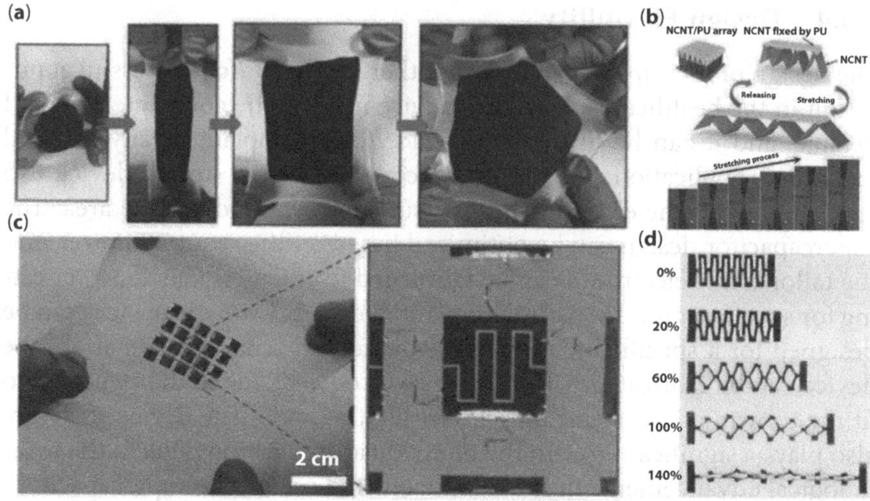


**Figure 7.5** (a) Stretchable supercapacitor fabrication techniques and real-time application in health monitoring [84]. (b) NCNT/PU composite array for stretchable supercapacitor application [85]. (c) IDT electrodes for energy storage application [87]. (d) Highly stretchable supercapacitor [86].

excellent mechanical strength under extreme strain. The device was gradually extended up to 140% without any failure and retained its original shape upon strain removal, as shown in Figure 7.5d [86]. The conductive stretchable devices provide numerous options to fabricate biocompatible supercapacitors.

### 7.3.5 Modification Strategies

There are various modification strategies to enhance the biocompatibility or electrochemical properties of the materials for implantable energy storage applications. Firstly, multifunctional biomaterials are biocompatible, degradable, and edible, compatible with skin, tissue, and blood vessels without toxic reactions. Gold, silver, and platinum are widely used biomaterials for implantable applications. Inorganic materials like hydroxyapatite and tricalcium phosphate are versatile biomaterials that can be modified chemically for energy storage applications and grafted with conductive polymers for the redox active energy storage application. Additionally, peptides and polysaccharides are reliable candidates for micro-energy storage devices due to their excellent biocompatibility and anti-inflammatory properties. Secondly, bioincompatible energy materials can be treated chemically, grafting the biocompatible functional group for implantable energy storage devices. The chemical treatment will make the parent material compatible with body fluids, tissues, and skin. The surface layer of the treated material can facilitate the material and body cell interaction to avoid any infection. The biocompatibility enhancement is generally carried out by the plasma treatments of various gases that lead to improved hydrophilicity. Introducing various functional groups over the material provides excellent biocompatibility and electrochemical properties. Biomolecule immobilization is another versatile strategy to enhance material biocompatibility. Anchoring of biomolecules to the material surface provides functional groups that enhance the cycle life of the device and create a localized immune system between the tissue and the device.

## 7.4 External Power Supply for Health Monitoring

The implementation of SCs in the healthcare sector to power external medical equipment is very crucial as the replacement of existing battery-based systems can provide a relocation in E-waste generation and can provide a reliable solution to the existing energy crisis. The advantage of SCs is the ability to charge fast and have higher power density. High-power medical equipment needs instantaneous power to run the diagnosis instrument or medical equipment inside the hospital, like patient chambers, operation theatres, or first air equipment. To provide high power and enhanced performance by charging the devices with nonconventional sources like solar or wind, the SCs can be a game changer in establishing a reliable, self-sustainable healthcare sector. The strategy can be useful for establishing quality healthcare infrastructure in underprivileged countries or remote locations where people are deprived of quality healthcare due to the non-availability of electricity or remote locations like hilltop, deserts, or any interior geographical location far from the limelight. Below is a detailed explanation of the various charging strategies utilizing an external power supply.

### 7.4.1 Health Monitoring

Smart bands and activity trackers have set the trend in miniature wearable electronics that continuously track health and alert any health disorder. However, the miniaturization of wearable sensors needs a micropower supply. Micro-supercapacitors can power these sensors effectively for long-term applications among micropower supply devices. However, the fabrication technique for MSC is still challenging with higher capacity. Here, 3D printing comes to solve the prevailing bottleneck for high-energy-density supercapacitors. Among the 3D printing techniques, inkjet and extrusion-based printing provide a huge scope for directly printing electronics utilizing conductive inks. The ink fabrication and rheological property optimization with higher ionic conductivity provide an opportunity to achieve high-performance energy storage devices. The devices can be printed according to the programmed structure and the capacity can be tailored as per need. As the miniaturization of bioimplants needs less energy to run, the customizable energy power source can be printed as per the 3D model designed or simulated. Researchers have developed conductive polymer-based MSCs for body health monitoring [88]. The PEDOT:PSS-based ink is directly printed into a designated design, as

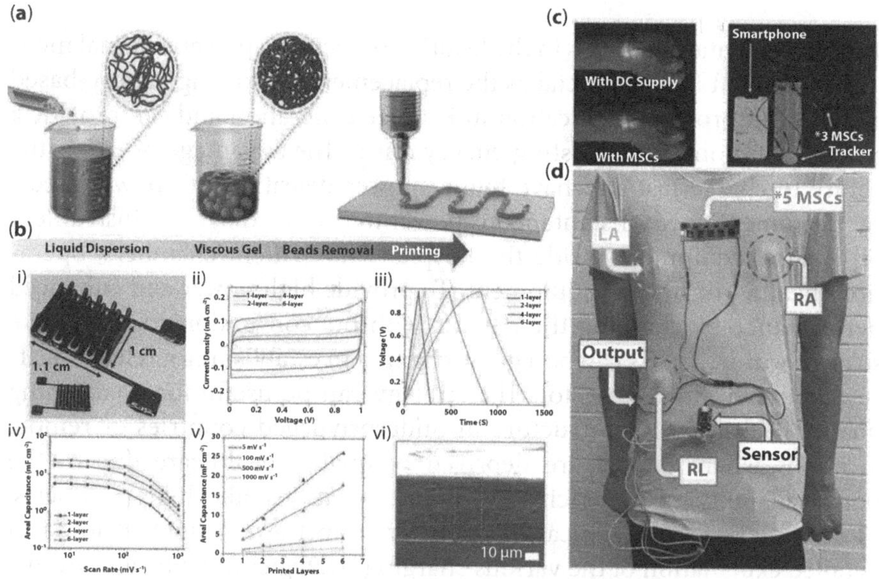


**Figure 7.6** Advanced manufacturing technique for the fabrication of biocompatible and wearable supercapacitor applications. (a) Illustration of 3D printable viscous gel preparation and extrusion techniques for direct printing process. (b) (i) Optical image of 3D printed MSC. (ii) CV curves of MSC with different printing layers from 1 to 6. (iii) GCD curves of different printing layers from 1 to 6. (iv) Areal capacitance of MSC as a function of scan rate. (v) Areal capacitance comparison with various printing layers and scan rates from 5 to 1000 mV $s^{-1}$. (vi) Morphology of the cross-section area of the MSC. (c) Comparison of luminous intensity of LED wristband. (d) 5-MSCs connected in series to power the ECG sensor for heartbeat monitoring. Reproduced with permission [88].

shown in Figure 7.6a. The aqueous conductive ink is treated with superabsorbent polymer beads to formulate viscous gel to achieve the required cross-linking for printing. The optimized ink is printed into a multilayer IDT electrode, and electrochemical analysis is carried out (Figure 7.6b (i)) As the obtained structure has undergone CV analysis, the current range was observed to increase the printed layers, which provides insight into controlling the current rating for the supercapacitor by customizing the printing technique shown in Figure 7.6b (ii). Figure 7.6b (iii) represents the higher charging and discharging time of the MSCs on increasing the printed layers. This provides a strategy to enhance the capacity of the MSCs by controlling the printing parameters and device architecture. Areal capacitance in different scans is plotted in Figure 7.6b (iv), which shows the agreement of scan rate to decrease at higher scans, illustrating the ideal supercapacitor property. A similar trend for specific capacitance enhancement on increasing the printed layers is observed in Figure 7.6b (v). The cross-sectional morphology is evidenced by Figure 7.6b (vi). The seamless fusion of the printed layer is evidenced by the obtained morphology, which results in enhanced specific capacitance. Thus, this strategy to develop new architecture with tunable capacity has huge potential for energy storage applications. Most importantly, the printed ink is biocompatible, which provides an endless opportunity for use in real-time medical applications. Further, the author has demonstrated the actual time application of the printed MSC for real-time health monitoring. A series of connected 3D-printed MSCs is employed to power the wristband, as shown in Figure 7.6c. The MSCs can power the band with abundant intensity. Further, a miniature tracking device is powered with a series of connected MSCs that contain various sensors. The device was successfully powered for 25 minutes, exhibiting excellent energy storage ability, and tracked remotely via mobile. The device was further employed for powering an electrocardiogram (ECG) sensor. The flexibility and architecture of the device allow it to be fixed over the body and act as a potential power source for the ECS sensor. An attenuation was observed for the ECG tracking shown in Figure 7.6d. However, the overall analysis was accurate. Along with real-time monitoring, the device was tested for different charging strategies, such as utilizing solar power and charging efficiently. The devices can be powered by body pulses and biomechanical motions with high- and low-frequency movements. This approach to fabricating customizable power sources with unique architecture leads to the implementation of biocompatible MSCs for implants and real-time health monitoring and establishes a self-powered energy system.

### 7.4.2 Remote Location Treatment

Numerous locations are hard to reach and away from the flat regions of the world. For instance, the mountain regions, deserts, and valleys, including some islands, are not powered by wired electricity. Those regions are powered mainly by wind and solar energy. Hence, when it comes to high-power electricity, conventional storage devices can be replaced with high-performance supercapacitors. A wide range of SCs is available to solve the existing energy crisis and run the devices in remote locations to provide quality healthcare. Implementation of such power sources will reduce the chemical-based power storage devices and can be used to avoid any explosion, e-waste generation or short-circuiting, and contamination of healthcare centers. Secondly, the fast capability of the devices is a breakthrough in harvesting and powering the devices with natural, non-conventional energy sources. Apart from this, the strategy to provide high-quality medical service to inpatients with the help of things is possible by establishing IoT-based healthcare.

### 7.4.3 Therapy

Therapy is necessary for cure in the healthcare sector, which may be active or passive based on the disease [89]. The medical instruments necessary for treatment need some power to run. Automation and advancements in the healthcare sector have miniaturized the devices that need a different power range. Some of the devices are wireless and contain sensors that also need some power to run. Thus, power sources have a huge potential to be explored and replaced with high-performance supercapacitors. Because the medical sector is very sensitive, large-scale trials are necessary for the real-time application of SCs. The targeted drug delivery and cancer cell treatments need a carrier equipped with miniature power sources. Smart band-aids have been developed for faster wound healing and tracking systems. Different sensors have been embedded in band-aids to excrete medication powered by micro-thin film batteries, which can be replaced with performance supercapacitors. Apart from this, miniature devices for *in vivo* imaging and treatment are developed, which need a biocompatible rechargeable power source. Those miniature implants have a vast scope to be powered by supercapacitors. Smart films and glasses in the optoelectronics sector need some portable power source. Such devices can be powered by supercapacitors.

### 7.4.4 Implantable Devices

The implantable devices are sensitive to foreign objects due to the sensitivity and inert surroundings inside the body part. The internal tissues and scaffolds continue to grow. Thus, adjusting the implantable to the sensitive internal part is a tremendous task. Biocompatibility, inertness, and size dependency are the major tasks for implantable devices. The internal injury or inflammation may cause tissue damage and localized infections inside the body. Growth of tissue may get damaged while coming in contact with the implants. Hence, overall concern about the implantable device, its design, and post-application treatment are the major factors that need to be considered. While designing an implantable biomedical device, the power management and the power supply device play major roles. The prime objective of implant design is the miniature device and energy storage unit with stable performance and long cycle life without compromising the capacity. The power source material and manufacturing techniques must be advanced and capable of high-resolution manufacturing. During targeted drug delivery, the carrier sometimes needs to be pushed in the passive path, which requires some power source to function inside the body. Apart from drug delivery, some implants are widely applied for live tracking, such as sensing and imaging of internal parts. The microdevices need miniature power sources to power the miniature camera and the wireless transmission equipment. Thus, here comes the challenge of powering the implantable power source to run for a long period of time. Wireless charging techniques are widely implemented for such charging applications. However, the efficiency and stability of the device are still challenging to match the industry requirements. Here, self-powered pacemakers and pulse-tracking equipment are largely implemented in the health sector, which can overcome the need for power sources for effective functioning. Apart from this, the post-life treatment for the implants is challenging as the implants need to be replaced after application. To overcome this challenge overall, bio-degradable implants with degradable power sources are designed and implemented successfully as implants in various animal bodies as trials. The prime requirement of the decomposable power source is biocompatible residue formation after decomposition, which will not further create any blockage inside the vessels of the body and can be excreted during purification cycles. Figure 7.7 represents the sericin-based implantable supercapacitor fabrication technique along with the biodegradability study for the same [30]. By sequentially cross-linking

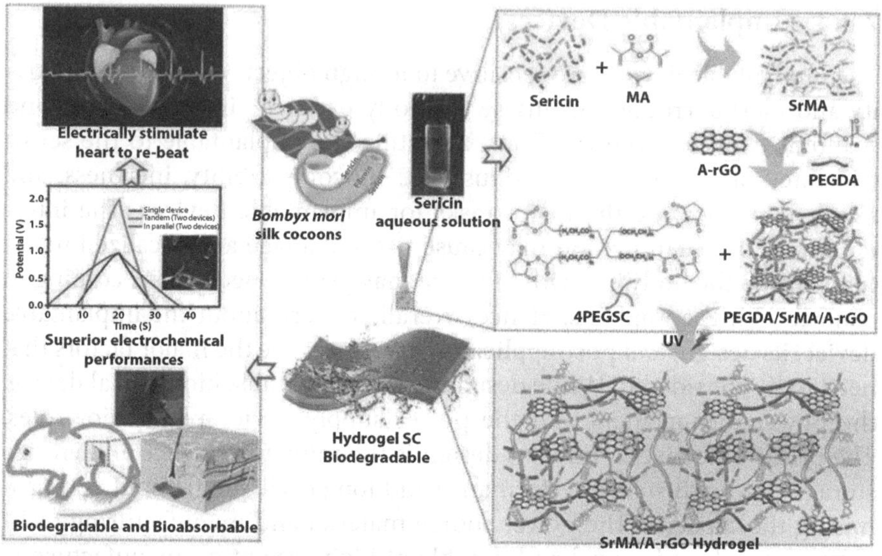


**Figure 7.7** Hydrogel-based biocompatible supercapacitor, fabrication technique, and implantable application for internal energy storage [30].

MA-and-A-rGO-modified sericin with four-arm polyethylene glycol succinimide carbonate (4PEGSC) and polyethylene glycol acrylate (PEGDA) in a weakly alkaline (pH 8.0) environment, the SrMA/A-rGO hydrogel electrode was created. Before gelling with 4PEGSC through amide bonds to form a network and further gelling with PEGDA through photoinitiated radical polymerization to form interpenetrating multiple networks, the SrMA was first conjugated with A-rGO by an amino acid with an aromatic ring, then gelled via amide bonds with 4PEGSC to create a network, and then gelated again with PEGDA via photoinitiated radical polymerization to create multiple interpenetrating networks.

The mechanical endurance of the material is proven by the tensile strength increment from 12.6% to 36.4% by the addition of 4PEGSC, as shown in Figure 7.8a and b. Further, the *in vivo* degradability study of the implantable material is studied in SD rats for 35 days (Figure 7.8c). Figure 7.8d shows the CV curves of the SC with different scan rates, and GCD profiles exhibit EDLC-type behavior (Figure 7.8e). The as-fabricated supercapacitor acquires a specific capacitance of 103.5 μF $cm^{-2}$ and attains a cyclic stability of 90.3 % (Figure 7.8d). The bio-based implantable SC attains excellent flexibility and performs uniformly under different bending conditions (Figure 7.8d, g–i). Capacitance retention under various bending movements proves the excellent capacity of the SC for energy

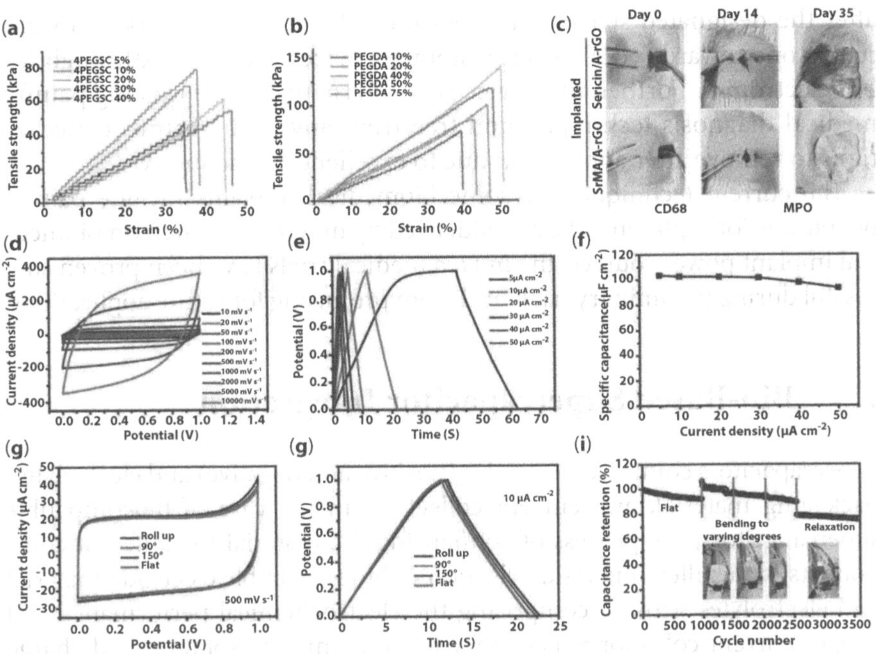


**Figure 7.8** Mechanical of electrochemical and bicompatability properties analysis of the implantable supercapacitors (a) Stress–strain curves of SrMA/A-rGO hydrogel film with different % of 4PEGSC and (b) stress–strain curves of SrMA/A-rGO hydrogel film with different % of PEGDA. (c) *In vivo* degradation study of the hydrogel electrode implant. (d) CV curves of SrMA/A-rGO based SC at different scan rates from 10 to 1000 mV $s^{-1}$. (e) GCD curves of SrMA/A-rGO at different current densities. (f) Specific capacitance as a function of current densities. (g) CV curves of SrMA/A-rGO-based SC under various bending angles. (h) GCD curves of SrMA/A-rGO-based SC under various bending angles. (i) Capacitance retention at various mechanical stimuli at a current density of 25 μA $cm^{-2}$ [30].

storage applications under extreme conditions. This work proposes an implantable hydrogel-based SC by utilizing bio-based protein as the electrode and the implantable application analysis without further compromising the mechanical and electrochemical properties. The SC attains a specific capacitance of 3.75 $Fg^{-1}$. The presence of zwitterionic structures in soy sauce makes it highly resistant to adverse temperature conditions and helps to exhibit uniform performance. The degradation and digestion study of edible SC is carried out under rocky soil and gastric fluids and has proven excellent degradability within 5 minutes. Further real-time demonstration was carried out by powering the LED. According to the need, the edible SC can be ingested, digested, and decomposed inside the body after the designated self-life or application. In short, this work is a stepping stone towards edible energy storage devices and a breakthrough for bio-electronics. Further, the device is an alternative to the existing painful medical diagnosis techniques and free from any post-treatment precautions to remove after application due to excellent degradability.

The current techniques for some biomedical implants include painful operations for replacements. As evidenced by different research on biomedical implant power sources, the *in vivo* medical trials have been proven successful during preliminary trials and seem promising for future applications.

## 7.5 Bio-Based Supercapacitor Integration

Supercapacitors comprise electrodes (positive and negative) and electrolytes, packaging material, and current collectors. In the case of biocompatible supercapacitors, the process of configuring the material for different components is a challenging task. Selection of biocompatible electrode material and electrolytes without comprising the electrochemical performance and proper current collector is necessary to overcome the solution and charge transfer resistance. The bio-based materials for various components provide an insight into bio-based energy storage devices; it will not only boost the energy storage devices in the biomedical sector overall but also reduce the generation of e-waste due to the degradability and biocompatibility. Here, various suitable materials for SC components are explained. For instance, the edible supercapacitor using cellulose-based material has demonstrated the potential application of such material in real-time application [90, 91].

Figure 7.9a shows the edible device electrode, electrolyte material, and fabrication technique [92]. Zwitterionic soy sauce and carbon-based electrodes are the prime components of the edible SC. Carbon-based electrode materials are prepared by mixing with glycerol and hydroxyethyl cellulose and cast using the doctor's blade method. Amino acid molecules in shoyu along with gelatin, glycerol, and hydroxy ethyl cellulose are formed into the edible electrolyte [93, 94]. The gel properties are studied extensively to be utilized successfully as an edible viscous medium for the edible SC. The electrochemical performance of the medical SC is measured by which an EDLC-type behavior is seen in both lower and higher scan rates [Figure 7.9b (i, ii)]. Figure 7.9b (iii) shows the lower solution and charge transfer resistance, proving excellent conductivity that can attain higher energy storage ability. Charge–discharge studies carried out at high and low currents agree with the EDLC behavior and show excellent electrochemical

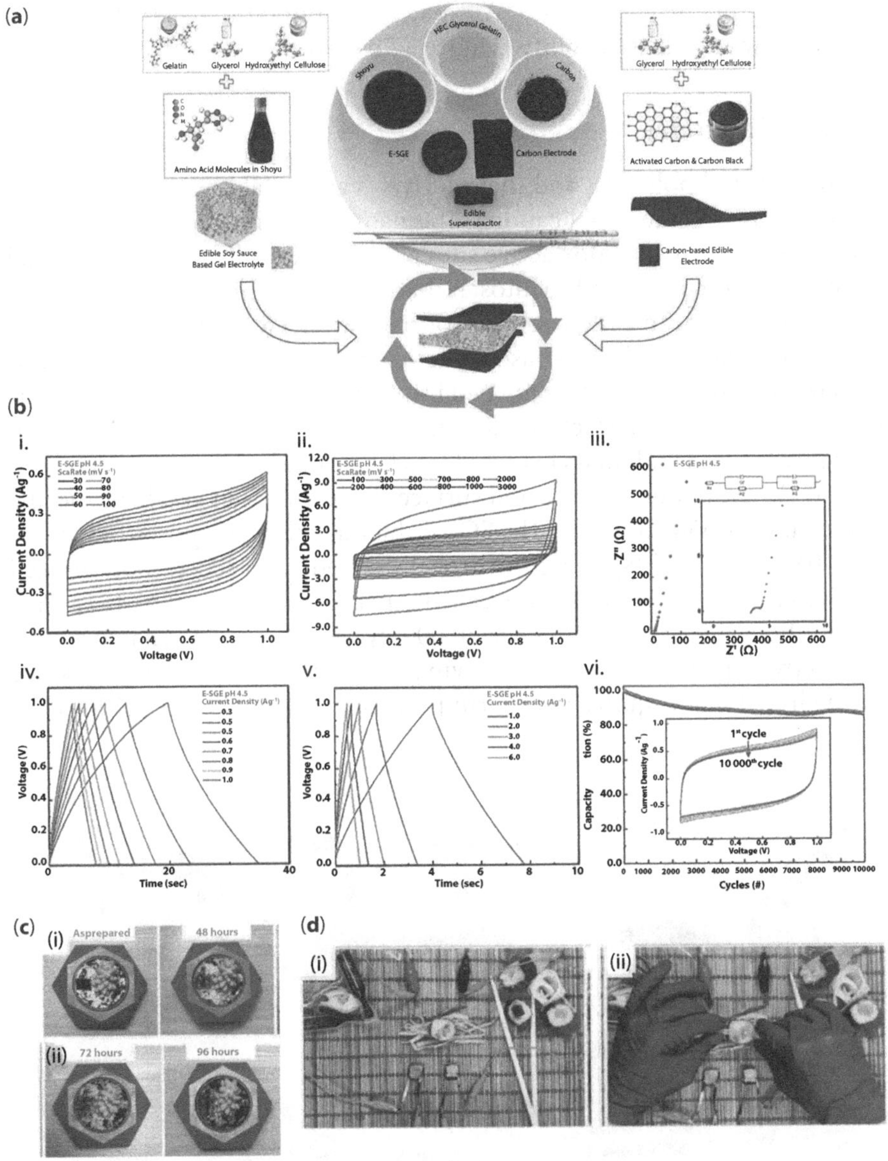


**Figure 7.9** Electrochemical performance analysis and the self-life analysis for the implantable and edible SCs. (a) Schematic representation of used materials for the fabrication of edible supercapacitors. (b) CV analysis of E-SGEs at (i) low and (ii) high scans; (iii) Nyquist plot obtained from the electrochemical impedance spectroscopy; (iv) GCD curves obtained at various low currents; (v) GCD curves obtained from high scans; (vi) capacity retention study from CV analysis for 10,000 cycles at a scan rate of 200 mV $s^{-1}$. (c) Degradation analysis on soil that is watered down. (d) Edible supercapacitor charging and real-time application. Reproduced with permission [92].

agreement [Figure 7.9b (iv, v)]. A capacity retention of 10,000 cycles with 86.5% shows excellent stability [Figure 7.9b (vi)].

### 7.5.1 Electrodes

Various natural polymers and their derivates are suitable electrode materials for energy storage applications. The natural polymers of various functional groups acquire redox active sites and show excellent electrochemical activities. Cellulose, starch, chitosan, bacterial cellulose, sodium alginate, and lignin have been widely used as source material for bio-based electrode fabrication [95–97]. The bio-based precursor can be carbonized, hybridized with redox active material, or formed into hydrogel with a conductive additive to form the active electrode for energy storage applications. As shown in Figure 7.10a, a high mass loading SC is fabricated by electrodeposition of Ni and Co hydroxide over a wood surface that exhibits an excellent specific capacitance of 5308 mF cm$^{-2}$. The SC exhibits an excellent cycle life of 90.3% over 10,000 cycles [98]. Starch hydrogel-derived activated carbon utilized for SC fabrication exhibits an excellent specific capacitance of 1314 Fg$^{-1}$. The cyclic stability of 93.5% over 10,000 cycles provides a reliable source for electrode material. The freeze-drying technique is adopted to synthesize the hydrogel that acquires a uniform porous structure (Figure 7.10b) [99].

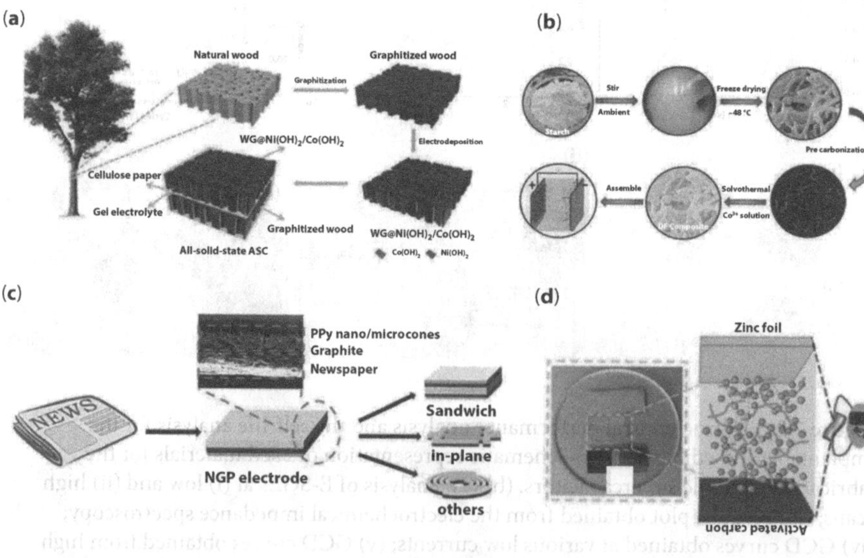


Figure 7.10 Bio-based electrode materials. (a) Wood-based composite for SC electrode [98]. (b) Starch-$Co_3O_4$-based supercapacitor [99]. (c) Cellulose-based newspaper-graphite electrode for energy storage application [100]. (d) Zn foil-hydrogel based supercapacitor [102].

Bio-derived carbon is another class of EDLC material self-doped with various elements that provide inherent enhanced electrochemical properties. Paper, a biocompatible material, attracts attention to the fabrication of biocompatible SCs. A paper-based energy storage system is developed using a pencil-drawn graphite layer, as shown in Figure 7.10c. Further, the electrodeposition of PPy over graphite creates a strong hydroxyl group bond with cellulose and makes a conductive layer with excellent electrochemical properties [100]. The carbon-based electrode materials are synthesized by single-step carbonization, laser treatment, microwave treatment, hydrothermal, etc. The specific capacitance obtained is close to 250 $Fg^{-1}$ and a retention of 50%, as reported by researchers from cellulose-derived carbon [101]. Researchers have proposed a strategy to derive bio-based carbon with different porosity and electrochemical properties. Figure 7.10d represents the cellulose-Zn-based hydrogel for SC application. The SC acquires a high potential window of 2 V and a high energy density of 192 Wh $kg^{-1}$. The hydrogel-based SC retains capacitance in very low temperatures below freezing point due to the presence of $ZnCl_2$ and seems promising for biomedical applications due to wide temperature tolerance.

### 7.5.2 Separator

Separators are an important component for planar-type supercapacitors, which prevent contact between two electrodes and allow the ions to migrate across via an electrolyte medium. However, cellulose, chitosan-based separators are widely used in supercapacitors [103, 104]. The addition of ionic additives can be used to make solid-state ionic conductivity separators for solid-state supercapacitors, flexible supercapacitors, etc. The separator fabrication can be cast, extruded, or electrospun. The customized thickness and porosity of the separator sometimes influence the electrochemical property of the supercapacitor.

### 7.5.3 Electrolyte

Bio-based electrolytes play a significant role in the overall electrochemical properties of the supercapacitor. The ionic conductivity of the supercapacitor depends on the electrolyte, which promotes the charge–discharge process and decides the energy storage capacity of the supercapacitor. Cellulose, gelatin, carboxyl chitosan, agar, starch, guar gum, and lignin are essential bio-precursors for supercapacitor electrolytes [97, 103]. Figure 7.11a shows all flour-based electrodes and electrolytes for SC application. The bio-based precursor can be a breakthrough for edible energy

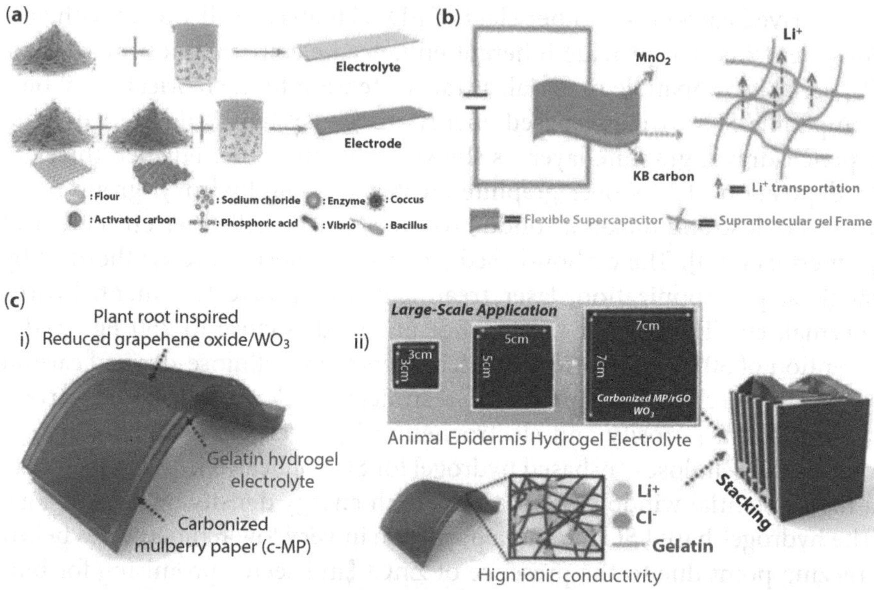


**Figure 7.11** Bio-based electrolyte for supercapacitor application. (a) Flour-derived carbon electrode and flour-based electrolyte for supercapacitor application [105]. (b) Chitosan-based hydrogel electrolyte for flexible supercapacitor [106]. (c) (i) Gelatin-based hydrogel electrolyte for energy storage application. (ii) Schematic of stacked gelatin-mulberry paper SCs [108].

storage devices. The flour electrolyte with NaCl phosphate solution exhibits self-healing properties and regains its shape, exhibiting excellent stability [105]. The diverse chemical structure of these bio-degradable polymers makes them versatile electrolytes due to functional polar chains like OH, $-NH_2$,-CONH, etc. These functional groups enhance the viscosity of the electrolyte and provide active sites for interaction with the salt anion and cations. These biopolymers are also utilized for the hydrogel-based electrolyte synthesis with higher ionic conductivity that facilitates enhanced electrochemical properties. Freestanding hydrogel based on $Ag^+$ ion and chitosan is utilized as an electrolyte with LiCl, as shown in Figure 7.11b. Chitogen with $Ag^+$ provides a faster gel time of 2 s that leads to excellent electrochemical performance while sandwiched between $MnO_2$ and carbon black as positrode and negatrode, respectively [106]. The addition of various amounts of salt can result in different concentrations of hydrogels for electrolyte application. Quasi-solid SC with sodium alginate-based hydrogen acquires a specific capacitance of 405 $Fg^{-1}$ with a cyclic stability of 74.9%, showing the ability of bio-based electrolytes for excellent energy storage application [107]. Researchers have reported that various hydrogels based on polymer electrolytes with hybridized additives lead to enhanced electrochemical performance. Figure 7.11c shows that high ionic conductivity pig skin-derived gelatin is utilized as an efficient electrolyte that is not only biocompatible but also highly ionic conductive. The schematic shows that the mulberry paper and gelatin act as electrodes and electrolytes, respectively. The device is further stacked for real-time application with an extended potential window of 12 V and efficiently lights an LED [108].

### 7.5.4 Current Collector/Packaging

The current collectors for SCs are mostly conductive metal sheets or wires based on the device configuration. However, as polymers lack conductivity, most of the bio-based current collectors are carbon-based materials like CNT, graphene, or graphite [104, 109]. However, the packaging and substrate material are replaced by polymeric materials like polylactic acid (PLA) and polycaprolactone (PCL) for the bio-based supercapacitor. PLA, a commercial biopolymer derived from starch, has excellent mechanical stability and flexibility. Ultrathin SC of 1 mm is fabricated by spraying SWNTs

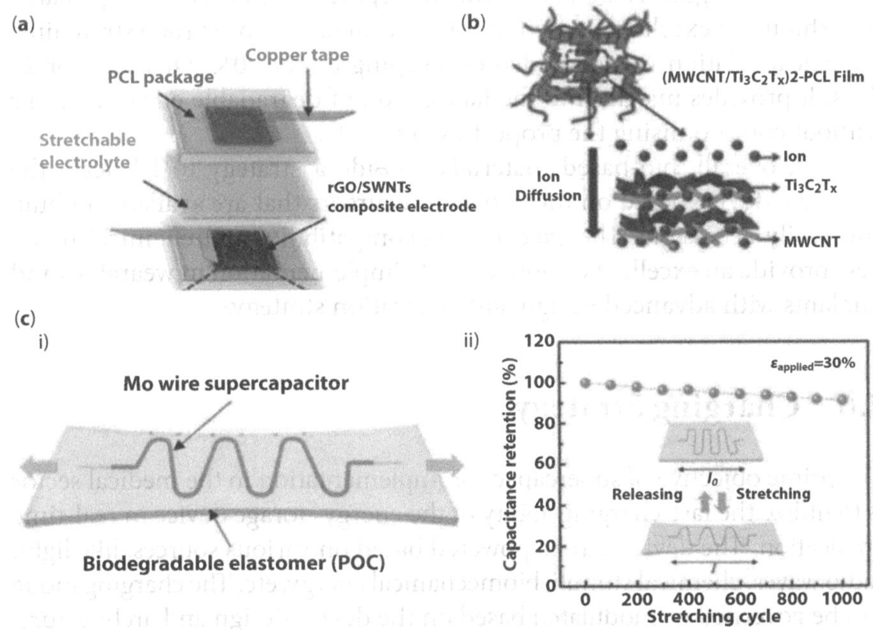


**Figure 7.12** Bio-based supercapacitor encapsulation. (a) PCL-based supercapacitor [110]. (b) PCL film-based supercapacitor illustration in MWCNT/$Ti_3C_2T_x$ SCs [111]. (c) (i) Biodegradable elastomer encapsulated supercapacitor (ii) Tensile stability of POC encapsulated SC [113].

over the PCL substate. The as-fabricated device exhibits excellent stretchability and ideal resistance to bending (Figure 7.12a) [110]. The porous PCL structure may hinder the ionic conductivity, which reduces the performance of the SC, leading to a decrease in electrochemical performance. To overcome the issue, researchers have developed a PCL-based thin film substrate by electrospinning, and further MXene and MWCNT are deposited over the electrospun substrate, as shown in Figure 7.12b [111]. The electrospun substrate provides abundant surface area and leads to a high rate of ion diffusion, resulting in faster electrochemical kinetics. The substrates can be electrospun or 3D printed based on the design requirement and further employed for the supercapacitor device fabrication. PLA/PBAT composites are versatile composites for packaging applications that provide excellent mechanical strength and flexibility to the supercapacitor [112]. The materials are not only biocompatible and degradable, but they are also chemically inert to the human body, which provides excellent stability for implantable or wearable applications. Owing to the development of new fabrication techniques and advancements in biocompatible materials, a POC-encapsulated molybdenum wire-based SC is fabricated, as shown in Figure 7.12c [113]. The biodegradable material encapsulated SC exhibits an excellent capacitance retention of 90% over 1000 stretching. The encapsulation was degraded on keeping under 90°C DI water for 21 days. It provides insight into the fabrication of degradable encapsulations without compromising the properties of the SC.

Thus, overall, bio-based materials provide a strategy to fabricate the complete device based on the natural precursors that are available in bulk and easily accessible. The excellent biocompatibility and chemical inertness provide an excellent option for wide implementation in wearables and implants with advanced design and fabrication strategy.

## 7.6 Charging Strategy

The prime objective of supercapacitor implementation in the medical sector is to utilize the fast-charging ability of the energy storage device in real-time application. The device can be powered based on various sources, like light, radio waves, chemical stimuli, biomechanical energy, etc. The charging mode can be governed or modulated based on the device design and architecture. The external SCs can be powered effectively utilizing the existing sources, but the inflamed SCs need to be charged wirelessly, via chemical gradients and internal pulses. Based on various charging strategies, SC charging in the medical sector is discussed in detail utilizing various energy sources.

### 7.6.1 Photovoltaic

Photovoltaic devices are widely used as portable power sources for supercapacitor recharging. Photovoltaics are semiconducting devices that can power the MSC effectively up to total capacity [114]. The PV cells encapsulated in biological tissues have been demonstrated to convert the near-IR to electrical energy. Initially, the potential PV to power the pacemaker was demonstrated, which broadened the scope for application in implantable devices. A photo-rechargeable supercapacitor with excellent electrochemical performance is fabricated by integrating an organic photovoltaic cell and CNT/PEDOT:PSS SC. An excellent photo-charging efficiency of 6% is achieved with a thickness below 50 μm. The biocompatibility with the ability to convert light energy makes it a sustainable source to empower medical implants. Organic photovoltaics with a thickness of 3 μm embedded into a 40-μm-wide electrode exhibit excellent capacity over 5000 times [115]. The epidermal implants with cellulose isolation membrane are used to hold the solar panel for supercapacitor recharging, which exhibits 99% capacity over 1000 cycles [116]. The potential for wireless recharging opens up for self-powered energy systems. Figure 7.13a shows the $MnO_2$ and carbon-based implantable electronic medical devices (IEMDs) that are constructed with solar cells and supercapacitors for *in vivo* self-powered implant application [24]. Successful *in vivo* implantation is demonstrated by transplanting the device into a rat body. The implanted device provides cycling stability over 1000 cycles that facilitate the pathway to design an efficient, biocompatible, *in vivo* implantable supercapacitor.

### 7.6.2 Ultrasonic

Ultrasonic energy harvesting is a potential way to recharge SCs wirelessly. This makes the ultrasonic-based energy harvesting systems a potential power source for implantable devices. This charging system exhibits a lower attenuation impact over the living tissues [117]. The charging efficiency is higher as compared to other sources of wireless charging of energy storage devices. This charging system is safe for biological implants. Piezoelectric-based ultrasonic systems are integrated with ultrasonic elements for the wireless charging of energy storage devices. The ability of ultrasonic harvesters limits the complication of powering the internal organs without any operation or displacement. Ultrasonic energy harvester as a power source for implantable electronics is shown in Figure 7.13a. Figure 7.13b shows the biocompatible CNT fibers, which are super hydrophilic, can perform in phosphate buffer solution, blood, and other body fluids and acquire a specific capacitance of

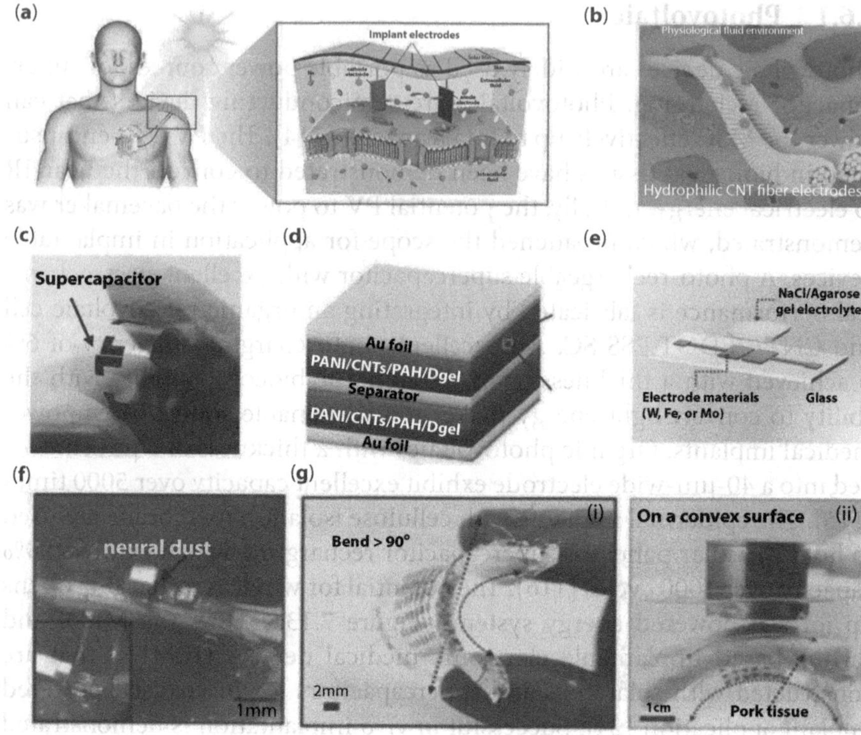


**Figure 7.13** Implantable supercapacitors. (a) $MnO_2$ and carbon-based implantable supercapacitors into the rat skin. Reproduced with permission [24] (b) Biocompatible hydrophilic CNT fibers. Reproduced with permission [23]. (c) Melanin-based supercapacitor [118]. (d) DNA hydrogel composite high-performance supercapacitor. Reproduced with permission [119]. (e) Biodegradable edible supercapacitor. Reproduced with permission [120]. (f) Ultrasonic energy harvesters for implantable bioelectronics, neural dust system of the rat. Reproduced with permission [121]. (g) (i) Flexibility analysis of ultrasonic energy harvester, (ii) *in vitro* study of the ultrasonic energy harvester over concave–convex surfaces. Reproduced with permission [122].

20.8 F $g^{-1}$. The SC provides excellent cyclic stability over 10,000 cycles with a rate capability of 98.3% [23]. Natural biomolecules also have potential in energy storage applications besides conductive polymers. Melanin-coated carbon paper-based SC exhibits a specific capacitance of 167 $Fg^{-1}$. The synergy between redox-active building blocks exhibits pseudocapacitive energy storage properties. The high-power density of 20 mW $cm^{-2}$ shows the excellent electrochemical properties of the melanin-based biomaterials that make it a suitable candidate for supercapacitor electrodes (Figure 7.13c) [118]. Polymer composite with biomolecules provides an excellent strategy for bio-based SC production that provides a potential way to fabricate high-performance SC, as shown in Figure 7.13d [119]. Further, the agarose-based SC is reported as an edible supercapacitor. Water-soluble SC with soluble metallic electrodes acts as a working electrode for the energy storage application, and agarose is the electrolyte shown in Figure 7.13e [120]. The neural dust system provides power to miniature electronics in the rats for recording the electroneurogram (ENG), as depicted in Figure 7.13f [121]. The neural dust system contains the piezoelectric crystal that converts the mechanical vibration induced due to the ultrasonic waves and generates electrical power. The rate of conversion depends on the frequency of the ultrasonic wave. Further, the flexible ultrasonic energy harvester is used for powering *in vivo* devices. The optimum output obtained is around 2 Vpp and 4 μA [122]. The *in vitro* experiment was carried out by mimicking the pork tissue. The device demonstrates excellent energy harvesting efficiency in *in vivo* concave–convex surfaces for harvesting energy to power the implantable SCs. The device acquires a significant decrease in the performance on increased pork skin that can be applied near skin applications. Further, for in-depth applications, the receiver and conversion device for ultrasonic applications need to be modified for highly efficient energy conversion and high-capacity powering of the SCs.

### 7.6.3 RF Energy/Inductive Coupling Charging

Radio frequency-based wireless charging systems are another reliable source to power implants efficiently. The energy transfer efficiency depends on the electromagnetic waves and the design of the transmitter and receiver antenna as shown in Figure 7.14a [123]. The bio-based materials like silicon nanomembranes and Mg, Mo-based materials are suitable for RF energy harvesting due to their desirable properties. Optogenetics is realized by connecting a soft silicon substrate-based thin film embedded with Ti/Au. The flexible antenna was designed with a small volume and can be operated at 2.34 GHz. The device was soft, flexible, and can be bendable as per the muscle movements. The mechanical properties of the device have excellent stability despite implant application. The ability of the device to adopt the anatomical shape makes it a versatile component for the modulation of internal pains caused by peripheral and spinal circuitry.

The biodegradable and implantable targeted drug delivery system is demonstrated by the researchers guided by the RF frequency. The release of the drug is induced on exposure to the electrical stimulus, and no leakage was observed after implantation. The device can be dissolved in PBD solution at 37°C and immersed for a week. The approach of drug delivery

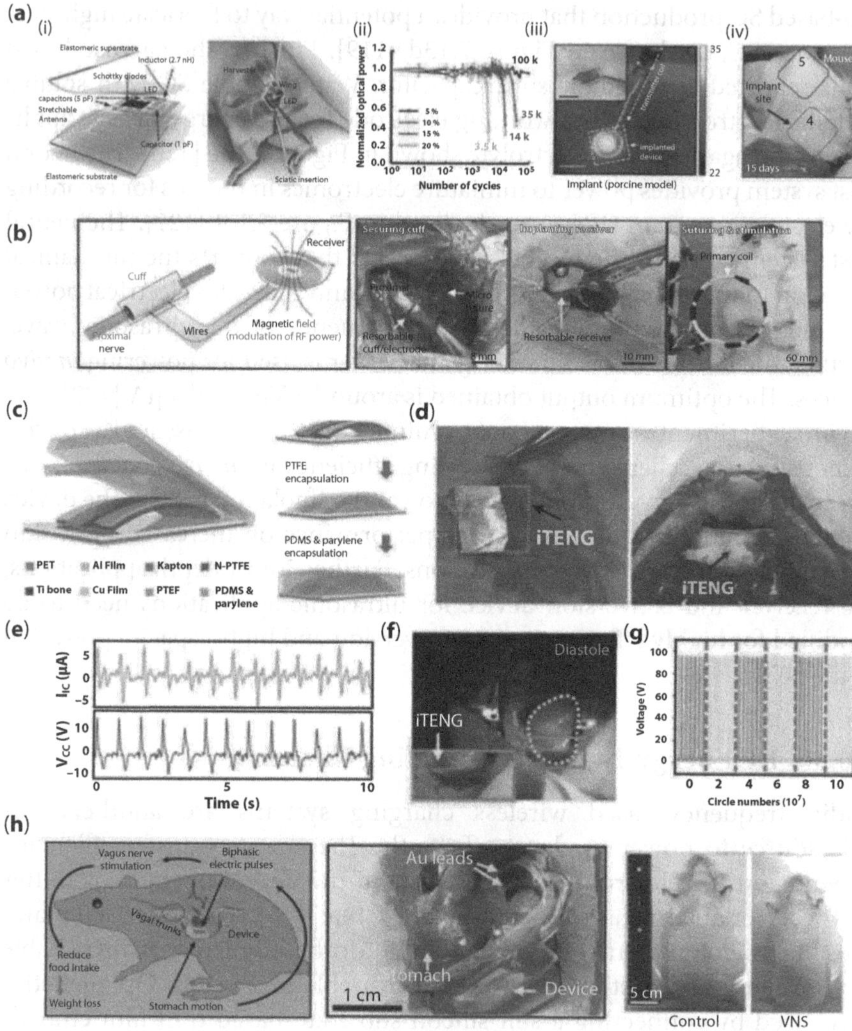


**Figure 7.14** External rechargeable supercapacitors for health monitoring and biological activity tracking. (a) (i) Illustration of flexible RF power transfer system; (ii) optical output analysis under repeated strain [123]; (iii) thermal image of the wirelessly activated device in porcine, reproduced with permission [129]; (iv) device analysis after 15 days of implantation, reproduced with permission [19]. (b) Device structure and implantation procedure to the sciatic nerve [124]. (c) A contact separation TENG. Reproduced with permission [20]. (d) TENG power implantable pacemaker [18]. (e) *In vivo* TENG performance analysis [20]. (f) Diastole and systole powered TENG for implantable device powering. (g) Stability study of the TENG. Reproduced with permission [130]. (h) Smart vagus nerve control system powered by TENG for next-generation weight control device. Reproduced with permission [131].

and bio-based RF systems leads toward developing sustainable biodegradable implants that can be powered wirelessly for a long-run application and disintegrated inside the body. Further, Figure 7.14b displays the design of an inductive couple system for the peripheral nervous system [124]. This strategy is employed for surgical procedures of sciatic nerve and attachment of the IC energy harvester to the subcutaneous tissue. The harvester comprises an antenna of Mg and a dielectric interlayer, diode, and capacitor. The system can be powered up to 100–300 mV from a distance of up to 80 mm. The application exhibits another strategy to power the implants wirelessly utilizing the inductive couple energy.

### 7.6.4 Chemical Energy

The biomechanical processes inside the body are induced by chemical reactions. The reactions continue to control the biological activities and catalytic reactions like metabolism and catabolism. Utilizing the chemical energy that is predominant in the body is a novel idea to incorporate the implants that can be powered by that energy efficiently. As a power source, batteries and supercapacitors have been proven to have potential applications in implants. Some fuel cells are powered by *in vivo* chemical reactions [18, 83, 124, 125]. Among the fuel cells, the glucose-based fuel cells are potential candidates for powering the bio-implants [126]. This inspires the strategy to power the supercapacitor utilizing the potential gradient in the chemical atmosphere of the body and the temperature gradient that induces chemical reactions inside the body. The glucose sensor for monitoring the health of diabetic conditions is developed to work in the long run and is powered by the chemical energy of the body [127]. The self-powered supercapacitor based on glucose sensing exhibits a specific capacitance of 1600 $Fg^{-1}$ and an excellent power density of 0.62 mW $cm^{-2}$. This is proposed for the self-powered body fluid sensors and bioelectronics. The glucose concentration gradient is the driving force for the supercapacitor power. The design of the device and the thin structure act as a reliable implantable power source.

Biomaterial-based SCs are a reliable energy storage material for medical applications and have the potential to be recharged from the body fluid gradient and chemical reactions. Along with glucose sensing, various biological signals can be recorded and utilized for real-time health monitoring. Apart from the glucose sensor, gastric juice concentration has the potential to power the microenergy storage devices efficiently by utilizing the salinity gradient. Capacitors with porous carbon electrodes can be powered continuously by switching electrolyte concentration from high to low. The temperature gradient in human skin can power the supercapacitor and fabricate a self-charging power cell. The Seebeck coefficient acts as a driving force for temperature-based self-powered energy systems. PVA gel solution with coupling agents like iron/ferrous chloride is a thermoelectric power source and can power the device effectively. The devices in the series can be integrated to broaden the potential window and power the device effectively. This method has not yet been established to power the implants effectively but has a huge potential to integrate with biomedical implants to structure a completely integrated self-powered energy system.

### 7.6.5 Mechanical Energy

Along with chemical reactions, the human body tends to vibrate along with the pulses and blood circulation. This helps monitor the body utilizing the sensors based on piezo or triboelectric mechanisms. There is a huge potential to power the implantable implants utilizing the internal and external body pulses [21, 128]. The piezoelectric nanogenerators are the potential energy sources for powering the micro-implants. Still, the circuit designs and deficiencies are significant factors for the real-time application of such power sources. Further, the extension of powering electronics utilizing triboelectric nanogenerator is explored by the researchers, which holds the potential to power bioelectronics efficiently. A miniature TENG is shown in Figure 7.14c, with PTFE, PDMS, and perylene for energy generation [129]. Figure 7.14d shows the real-time powering of the pacemaker of a rat using i-TENG [18]. The 1.2*1.2 cm TENG produces an electrical output of 12 V and 0.25 μA, respectively, with a power density of 8.44 mW $m^{-2}$. The lower output is inefficient for powering the device for large animals. However, the strategy can lead to power MSC efficiency for miniature bio-electronics [19].

Further implantable TENG is composed of a multilayer PTFE/Kapton and aluminum layer harness with a 14 V and 5 μA electrical output, respectively, that can power a pacemaker of Yorkshire porcine (Figure 7.14e) [20]. The *in vivo* energy generation from the TENG can be utilized to power implantable SCs and power micro bioimplants effectively. Figure 7.14f shows the energy generation from the diastole and systole of the heart [130]. The implantable TENG consists of PTFE, EVA spacer, titanium keel, and PDMS encapsulation. The corresponding electrical output of the device is 65.2 V and 5.9 μA, respectively. The excellent stability of the TENG device is achieved with 95 V over 100 million mechanical stimuli cycles, as shown in Figure 7.14g. A smart vagus nerve stimulation system is demonstrated utilizing the flexible TENG that can help to control the food intake and weight control utilizing the electrical signals induced from the peristalsis of the stomach as shown in Figure 7.14h [131]. This smart approach provides insight into the fabrication of self-powered sensory system power by a micro supercapacitor and integrated TENG device for supplying intermittent power.

## 7.7 Challenges and Future Prospects

***1. New biocompatible material:*** Balancing cycle life and specific capacitance with the existing energy material for SC application is a challenging task. Despite this drawback, the biocompatible SC material powers the implants or monitoring devices. The new strategy for biomaterial-based energy material is the need of the hour, using different hybridization techniques or compositing with existing biomaterials. The biocompatible ionic electrolytes can also enhance the electrochemical performance by enhancing the specific capacitance or widening the potential window to power the miniature electronics, bio-implants, etc. Noble surface treatments with biocompatible coating or grafting functional groups can enhance the charge storage mechanism, inducing redox active sites.

***2. Advanced fabrication techniques:*** The existing supercapacitor fabrication techniques need to be advanced with technological advancements. Current manufacturing techniques like spray coating, hydrothermal treatments, 3D printing or direct writing, and laser engravements need to be more precise with high-resolution electrode lining that can fulfill the need for miniaturized power sources for the healthcare sector. Femtosecond-laser writing, direct ink printing, and photolithography are advanced technologies for future energy storage device fabrication on a miniature scale with high resolution.

***3. Power management circuits:*** The implants and health monitoring devices need advanced power management circuits for fast charging, controlled power release during application, and auto-switching or sleep mode during the non-utilization time. The PMC can be a breakthrough in matching the energy needs of the device and the power supply source to tune and regulate power consumption. Apart from this, PMC can be a breakthrough for effectively charging the devices from ambient sources.

***4. Charging technique:*** The conventional charging technique of supercapacitors needs to be streamlined with advanced technologies like Wi-Fi and Li-fi techniques with high efficiency. The RF charging and wireless charging techniques need to be optimized for charging the implant sources to avoid any operation for replacements. The low pulse charging technique can open up countless opportunities to charge the implant from the internal pulses induced by the nerves or tissues inside the body.

***5. New designs:*** To match the needs of the miniatured biomedical devices, a new design strategy is required for the biomedical SCs. Though wire-type

IDT electrodes have been widely used for the application nowadays, quantum dot-based energy storage systems are required for high-performance SCs. The existing wire-type or planar IDT SC design needs to be upgraded for optimized electrochemical performance with higher specific capacitance and better cycle life.

***6. Simulation and modeling:*** Implementation of simulation and modeling can provide a faster way to formulate biocompatible material, design SC structure, and simulate the performance to reduce the experimental errors and get a ready-made solution to the existing challenges over material and designs. The software-assisted design and program can help the researchers find a faster approach for high-performance SCs to power the implants and a new strategy to mitigate operational failures. Material, methods, and new software-assisted controlled programs can power the devices effectively with the existing sources.

## 7.8 Conclusion

This chapter covers the implementation of supercapacitors in the healthcare sector and the material selection protocol for biomedical applications. Current trends and achievements have been discussed in detail in various healthcare applications, both *in vivo* and *ex vivo*. Different biocompatible, degradable, edible, and implantable supercapacitors and their performance and structure are discussed. Protocols for material selection and various design and fabrication strategies are described. The flexible architecture and material treatments for biocompatibility provide insight into developing new-generation power sources for medical applications for various implants and real-time integrated health monitoring systems. The focus is now on new material development, treatment strategy for existing materials for biocompatibility, and advanced manufacturing techniques for miniatured power source fabrication. Miniaturization, mass production, and intelligent supercapacitors with high cycles are the future.

## Acknowledgements

Authors thank Indo-German Project (Scheme No. DST/INT/DAAD/P-12/2022) and UGC-DAE-CSR, Indore (Ref. No: CRS/2021-22/04/614) for the financial research support. The Basic Science Research Program supported this research work through the National Research Foundation of Korea (NRF) grant funded by the Korea government (MSIT) (2021R1A4A2000934, & 2023R1A2C3004336).

## References


1. Molefi, M., Markus, E.D., Abu-Mahfouz, A., Wireless power transfer for IoT devices-a review, in: *2019 International multidisciplinary information technology and engineering conference (IMITEC)*, IEEE, pp. 1–8, 2019.
2. Jayakumar, H. *et al.*, Energy-efficient system design for IoT devices, in: *2016 21st Asia and South Pacific design automation conference (ASP-DAC)*, IEEE, pp. 298–301, 2016.
3. Singh, S., Sharma, P.K., Moon, S.Y., Park, J.H., Advanced lightweight encryption algorithms for IoT devices: survey, challenges and solutions. *J. Ambient Intell. Humaniz Comput.*, 15, 1–18, 2017.
4. Su, Y. *et al.*, Stretchable Transparent Supercapacitors for Wearable and Implantable Medical Devices. *Adv. Mater. Technol.*, 7, 2100608, 2022.
5. Meng, C., Gall, O.Z., Irazoqui, P.P., A flexible super-capacitive solid-state power supply for miniature implantable medical devices. *Biomed. Microdevices*, 15, 973–983, 2013.
6. Naveen, M.H., Gurudatt, N.G., Shim, Y.B., Applications of conducting polymer composites to electrochemical sensors: A review. *Appl. Mater. Today*, 9, 419–433, 2017.
7. Zou, J. *et al.*, Coupled Supercapacitor and Triboelectric Nanogenerator Boost Biomimetic Pressure Sensor. *Adv. Energy Mater.*, 8, 1702671, 2018.
8. Khan, A., Khan, A.A.P., Rahman, M.M., Asiri, A.M., Al-Youbi, A.O., Toward designing efficient rice-shaped polyaniline@bismuth oxide nanocomposites for sensor application. *J. Solgel Sci. Technol.*, 76, 519–528, 2015.
9. Jia, J. *et al.*, Advancing MXene-based integrated microsystems with micro-supercapacitors and/or sensors: Rational design, key progress, and challenging perspectives. *J. Materiomics*, 9, 1242–1262, 2023.
10. de Oliveira, F.M., de Melo, E., II, da Silva, R.A.B., 3D Pen: A low-cost and portable tool for manufacture of 3D-printed sensors. *Sens. Actuators B Chem.*, 321, 128528, 2020.
11. Li, Y., Li, N., Oliveira, N.D., Wang, S., Implantable Bioelectronics Toward Long-Term Stability and Sustainability. *Matter*, 4, 1125–1141, 2021.
12. Vaghasiya, J.V., Mayorga-Martinez, C.C., Vyskočil, J., Sofer, Z., Pumera, M., Integrated Biomonitoring Sensing with Wearable Asymmetric Supercapacitors Based on Ti3C2 MXene and 1T-Phase WS2 Nanosheets. *Adv. Funct. Mater.*, 30, 1–10, 2020.
13. Wang, X. *et al.*, Bioinspired stretchable triboelectric nanogenerator as energy-harvesting skin for self-powered electronics. *Nano Energy*, 39, 429–436, 2017.
14. Chen, X. *et al.*, Stretchable Supercapacitors as Emergent Energy Storage Units for Health Monitoring Bioelectronics. *Adv. Energy Mater.*, 10, 1902769, 2020.
15. Beker, P., Koren, I., Amdursky, N., Gazit, E., Rosenman, G., Bioinspired peptide nanotubes as supercapacitor electrodes. *J. Mater. Sci.*, 45, 6374–6378, 2010.

16. Dhirendra Prasad, R. *et al.*, In Memory of Prin. RCP Yadav, BVC Patna: A Critical Review on Recent Developments in Advanced Super-capacitors for Veterinary Medicine. *ES Food Agrofor.*, 11, 805, 2023.
17. Rarotra, S. *et al.*, Progress and challenges on Battery Waste Management: A Critical Review. *ChemistrySelect*, 5, 6182–6193, 2020.
18. Zheng, Q. *et al.*, *In Vivo* Powering of Pacemaker by Breathing-Driven Implanted Triboelectric Nanogenerator. *Adv. Mater.*, 26, 5851–5856, 2014.
19. Tao, H. *et al.*, Silk-based resorbable electronic devices for remotely controlled therapy and *in vivo* infection abatement. *Proc. Natl. Acad. Sci.*, 111, 17385–17389, 2014.
20. Zheng, Q. *et al.*, *In Vivo* Self-Powered Wireless Cardiac Monitoring *via* Implantable Triboelectric Nanogenerator. *ACS Nano*, 10, 6510–6518, 2016.
21. Wang, Z.L., Triboelectric Nanogenerators as New Energy Technology for Self-Powered Systems and as Active Mechanical and Chemical Sensors. *ACS Nano*, 7, 9533–9557, 2013.
22. Stauss, S. and Honma, I., Biocompatible Batteries—Materials and Chemistry, Fabrication, Applications, and Future Prospects. *Bull. Chem. Soc. Jpn.*, 91, 492–505, 2018.
23. He, S. *et al.*, Biocompatible carbon nanotube fibers for implantable supercapacitors. *Carbon N Y*, 122, 162–167, 2017.
24. Ralls, A., Kumar, P., Misra, M. *et al.*, Material Design and Surface Engineering for Bio-Implants. *JOM*, 72, 684–696, 2020.
25. Boddupalli, A., Zhu, L., Bratlie, K.M., Methods for Implant Acceptance and Wound Healing: Material Selection and Implant Location Modulate Macrophage and Fibroblast Phenotypes. *Adv. Healthc. Mater.*, 5, 2575–2594, 2016.
26. Jiang, J. *et al.*, Uniform implantation of CNTs on total activated carbon surfaces: a smart engineering protocol for commercial supercapacitor applications. *Nanotechnology*, 28, 145402, 2017.
27. Bhat, M.A., Rather, R.A., Shalla, A.H., PEDOT and PEDOT:PSS conducting polymeric hydrogels: A report on their emerging applications. *Synth. Met.*, 273, 116709, 2021.
28. Valero, A., Mery, A., Gaboriau, D., Gentile, P., Sadki, S., One Step Deposition of PEDOT–PSS on ALD Protected Silicon Nanowires: Toward Ultrarobust Aqueous Microsupercapacitors. *ACS Appl. Energy Mater.*, 2, 436–447, 2019.
29. Ur Rehman, H., Shuja, A., Ali, M., Murtaza, I., Meng, H., Evaluation of defects and current kinetics for aging analysis of PEDOT:PSS based supercapacitors. *J. Energy Storage*, 28, 101243, 2020.
30. Lv, Q. *et al.*, A Degradable and Biocompatible Supercapacitor Implant Based on Functional Sericin Hydrogel Electrode. *Adv. Energy Mater.*, 13, 2208034, 2023.
31. Gerritsen, M., Performance of subcutaneously implanted glucose sensors for continuous monitoring. *Neth. J. Med.*, 54, 167–179, 1999.
32. Shaw, G.W., Claremont, D.J., Pickup, J.C., *In vitro* testing of a simply constructed, highly stable glucose sensor suitable for implantation in diabetic patients. *Biosens. Bioelectron.*, 6, 401–406, 1991.

33. Lu, X., Wang, C., Favier, F., Pinna, N., Electrospun nanomaterials for supercapacitor electrodes: designed architectures and electrochemical performance. *Adv. Energy Mater*, 7, 1601301, 2017.
34. Corker, A., Ng, H.C.H., Poole, R.J., García-Tuñón, E., 3D printing with 2D colloids: Designing rheology protocols to predict 'printability' of soft-materials. *Soft Matter*, 15, 1444–1456, 2019.
35. Tanwilaisiri, A. *et al.*, Design and fabrication of modular supercapacitors using 3D printing. *J. Energy Storage*, 16, 1–7, 2018.
36. Das, J.P., Kim, S.J., Ramadoss, A., Additive manufacturing for functionalized nanomaterials dedicated to supercapacitors, in: *Functionalized Nanomaterials Based Supercapacitor: Design, Performance and Industrial Applications*, pp. 131–160. Springer Nature, Singapore, 2023.
37. Naoi, K., Naoi, W., Aoyagi, S., Miyamoto, J., Kamino, T., New generation "nanohybrid supercapacitor". *Acc. Chem. Res.*, 46, 1075–1083, 2013.
38. Li, M. *et al.*, 3D Printed Supercapacitor: Techniques, Materials, Designs, and Applications. *Adv. Funct. Mater.*, 33, 2002180, 2023.
39. Sharma, P. and Kumar, V., Current technology of supercapacitors: A review. *J. Electron. Mater.*, 49, 3520–3532, 2020.
40. Yan, X., Tong, Y., Wang, X., Hou, F., Liang, J., Extrusion-Based 3D-Printed Supercapacitors: Recent Progress and Challenges. *Energy Environ. Mater.*, 5, 800–822, 2022.
41. Horn, M., MacLeod, J., Liu, M., Webb, J., Motta, N., Supercapacitors: A new source of power for electric cars? *Econ. Anal. Policy*, 61, 93–103, 2019.
42. Delbari, S.A. *et al.*, Transition metal oxide-based electrode materials for flexible supercapacitors: A review. *J. Alloys Compd.*, 857, 158281, 2021.
43. Vaghasiya, J.V., Mayorga-Martinez, C.C., Sofer, Z., Pumera, M., MXene-Based Flexible Supercapacitors: Influence of an Organic Ionic Conductor Electrolyte on the Performance. *ACS Appl. Mater. Interfaces*, 12, 53039–53048, 2020.
44. Xu, M. *et al.*, Green conversion of Ganoderma lucidum residues to electrode materials for supercapacitors. *Adv. Compos Hybrid Mater*, 4, 1270–1280, 2021.
45. Miller, E.E., Hua, Y., Tezel, F.H., Materials for energy storage: Review of electrode materials and methods of increasing capacitance for supercapacitors. *J. Energy Storage*, 20, 30–40, 2018.
46. Tan, Y.B. and Lee, J.-M., Graphene for supercapacitor applications. *J. Mater. Chem. A Mater.*, 1, 14814–14843, 2013.
47. Chavhan, M.P., Slovak, V., Lu, C.-H., Thickness-controlled porous hexagonal NiO nanodiscs electrodes for use in supercapacitors: How nanodiscs thickness influences electrochemical performance. *J. Energy Storage*, 50, 104329, 2022.
48. Shown, I., Ganguly, A., Chen, L., Chen, K., Conducting polymer-based flexible supercapacitor. *Energy Sci. Eng.*, 3, 2–26, 2015.

49. Wang, Y. *et al.*, Supercapacitor devices based on graphene materials. *J. Phys. Chem. C*, 113, 13103–13107, 2009.
50. Devillers, N., Jemei, S., Péra, M.-C., Bienaimé, D., Gustin, F., Review of characterization methods for supercapacitor modelling. *J. Power Sources*, 246, 596–608, 2014.
51. Snook, G.A., Kao, P., Best, A.S., Conducting-polymer-based supercapacitor devices and electrodes. *J. Power Sources*, 196, 1–12, 2011.
52. Liu, C., Yu, Z., Neff, D., Zhamu, A., Jang, B.Z., Graphene-based supercapacitor with an ultrahigh energy density. *Nano Lett.*, 10, 4863–4868, 2010.
53. Kang, J. *et al.*, Extraordinary supercapacitor performance of a multicomponent and mixed-valence oxyhydroxide. *Angew. Chem.*, 127, 8218–8222, 2015.
54. Idrees, M., Ahmed, S., Mohammed, Z., Korivi, N.S., Rangari, V., 3D printed supercapacitor using porous carbon derived from packaging waste. *Addit. Manuf.*, 36, 101525, 2020.
55. Li, L. *et al.*, Highly Stretchable Micro-Supercapacitor Arrays with Hybrid MWCNT/PANI Electrodes. *Adv. Mater. Technol.*, 2, 1600282, 2017.
56. Ravichandran, V. et al., High-Performance Redox-Active Organic Molecule Grafted Graphene Based On-Chip Micro-Supercapacitor Towards Self-Powered Environmental Monitoring Station. *Chem. Eng. J.*, 482, 148822, 2024.
57. Qin, S. *et al.*, Hybrid Piezo/Triboelectric-Driven Self-Charging Electrochromic Supercapacitor Power Package. *Adv. Energy Mater.*, 8, 1800069, 2018.
58. Huang, X. and Gou, L., High performance asymmetric supercapacitor based on hierarchical flower-like NiCo2S4@polyaniline. *Appl. Surf. Sci.*, 487, 68–76, 2019.
59. Dubey, R. and Guruviah, V., Review of carbon-based electrode materials for supercapacitor energy storage. *Ionics (Kiel)*, 25, 1419–1445, 2019.
60. Sajedi-Moghaddam, A., Rahmanian, E., Naseri, N., Inkjet-printing technology for supercapacitor application: Current state and perspectives. *ACS Appl. Mater. Interfaces*, 12, 34487–34504, 2020.
61. Siddique, S. *et al.*, Scalable Synthesis of Atomically Thin Gallium Telluride Nanosheets for Supercapacitor Applications. *ACS Appl. Nano Mater.*, 4, 4829–4838, 2021.
62. Lu, X. *et al.*, 3D printing well organized porous iron-nickel/polyaniline nanocages multiscale supercapacitor. *J. Alloys Compd.*, 760, 78–83, 2018.
63. Guo, Y. *et al.*, Multicomponent hierarchical Cu-doped NiCo-LDH/CuO double arrays for ultralong-life hybrid fiber supercapacitor. *Adv. Funct. Mater.*, 29, 1809004, 2019.
64. Li, X., Li, H., Fan, X., Shi, X., Liang, J., 3D-Printed Stretchable Micro-Supercapacitor with Remarkable Areal Performance. *Adv. Energy Mater.*, 10, 1–12, 2020.
65. Zhang, H. *et al.*, A general strategy toward transition metal carbide/carbon core/shell nanospheres and their application for supercapacitor electrode. *Carbon N Y*, 100, 590–599, 2016.

66. Ghasem Hosseini, M. and Shahryari, E., A Novel High-Performance Supercapacitor based on Chitosan/Graphene Oxide-MWCNT/Polyaniline. *J. Colloid Interface Sci.*, 496, 371–381, 2017.
67. Song, Y. *et al.*, Integrated self-charging power unit with flexible supercapacitor and triboelectric nanogenerator. *J. Mater Chem. A Mater.*, 4, 14298–14306, 2016.
68. Wang, X. *et al.*, Controllable Synthesis of NiCo LDH Nanosheets for Fabrication of High-Performance Supercapacitor Electrodes. *Electroanalysis*, 29, 1286–1293, 2017.
69. Zhang, J. *et al.*, MXene: A potential candidate for yarn supercapacitors. *Nanoscale*, 9, 18604–18608, 2017.
70. Guan, F. *et al.*, Mechanically robust reduced graphene oxide/bacterial cellulose film obtained via biosynthesis for flexible supercapacitor. *Chem. Eng. J.*, 360, 829–837, 2019.
71. Li, B. *et al.*, Self-supporting graphene aerogel electrode intensified by NiCo2S4 nanoparticles for asymmetric supercapacitor. *Electrochim. Acta*, 314, 32–39, 2019.
72. Das, J.P., Divakaran, N., Y, A., T P, M., Mohanty, S., Fabrication of 3D printing extrusion-based Nylon 12/Bi2O3 composite filaments for thermal interface device application. *J. Manuf. Process.*, 85, 885–893, 2023.
73. Singh, G. and Chanda, A., Mechanical properties of whole-body soft human tissues: a review. *Biomed. Mater.*, 16, 062004, 2021.
74. Keum, K. *et al.*, Flexible/Stretchable Supercapacitors with Novel Functionality for Wearable Electronics. *Adv. Mater.*, 32, 15261–15272, 2020.
75. Yu, M., Peng, Y., Wang, X., Ran, F., Emerging Design Strategies Toward Developing Next-Generation Implantable Batteries and Supercapacitors. *Adv. Funct. Mater.*, 33, 2301877, 2023.
76. Li, H. *et al.*, Fully Bioabsorbable Capacitor as an Energy Storage Unit for Implantable Medical Electronics. *Adv. Sci.*, 6, 1801625, 2019.
77. Chen, K. *et al.*, An Edible and Nutritive Zinc-Ion Micro-supercapacitor in the Stomach with Ultrahigh Energy Density. *ACS Nano*, 16, 15261–15272, 2022.
78. Sheng, H. *et al.*, A thin, deformable, high-performance supercapacitor implant that can be biodegraded and bioabsorbed within an animal body. *Sci. Adv.*, 7, 1–11, 2021.
79. Lamberti, A., Clerici, F., Fontana, M., Scaltrito, L., A Highly Stretchable Supercapacitor Using Laser-Induced Graphene Electrodes onto Elastomeric Substrate. *Adv. Energy Mater.*, 6, 1600050, 2016.
80. Li, P. *et al.*, Stretchable All-Gel-State Fiber-Shaped Supercapacitors Enabled by Macromolecularly Interconnected 3D Graphene/Nanostructured Conductive Polymer Hydrogels. *Adv. Mater.*, 30, 1800124, 2018.
81. Choi, C. *et al.*, Elastomeric and Dynamic $MnO_2$ /CNT Core–Shell Structure Coiled Yarn Supercapacitor. *Adv. Energy Mater.*, 6, 1502119, 2016.

82. Wang, X. *et al.*, High-performance stretchable supercapacitors based on intrinsically stretchable acrylate rubber/MWCNTs@conductive polymer composite electrodes. *J. Mater Chem. A Mater*, 6, 4432–4442, 2018.
83. Yu, J. *et al.*, A High Performance Stretchable Asymmetric Fiber-Shaped Supercapacitor with a Core-Sheath Helical Structure. *Adv. Energy Mater.*, 7, 1600976, 2017.
84. Yu, J. *et al.*, Omnidirectionally Stretchable High-Performance Supercapacitor Based on Isotropic Buckled Carbon Nanotube Films. *ACS Nano*, 10, 5204–5211, 2016.
85. Zhang, Z. *et al.*, Nitrogen-Doped Core-Sheath Carbon Nanotube Array for Highly Stretchable Supercapacitor. *Adv. Energy Mater.*, 7, 1601814, 2017.
86. He, S. *et al.*, Stretchable supercapacitor based on a cellular structure. *J. Mater. Chem. A Mater.*, 4, 10124–10129, 2016.
87. Yun, J. *et al.*, Stretchable patterned graphene gas sensor driven by integrated micro-supercapacitor array. *Nano Energy*, 19, 401–414, 2016.
88. Li, T. *et al.*, Facile material extrusion of 3D wearable conductive-polymer micro-super-capacitors. *Addit. Manuf.*, 74, 103714, 2023.
89. Tian, W. *et al.*, Implantable and Biodegradable Micro-Supercapacitor Based on a Superassembled Three-Dimensional Network Zn@PPy Hybrid Electrode. *ACS Appl. Mater. Interfaces*, 13, 8285–8293, 2021.
90. Wang, H. *et al.*, *In situ* formation of a renewable cellulose hydrogel electrolyte for high-performance flexible all-solid-state asymmetric supercapacitors. *Sustain. Energy Fuels*, 3, 3109–3115, 2019.
91. Gui, Z. *et al.*, Natural Cellulose Fiber as Substrate for Supercapacitor. *ACS Nano*, 7, 6037–6046, 2013.
92. Durukan, M.B. *et al.*, An Edible Supercapacitor Based on Zwitterionic Soy Sauce-Based Gel Electrolyte. *Adv. Funct. Mater.*, 34, 6, 2023.
93. Kim, E.-K. *et al.*, KCl-assisted synthesis of hierarchically porous carbon materials from water-soluble 2-hydroxyethyl cellulose for high-performance green supercapacitors. *Sustain. Mater. Technol.*, 34, e00520, 2022.
94. Kadir, M.F.Z., Non-Faradaic-based supercapacitor fabricated with fish skin gelatin biopolymer electrolyte. *Ionics (Kiel)*, 27, 2219–2229, 2021.
95. Feng, L. *et al.*, Phosphoric acid/glycerol-based anti-freezing organohydrogel electrolyte for flexible supercapacitor with excellent low-temperature stability. *J. Power Sources*, 580, 233453, 2023.
96. Peng, S. *et al.*, High-performance and flexible solid-state supercapacitors based on high toughness and thermoplastic poly(vinyl alcohol)/NaCl/glycerol supramolecular gel polymer electrolyte. *Electrochim. Acta*, 324, 134874, 2019.
97. Rosi, M., Iskandar, F., Abdullah, M., Khairurrijal, Hydrogel-Polymer Electrolytes Based on Polyvinyl Alcohol and Hydroxyethylcellulose for Supercapacitor Applications. *Int. J. Electrochem. Sci.*, 9, 4251–4256, 2014.
98. Liu, Z. *et al.*, Redox active coating on graphite surface of hierarchically porous wood electrodes for supercapacitor application. *Mater. Today Commun.*, 24, 101045, 2020.

99. He, L. *et al.*, A new strategy for porous carbon synthesis: Starch hydrogel as a carbon source for Co3O4@C supercapacitor electrodes. *Ceram. Int.*, 48, 8104–8111, 2022.
100. Huang, Q., Yang, Y., Chen, R., Wang, X., High performance fully paper-based all-solid-state supercapacitor fabricated by a papermaking process with silver nanoparticles and reduced graphene oxide-modified pulp fibers. *EcoMat*, 3, 1–11, 2021.
101. Titirici, M.-M., Robin., J., Falco, C., & Sevilla, M. Black perspectives for a green future: hydrothermal carbons for environment protection and energy storage. *Energy Environ. Sci.*, 5, 6796, 2012.
102. Yang, L. *et al.*, Zinc ion trapping in a cellulose hydrogel as a solid electrolyte for a safe and flexible supercapacitor. *J. Mater. Chem. A Mater.*, 8, 12314–12318, 2020.
103. Sudhakar, Y.N., Selvakumar, M., Bhat, D.K., Preparation and characterization of phosphoric acid-doped hydroxyethyl cellulose electrolyte for use in supercapacitor. *Mater. Renew. Sustain. Energy*, 4, 10, 2015.
104. Fan, H. and Shen, W., Gelatin-Based Microporous Carbon Nanosheets as High Performance Supercapacitor Electrodes. *ACS Sustain. Chem. Eng.*, 4, 1328–1337, 2016.
105. Hu, M. *et al.*, A flour-based one-stop supercapacitor with intrinsic self-healability and stretchability after self-healing and biodegradability. *Energy Storage Mater*, 21, 174–179, 2019.
106. Cao, L. *et al.*, Biopolymer-chitosan based supramolecular hydrogels as solid state electrolytes for electrochemical energy storage. *Chem. Commun.*, 53, 1615–1618, 2017.
107. Zhang, X. *et al.*, Inorganic Salts Induce Thermally Reversible and Anti-Freezing Cellulose Hydrogels. *Angew. Chem. Int. Ed.*, 58, 7366–7370, 2019.
108. Yun, T.G., Jang, J.-S., Cheong, J.Y., Kim, I.-D., Organism epidermis/plant-root inspired ultra-stable supercapacitor for large-scale wearable energy storage applications. *Nano Energy*, 82, 105776, 2021.
109. Pérez-Madrigal, M.M., Edo, M.G., Alemán, C., Powering the future: application of cellulose-based materials for supercapacitors. *Green Chem.*, 18, 5930–5956, 2016.
110. Jeong, H.T., Kim, Y.R., Kim, B.C., Flexible polycaprolactone (PCL) supercapacitor based on reduced graphene oxide (rGO)/single-wall carbon nanotubes (SWNTs) composite electrodes. *J. Alloys Compd.*, 727, 721–727, 2017.
111. Zhou, Z. *et al.*, Layer-by-layer assembly of MXene and carbon nanotubes on electrospun polymer films for flexible energy storage. *Nanoscale*, 10, 6005–6013, 2018.
112. Mathew, J., Das, J.P., TP, M., Kumar, S., Development of poly (butylene adipate-co-terephthalate) PBAT toughened poly (lactic acid) blends 3D printing filament. *J. Polym. Res.*, 29, 474, 2022.
113. Lee, H. *et al.*, Facile fabrication of a fully biodegradable and stretchable serpentine-shaped wire supercapacitor. *Chem. Eng. J.*, 366, 62–71, 2019.

114. Bhatnagar, P. *et al.*, Transparent photovoltaic memory for neuromorphic device. *Nanoscale*, 13, 5243–5250, 2021.
115. Liu, R. *et al.*, An Efficient Ultra-Flexible Photo-Charging System Integrating Organic Photovoltaics and Supercapacitors. *Adv. Energy Mater.*, 10, 2000523, 2020.
116. Chae, J.S. *et al.*, A biocompatible implant electrode capable of operating in body fluids for energy storage devices. *Nano Energy*, 34, 86–92, 2017.
117. Ouyang, H., Jiang, D., Fan, Y., Wang, Z.L., Li, Z., Self-powered technology for next-generation biosensor. *Sci. Bull. (Beijing)*, 66, 1709–1712, 2021.
118. Kumar, P. *et al.*, Melanin-based flexible supercapacitors. *J. Mater. Chem. C Mater.*, 4, 9516–9525, 2016.
119. Hur, J. *et al.*, DNA hydrogel templated carbon nanotube and polyaniline assembly and its applications for electrochemical energy storage devices. *J. Mater. Chem. A Mater.*, 1, 14460, 2013.
120. Lee, G. *et al.*, Fully Biodegradable Microsupercapacitor for Power Storage in Transient Electronics. *Adv. Energy Mater.*, 7, 1700157, 2017.
121. Seo, D. *et al.*, Wireless Recording in the Peripheral Nervous System with Ultrasonic Neural Dust. *Neuron*, 91, 529–539, 2016.
122. Jiang, L. *et al.*, Flexible piezoelectric ultrasonic energy harvester array for bio-implantable wireless generator. *Nano Energy*, 56, 216–224, 2019.
123. Park, S., II *et al.*, Soft, stretchable, fully implantable miniaturized optoelectronic systems for wireless optogenetics. *Nat. Biotechnol.*, 33, 1280–1286, 2015.
124. Koo, J. *et al.*, Wireless bioresorbable electronic system enables sustained non-pharmacological neuroregenerative therapy. *Nat. Med.*, 24, 1830–1836, 2018.
125. Bhunia, K. *et al.*, Cobalt Nanoparticle-integrated Nitrogen-doped Carbon Nanotube as an Efficient Bifunctional Electrocatalyst for Direct Methanol Fuel Cells. *Chem. Eng. J.*, 473, 145028, 2107160, 2023.
126. Wang, L. *et al.*, Designing Porous Antifouling Interfaces for High-Power Implantable Biofuel Cell. *Adv. Funct. Mater.*, 31, 2021.
127. Xiao, X., Conghaile, P. Ó., Leech, D., Ludwig, R., Magner, E., An oxygen-independent and membrane-less glucose biobattery/supercapacitor hybrid device. *Biosens. Bioelectron.*, 98, 421–427, 2017.
128. Jung, S., Lee, J., Hyeon, T., Lee, M., Kim, D., Fabric-Based Integrated Energy Devices for Wearable Activity Monitors. *Adv. Mater.*, 26, 6329–6334, 2014.
129. Lee, C.H. *et al.*, Biological lipid membranes for on-demand, wireless drug delivery from thin, bioresorbable electronic implants. *NPG Asia Mater.*, 7, e227–e227, 2015.
130. Ouyang, H. *et al.*, Symbiotic cardiac pacemaker. *Nat. Commun.*, 10, 1821, 2019.
131. Yao, G. *et al.*, Effective weight control via an implanted self-powered vagus nerve stimulation device. *Nat. Commun.*, 9, 5349, 2018.

# 8

# Recent Trends in the Development of Sustainable Supercapacitors

**Sandeep B. Wategaonkar[1,2], Umesh V. Shembade[3], Mayuri G. Magadum[3], Jaywant V. Mane[1], Prathamesh B. Patil[1], Tushar T. Bhosale[3], Nishigandha B. Chougale[3] and Annasaheb V. Moholkar[3]***

[1]*Department of Chemistry, D. Y. Patil Agriculture and Technical University, Talsande, (M.S), India*
[2]*Department of Chemistry, Kisanveer Mahavidyalaya, Wai, (M. S), India*
[3]*Thin Film Nanomaterials Laboratory, Department of Physics, Shivaji University, Kolhapur, (M.S), India*

## *Abstract*
In the current state, energy storage systems are becoming more and more popular day by day as a result of the increasing demand for electrical power supply for portable devices like mobile phones, electric vehicles, and domestic equipment. The fast advancement of portable and wearable electronics raises new needs for energy storage technologies, including flexibility, intelligent features, and extended outside operation. Supercapacitors (SCs) are regarded as the most promising energy storage technologies because of their superior qualities over conventional capacitors, as well as their better power density than batteries and environmentally free-to-design devices. To increase the efficiency of SCs, we have tried to cover a wide range of electrode materials in this chapter, including carbon-based materials, metal oxides, conducting polymers, and their nanocomposites, which have demonstrated excellent potential for energy storage. This chapter will provide a thorough introduction to the various electrolytes and electrode materials used in the most recent supercapacitor technologies. The highlight of this chapter is how the supercapacitive properties enhance the social impact of the most advanced SCs technologies in their output, cycle stability, cost analysis, and environmental impact. The conclusion addresses the current issues and exciting opportunities for the growing industry. Finally, the future of SC technology is examined along with suggestions for developing high-performance, low-cost, and environmentally friendly supercapacitors.

**Corresponding author*: avmoholkar@gmail.com

Basheer Ahamed and Chaudhery Mustansar Hussain (eds.) Sustainable Supercapacitors: Next-Generation of Green Energy Storage Devices, (227–264) © 2025 Scrivener Publishing LLC

***Keywords***: Sustainable SCs, electrodes, electrolytes, hybrid nanocomposites, energy storage

## 8.1 Introduction

As the need for significantly improved electrochemical energy storage devices grows, an inventive new material architecture for energy storage is being created. Batteries can be charged and discharged more quickly, have a higher power density, and have longer life cycles than electrochemical capacitors (ECs), a form of capacitive energy storage device. In Ragone's specific power vs. specific energy diagram, these components are positioned in between batteries and dielectric capacitors. ECs can serve as backup power supplies, memory backup systems, or peak power sources in electric vehicles. Therefore, there will always be a need for equally adaptable energy sources and energy storage systems that can be stretched while still sustaining their electrochemical processes. One of the best solutions for powering electronic devices is LIBs because of their high energy density, lengthy cycle life, and power density. SCs have a higher power density compared to LIBs, and intermediate energy density has also attracted a lot of attention lately. LIBs and SCs have historically been since they rely on the traditional configuration and slurry-casting approach, and the current fabrication techniques are often inflexible and hefty [1, 2].

A carbon substance with a huge surface area serves as an illustration of an EDLC. The benefits of carbon materials include their improved electrical conductivity, appropriate physical–chemical characteristics (such as their porous nature, etc.), and commercial viability. Utilizing the electro-adsorption or the pseudocapacitors charge due to a redox reaction that occurs on the electrode's surface, carbon-based materials have the disadvantage that they have a lower energy density (in comparison to pseudo-capacitors, which have a higher energy EDLCs) [3, 4]. Recently, polymeric materials have drawn a lot of interest for supercapacitor applications because of their advantageous qualities, including flexibility, lightweight, and steady cycling behavior. Additionally, replacing heavy metals in battery electrodes with redox-active polymers is very appealing in terms of sustainability and recyclable materials. Electroactive conducting polymers (ECPs), even though many other polymers have been used as redox-active components in energy storage devices, offer promising electrodes in supercapacitors because of their affordability, low weight, versatility, and high charge/discharge rate [5, 6].

The ongoing search for safe and effective energy sources is prompted by the escalating environmental issues as well as the rising energy demand. High-efficiency energy storing systems are necessary because the majority of energy sources are not reliable. Due to its extremely large capacity for storing energy, supercapacitors attract the attention of researchers. The electrode materials have a direct impact on the supercapacitor's effectiveness and efficiency. The potential of transition metal oxides as a material for phony capacitors is extensively investigated. Significant research has been paid to the various electrode-based materials that exhibit extremely high theoretical capacitance [7, 8].

The prospective uses of various nanostructured materials in supercapacitors have attracted increasing amounts of scientific interest over the last few decades. These nanostructured materials often display better capacitance behavior as compared to their bulk counterparts. The materials with nanostructures have improved chemical kinetics and activity, shorter ionic diffusion channel lengths, larger surfaces, and a large number of active sites for electrochemical processes. It involves topics like materials, energy, chemistry, electronics, and other multidisciplinary study fields. It can be used as a novel type of energy storage device with excellent performance and environmental friendliness in numerous fields, such as industrial control, power, transportation, intelligent instruments, consumer electronics, national defense, communications, medical equipment like defibrillators, pulsed lasers, new energy vehicles, and so on. Despite having boundless potential and opportunity, the supercapacitor faces significant obstacles [9, 10].

This book chapter compiles the potential, challenges, and development patterns of supercapacitors based on the level of research at the time.

## 8.2 Recent Trends in Electrode Materials

A current collector is covered with a slurry containing active chemicals, conductive materials (carbon black), and polymer binder before being dried and pressed to create the conventional electrode. Due to the ease with which the electrode materials can be removed from the current collector and the difficulty with which metallic current collectors can be bent and then released repeatedly, the electrochemical performance degrades. The materials of the delaminated electrodes are capable of penetrating even the separator, resulting in short circuits and thermal runaway. The main parts of LIBs and SCs include cathode and anode, separator/

electrolyte, current collector, and packaging materials. Every component must have the ability to deform with a compatible feature for flexible and stretchable LIBs and SCs to function [11, 12]. As the current components are not flexible enough, it would be ideal if they were entirely replaced with new ones that could be deformed. The study of flexible and stretchy LIBs and SCs now faces three major challenges: electrochemical performance stability under dynamic settings; high energy and power density; and the design and manufacture of stretchable, flexible electrodes. The development of flexible and stretchable energy storage technologies is still being studied in laboratories. Consequently, creating new techniques for flexible and stretchy energy [13].

A variety of materials are used depending on the three basic types of supercapacitors (SCs). Supercapacitors come in three primary varieties: EDLC, pseudocapacitor, and hybrid supercapacitor, as was previously mentioned. Different electrode materials are used for various types. The most common electrode types used in EDLC are carbon-based electrodes, metal oxide electrodes, polymer electrodes, asymmetric electrodes utilized in hybrid supercapacitors, and electrodes based on polymers, metal oxide, and polymer. supercapacitors' electrode materials, which are utilized to hold charge [14].

The electrode material of the supercapacitor needs to have pores of a certain size, because the pore size of the electrode material has a substantial impact on the electrochemically active surface area. The capacitance of SCs is significantly influenced by the specific surface area of the electrode materials. The electrodes for the SCs should also be economical, environmentally friendly, and have great electrical and thermal stability and conductivity. These characteristics could be perfect for improving capacitance performance. High surface area electrode materials and electrode materials with small pore sizes are the two most crucial conditions for creating an SC device. The materials used to make electrodes with a lot of surface area have the highest specific capacitance. The materials employed to make the large pore-sized electrodes, however, significantly reduced capacitance. Additionally, by modifying the size dispersion, it is possible to increase the storage capacity, a sign of a significant power density in SCs. The efficient dispersion of micropores and mesopores in the electrode material can enable rapid mass and ion transport along an unbroken process, increasing the availability of the electrolyte and making an excellent electrode material for SC applications [2, 15].

### 8.2.1 Carbon-Based Electrodes

There are numerous domains where carbon-based electrode materials are used. The most often utilized electrode materials in supercapacitors are those that contain carbon, such as graphene, reduced graphene oxide (rGO), carbon aerogel, and carbon nanotube (CNT). because of their easy availability, high thermal and chemical stability, huge surface area, consistent pore size distribution, and good electrical conductivity. The electrodes used in EDLC devices most frequently are those made of carbon-based materials. EDLC does not involve any chemical reactions; instead, the charge storage mechanism occurs at the interface between the electrodes and the electrolyte. As a result, the electrode materials' large surface area, homogeneous pore distribution, and large pore sizes significantly affect the performance of the EDLC device. Creating the Helmholtz layer, a thin layer on the electrode surface is the initial stage in storing the charge in EDLC using carbon materials. Contact of this layer with electrolyte ions that are not dissolved allows it to store charges. A thicker outer Helmholtz layer is created as a result of the dissolved ions' real presence, which is triggered by strong electrostatic forces. This larger outer Helmholtz layer then contributes to the charge storage. In studies using cyclic voltammetry (CV) and galvanotactic charge–discharge profiles (GCD), carbon-based electrode materials had triangular and rectangular forms [7, 16, 17]. Details on the various carbon-based electrode materials used in supercapacitors are provided in this section (Figure 8.1).

**a) Graphene-Based Electrodes**

The carbon atoms in the $sp^2$ hybrid material known as graphene are organized in a honeycomb-like crystal structure. Graphene is a crucial component of a specific carbon substance with exceptional electrical conductivity. A simple source of this one-atom-thick substance is graphite. Because of its substantial surface area, chemical and thermal stability, and great mechanical strength, carbon is a material that can be used in energy storage systems. Additionally, the 2D structure of graphene reduces the electrodes' surface area, opening up a broader potential window, improving mobility, and creating distinctive porous patterns that increase the device's specific capacitance. The proximity of the electrolyte to both of the graphene sheet's exterior surfaces makes using graphene as an electrode material advantageous [18].

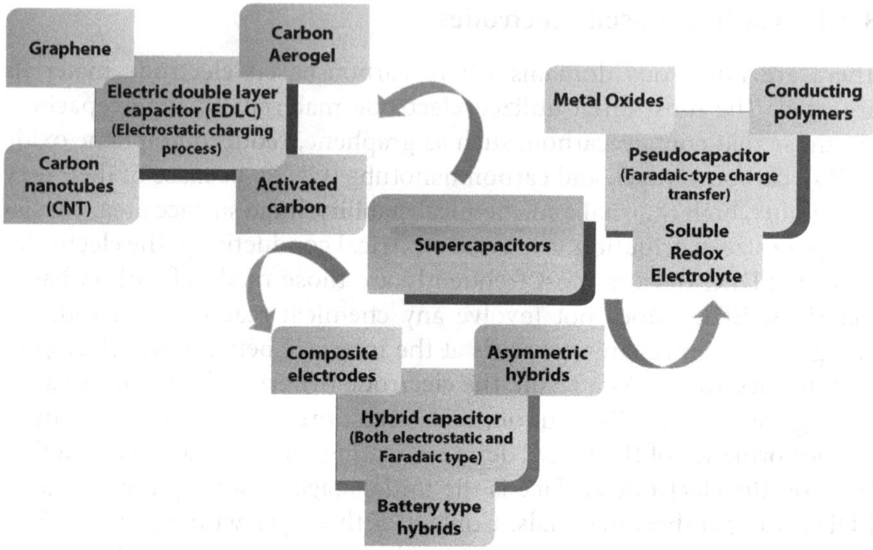


**Figure 8.1** The classification of the supercapacitors based on the electrode materials.

There are numerous ways to create graphene, and the technique chosen will depend on the specific purpose that graphene must perform. Mechanical exfoliation and chemical vapor deposition (CVD), two of the various ways to make graphene, are capable of producing high-quality single-layer graphene films but have low yields, whereas liquid-phase exfoliation shows significant potential since the graphene made from reduced graphene oxide (GO) exhibits a good connection with the materials used to make the electrodes and the electrolyte [19]. Graphene-nano porous carbon material is one of many graphene-based electrodes that are now on the market to study how porosity affects the performance of supercapacitors, which were successfully coupled with graphene. Microporous and mesoporous materials can be created to varying degrees depending on the formulation technique, and the pore quantity, dimension, and area can all be precisely managed [20]. By combining carbonization and activation of phytic acid (PA) on the surface of graphene oxide (GO) in the presence of ethylenediamine (EDA), Zheng *et al.* [21] developed the porous carbon-coated graphene nanohybrid with nitrogen/phosphorus co-doping. The created material comprises a unique 3D nanohybrid structure composed of graphene sheets sandwiched between porous carbon nanosheets, a hierarchically micro/mesoporous structure, a 596 $m^2\ g^{-1}$ high specific surface area, and an efficient N/P co-doping. With a capacitance retention ratio of 75% at 20 A $g^{-1}$ as a supercapacitor electrode and a gravimetric capacitance of 201 F $g^{-1}$ at a current density of 0.5 A $g^{-1}$, the produced material performs admirably. The developed symmetric supercapacitor in 6 M KOH also features an extraordinary cycle stability with 84.6% retention after 20,000 cycles, a high gravimetric energy density of 9.10 Wh $kg^{-1}$, a significant volumetric energy density of 9.07 Wh $L^{-1}$, and other noteworthy features. The modified graphene material increases the effective surface area and minimizes graphene formation. These modified cations frequently have the purpose of enhancing the capacitance of electrodes.

By using the electrodeposition approach, Cheng *et al.* [22] created graphene with $MnO_2$ nanostructures that are like flowers. The specific capacitance of the graphene electrode reached 245 F $g^{-1}$ at a charging current of 1 mA and 328 F $g^{-1}$ for $MnO_2$-coated graphene with an energy density of 11.4 Wh/kg and a power density of 25.8 kW/kg. According to this research, for high-performance energy storage systems, the graphene-based electrodes are a strong possibility.

The G/Ni-Al-LDH composition was hydrothermally produced by Wimalasiri *et al.* [23], the prepared Ni-Al-LDH has been improved by graphene, and the electrochemical activity of the composite materials in supercapacitor applications has been improved. It was possible to realize the electrically charged materials under two advantageous conditions because using combined materials in their layered form increased contact between electrochemically active areas.

**b) rGO-Based Electrodes**

Reduced graphene oxide (rGO) is the material utilized as an electrode in supercapacitors. Graphene oxide (GO) has many functional groups, such as hydroxyl, epoxide, and carbonyl. The groups cause the $sp^2$ hybrid carbon atoms to change into $sp^3$ carbon atoms, causing structural changes, lower conductivity, and worse performance. The higher electrical conductivity of the rGO consequently enhances the performance of the supercapacitor. The rGO is a promising electrode material for supercapacitors because of its many properties, which include high electrical conductivity, high surface-to-volume ratio, chemical stability, high specific capacitance, low cost, and ease of handling. There are many ways to convert GO to rGO, including thermal reduction, electrochemical reduction, laser reduction, and chemical reduction of GO [24].

Akturk *et al.* [25] created a hybrid nanocomposite of reduced graphene oxide and zinc oxide (rGO/ZnO), which they used as an electrode material for supercapacitor applications. The electrochemical performance of a two-electrode cell was used to study a hybrid rGO/ZnO nanocomposite in 6 M KOH. After 30 cycles at 0.30 A $g^{-1}$, it was discovered that the capacitance retention of all electrodes was 100%. The rGO electrode's ability to conduct current and diffuse ions was enhanced by ZnO, which is abundant in core defects.

Rajesh Kumar and colleagues created the $Co_3O_4$/CoO nanoparticles using microwave irradiation. Following that, rGO nanosheets were used to hold these nanoparticles in place in thin layers. The specific capacitance measured at a scan rate of 5 mV $s^{-1}$ in 1 M KOH for the mesoporous structure of rGO@$Co_3O_4$/CoO shows Cs of 276.1 F $g^{-1}$ and reveals the long-term cycling stability of the 82.37% over 10,000 cycles [26].

**c) Carbon Aerogel-Based Electrodes**

The most often used electrode material in EDLC is carbon aerogel. It boosts the device's energy and power density as well as capacitance. The pore structure of the electrode determines power and energy densities and the carbon aerogel structure's unique tunability provides a tool for enhancing both. The area-specific capacity of the electrode material is crucial for an EDLC to operate effectively. Aerogel is a synthetic material that is created from a gel in which gas has been substituted for the liquid component. It is a porous solid substance. A 3D network of connected nanoparticles makes up the carbon aerogel. Carbonization produces nanostructured carbons with a variety of characteristics from organic aerogels created by the sol-gel reaction of particular organic monomers. Among the common organic types of carbon aerogels are polyurethane, phenolic, and resorcinol-furfural aerogels. The use of carbon aerogel electrodes is receiving significantly more attention. Thus, in energy-related applications including catalysis, capacitive deionization, supercapacitors, batteries, and hydrogen generation, aerogel is used as an electrode material. Porosity, a huge surface area, low mass densities, high-temperature resistance, superior electrical conductivity, rapidly moving masses, and a sizable energy storage capacity are only a few of the unique properties of carbon aerogel [27, 28].

Huang *et al.* [29] present a hydrothermal technique for the quick, inexpensive, and simple synthesis of carbon aerogel composite to three-dimensional flower-like $MoS_2$. Excellent electrochemical performance is produced by the 3D-layered nanostructure of the $MoS_2$/carbon aerogel, which is an efficient active material and reduces the desirable diffusion distance of charged particles. It demonstrates that the specific capacitance in a 1 M $Na_2SO_4$ aqueous solution is 260 F $g^{-1}$. It demonstrated exceptional electrochemical stability, retaining 92.4% of its capacitance over 1500 cycles. Additionally, the carbon aerogel exhibits a 15 F $g^{-1}$ specific capacitance.

Nitrogen-doped carbon aerogel was produced by Mirzaeian *et al.* [30] via the polycondensation of resorcinol, formaldehyde, and melamine followed by carbonization. The synthetic carbon materials exhibiting amorphous structure were employed as an electrode material for a supercapacitor that exhibits a 208 F $g^{-1}$ specific capacitance in a 6 M KOH electrolyte solution. The nitrogen-doped carbon aerogel has a surface area of 841 $m^2 g^{-1}$, which is determined by BET. This result demonstrates the material's high porosity, which increases conductivity and the active sites of the electrodes, improving the capacitive behavior of the device.

By pyrolyzing crosslinked polycarbosilane aerogels and treating them with chlorine at 700 and 1000°C, carbide-derived carbon (CDC) aerogels with hierarchical porosity were created by Oschatz *et al.* [31]. To add more micropores and remove surface functional structures, it is then processed with $CO_2$. In comparison to material prepared at 1000°C, the CDC aerogel prepared at 700°C has larger micropores and a higher surface area, which shows more capacitance. The obtained active material was employed as electrode material in various aqueous electrolytes, such as 1 M $H_2SO_4$, 1 M $Li_2SO_4$, 1 M LiCl, and 1 M HCl, as well as organic electrolytes and ionic liquid electrolytes, such as 1 M TEAB and 1 M EMIBF electrolytes, respectively. Among all the electrolytes, the $H_2SO_4$ was proven to have the best capacities for CDC-Aero-700 at low scan rates and current densities of all the 1 M aqueous electrolytes, at about 150 F $g^{-1}$. Even at current densities as high as 60 A/g in organic electrolyte, activated CDC aerogels operate steadily. The largest specific capacities, up to 170 F $g^{-1}$ in an ionic liquid electrolyte, are demonstrated by the CDCAero-700 sample [32].

**d) Carbon Nanotube-Based Electrodes**

A 3D carbon nanotube (CNT) is made up of several atomic layers of graphite. The CNTs considerably improve electrode performance by permitting a continuous charge distribution that virtually covers the entire accessible surface area thanks to their mesoporous and well-conducting networks. Depending on their size and shape, CNTs are used in a range of industries and applications. CNTs are employed in batteries and fuel cells and may support catalytic particles more effectively due to their mesoporous nature. It is utilized as an electrode material in supercapacitors if the diameter is small. Due to its unique characteristics, such as its high electrical conductivity, mesoporous structure, high resistance, vast surface area, and thermal chemical stability, CNTs are used as an electrode material in electrochemistry-based applications such as electrochemical sensors.

Ions can diffuse more readily into the mesoporous network, which results in a significant surface area. The ions are transformed significantly better, the cell resistance is reduced, and the large surface area is used to produce capacitance more effectively. Moreover, the elastic and flexible CNTs enhance the device's capacitance [27, 33].

In EDLC, CNTs are generally used as an electrode material. EDLC using carbon nanotubes in the electrode demonstrated relatively high capacitances because of the huge surface area that the electrolyte can reach. Carbon nanotubes' improved mechanical and electrical properties in this case allow for their usage as an electrically conductive addition to the electrode of an EDLC. CNTs and their composites are credited with supercapacitor devices' better performance. Specifically, CNT composites are made of materials that can act as pseudocapacitors, including CNTs and metal oxides or CNTs and conducting polymers. In this situation, CNTs function as a conducting mesoporous network that facilitates quick diffusion of components and reaction by-products while also accelerating charge transfer during fuel oxidation or oxygen reduction. CNTs are the optimal conducting channel to enable good charge propagation because they are the ideal strong backbone with stretchy properties. This greatly enhances the cyclability of the supercapacitor, especially if volumetric changes in the material happen during charging and discharging [34–36].

Li *et al.* [37] developed the hydrothermal and sulfuration process-derived hierarchically hybrid composite of $CuCo_2S_4$/CNTs. The produced material decreased series and charge transfer resistances. Additionally, the composite provided efficient pathways for electron transfer and ion diffusion. This resulted in a faradic reaction that increased the capacitance of the $CuCo_2S_4$/CNT electrode. The energy density is 23.2 Wh kg$^{-1}$, the power density is 402.7 W kg$^{-1}$, and the electrode has a significantly higher specific capacitance of 557.5 F g$^{-1}$ than the pure $CuCo_2S_4$ electrode (373.4 F g$^{-1}$). The composite material also has a very high degree of cyclic stability, with device retention at 85.7% after 10,000 cycles of testing.

The GN/MWCNT (graphene/multiwalled carbon nanotube) film was created by Lu *et al.* [38] using gas-based hydrazine to convert GO into GN. In this instance, the GN sheet and MWCNT were uniformly sandwiched together. The GN/MWCNT composite effectively increases the electrolyte/electrode contact area and improves the movement of the electrolyte ion and electron into the electrode. Therefore, the composite material shows the 265 F g$^{-1}$ specific capacitance at a current density of 0.1 A g$^{-1}$, and following 2000 nonstop charge/discharge cycles, the device exhibits good cyclic stability and a specific capacitance retention of 97%.

Polyaniline/carbon nanotube (PANI/CNT) nanofiber mats were made using the electrospinning process, as described by Simotwo *et al.* [39]. Here, PANI/CNT nanofiber was initially created, and then PEO (polyethylene oxide) was added in a small quantity. The constructed electrode performs well electrochemically in this instance because the CNTs increase electrical conductivity. Comparing the specific capacitance of the PANI/CNT electrode to that of the PANI/PEO electrode reveals a higher specific capacitance of 385 F $g^{-1}$. Additionally, after 1000 cycles, specific capacitance retention of 81.4% was noted for the PANI-CNT-based electrode. As a result of the excellent electrolyte penetration provided by the interconnected nanofiber network and the internal flexibility within the structure of the polymer, rapid ion transport to the active sites is made possible.

### 8.2.2 Metal Oxide-Based Electrode Materials

The supercapacitor electrode requires materials with high electrical conductivity, high thermal and chemical stability, pore structure, large surface areas, and materials that exhibit both the electric double-layer type response and the faradic reaction. As a result, metal oxides are the most promising alternatives for electrode materials in the supercapacitor. The cell's low cost, simplicity of handling, distinctive design, and large surface area boost its specific capacitance. Supercapacitors have used a variety of metal oxides, including those based on ruthenium (Ru), nickel (Ni), cobalt (Co), iron (Fe), and manganese (Mg). This section covers single metal oxides, binary metal oxides, ternary metal oxides, and quaternary metal oxides for supercapacitors.

**a) Single Metal Oxide-Based Electrodes**

Single metal oxides, such as NiO, MnO, and $WO_3$ have been shown to exhibit outstanding redox activity, and high ionic conductivity and exist in a significant natural abundance; as a result, a high theoretical energy storage capacity can be produced at a comparatively low cost. These materials are also environmentally friendly, inexpensive, and simple to synthesize. Chemical precipitation was used by Li *et al.* [40] to create $Ni(OH)_2$ nanosheets, which were then transformed into NiO nanosheets through calcination. The electrodes for the supercapacitor were made of single-crystalline $Ni(OH)_2$ and NiO nanosheets. The specific capacitance of NiO nanosheets is 388 F $g^{-1}$ at a current density of 5 A $g^{-1}$, demonstrating their ability to deliver excellent electrochemical capacitance. To achieve high electrochemical performance, Liu *et al.* [41] synthesized CoO rhombus nanopillar and nano brush array electrodes for supercapacitors. When compared to the CoO nano brush array-based electrode (169 F $g^{-1}$), the CoO rhombus nanopillar array exhibits a greater specific capacitance of 509 F $g^{-1}$ at a scan rate of 10 mV/s. The CoO rhombus nanopillar array's huge surface area, pore structure, and high performance show good cycle stability.

Unpretentious and environmentally friendly electrode material for supercapacitors was recently created by Shembade *et al.* [3]. The hydrothermal process was used to create the tungsten oxide nanosheets by varying the concentrated acids at the same temperature. The monoclinic $WO_3$ on nickel foam (M-$WO_3$-NF) exhibits outstanding electrochemical performance in 1 M KOH. The developed M-$WO_3$-NF electrode has a specific capacity of 72 mAh/g, a gravimetric capacitance of 600 F $g^{-1}$, and high energy and power densities of 17 Wh/kg and 321 W/kg, respectively. Furthermore, the electrode exhibits a superior Columbic efficiency of 96.9% at a current density of 5 mA/$cm^2$.

**b) Binary Metal Oxide-Based Electrodes**

There is a great deal of interest in binary metal oxides. One or more active and passive ions are present, as well as one transition metal ion. It has better electrical conductivity and more advantageous oxidation states, which when combined can result in superior electrochemical performance and a larger potential window. The most common electrode materials for supercapacitors are binary metal oxides, which have substantial redox activity. Examples include those based on Ni-Co-O, Co-Fe-O, and Ni-Fe-O. The binary oxides perform electrochemically more effectively than single metal oxides [42, 43].

For the flexible solid-state battery supercapacitor hybrid device, Hong *et al.* [44] research $NiCo_2O_4$ and $NiCo_2O_4$@$NiMoO_4$ composite materials. Different directing agents, such as urea, HMT, and NH4F, were used to synthesize $NiCo_2O_4$ and $NiCo_2O_4$@$NiMoO_4$; for the electrochemical performance, two distinct electrodes, Ni foam and carbon cloth, were employed. The high specific capacitance of 1.87 F/$cm^2$, which is equivalent to 0.68 mAh/$cm^2$, is demonstrated by the $NiCo_2O_4$@$NiMoO_4$/NF positive electrode produced using NH4F. The prepared electrode further demonstrates that the maximum energy density of 70.78 Wh/kg was likewise attained at the power density of 3.25 kW/kg for this device. Additionally, in the 5000 charging/discharging cycles, a CF retention of 80% and a Coulombic efficiency of greater than 95% were achieved. For use in supercapacitors, Yang *et al.* [45] produced an undoped MnOx film thin film and co-doped amorphous MnOx film, which exhibits a specific capacitance of 99 F $g^{-1}$ at a scan rate of 5 mV $s^{-1}$, which is higher than the undoped MnOx (47 F $g^{-1}$). Here, elemental doping is a useful technique for enhancing the functionality of metal oxide pseudo-capacitive. To achieve a high energy density for a supercapacitor device, Ishaq *et al.* [46] work with the binary material $CoFe_2O_4$ and its composite with rGO. When evaluated at 5 mV $s^{-1}$, the nanocomposite of the rGO/$CoFe_2O_4$ electrode showed a gravimetric capacitance of 97 F $g^{-1}$ and an aerial capacitance of 165 mF $cm^{-2}$.

The Mn-Ni binary metal oxides ($NiMnO_3$ and $NiMn_2O_4$) were produced hydrothermally by Dinesh *et al.* [47] and used as supercapacitor electrode materials. Due to the different oxidation states of Mn and Ni and their larger dimension, $NiMnO_3$ demonstrated better electrochemical characteristics than $NiMn_2O_4$. At 1.0 A $g^{-1}$ of current density, $NiMnO_3$ produced a capacitance of 230 F $g^{-1}$. The electrode exhibited better equilibrium during the GCD cycle and 67% of its capacitance after 2000 cycles. In addition, $NiMnO_3$ was modified to function as a catalyst for OER and showed an overpotential at 0.256 V vs. RHE.

Binary tungsten-molybdenum oxide nanoneedle arrays ($W_{0.4}Mo_{0.6}O_3$) were generated by Peng *et al.* [48] using a simple hydrothermal technique. The developed nanoneedle array electrode exhibits outstanding electrochemical performance with a large specific capacity of 115.7 mAh $g^{-1}$ at a current density of 1 A $g^{-1}$ and rate capacity, due to a large number of surface reaction sites that are active and diffusion of electrolyte ions. Additionally, they design an asymmetric supercapacitor device, with $W_{0.4}Mo_{0.6}O_3$ nanoneedle arrays as a negative electrode and PMCHC (polygni multiflora caulis-based honeycomb-like carbon) as a positive electrode. These electrodes have a superior output voltage of 1.5 V and excellent energy and power densities of 20.2 Wh $kg^{-1}$ and 750.6 W$kg^{-1}$, respectively. The retention rate after 2000 cycles of cyclic stability was 82.3%.

**c) Ternary and Quaternary Metal Oxide-Based Electrodes**

Supercapacitors can use a variety of transition metal oxides as their electrode material since more faradic redox processes are required to boost specific capacitance. Increased electrical conductivity and improved electrochemical performance may result from adding more metals to the oxide. The morphological substance may also have a major impact on the electrical conductivity and electrochemical performance [49].

RGO/-$Ni(OH)_2$/$CeO_2$ ternary nanocomposite was created by a hydrothermal technique by Godlaveeti *et al.* [50]. Additionally, they synthesized the quaternary nanocomposite by incorporating a small amount of $TiO_2$, $SnO_2$, and $MnO_2$ nanorods. As shown by employing the CV, GCD, and EIS, the produced quaternary material demonstrates its electrochemical performance. According to the results, the generated ternary composite, in addition to $MnO_2$ NRs, had a higher specific capacitance (1250 F $g^{-1}$) at current densities of 0.5 A $g^{-1}$ in 3 M KOH aqueous electrolyte solution than other metal oxides including $SnO_2$ NPs and $TiO_2$ NRs. In addition to $MnO_2$ NRs, ternary nanocomposite preserved 78.46% and showed improved cyclic stability after 3000 cycles.

Sahoo *et al.* [51] used hydrothermal synthesis to create innovative 3D ternary Fe-Ni-Co oxide nanoflake arrays on Ni foam. The supercapacitor electrode performed exceptionally well electrochemically, with a high specific capacitance of 867 F $g^{-1}$ and outstanding cycling stability, retaining 92.3% of its capacitance after 10,000 charge–discharge cycles. Additionally, a device made from this material that was an aqueous hybrid asymmetric supercapacitor showed outstanding capacitance retention after 5000 cycles and a high energy density of 40 Wh $kg^{-1}$.

High electrical conductivity and high supercapacitive performance were achieved with ternary metal oxides of manganese, nickel, and cobalt (MNC-GNP) by Malik *et al.* [52] using graphene nanoplatelets. As a result of the porous shape and synergistic impact, the composite of MNC-GNP demonstrated outstanding electrochemical performance. MNC-GNP exhibited an enhanced specific capacity of 605 mAh $g^{-1}$ which is higher compared to MNC, which exhibited 243 mAh $g^{-1}$ at a current density of 2 mV $s^{-1}$, with a capacity retention of 99.5% up to 1000 cycles.

## 8.3 Role of Different Electrolytes in the Field of Sustainable SCs

Supercapacitors are regarded as potential energy storage technologies because of their high Pd, quick charge–discharge, increased service life, wide working temperature range, and outstanding capacitance retention. Electrochemically productive supercapacitors are affected by the nature of the electrode materials, the kind of electrolyte, the thickness of the separator, etc. One of these qualities, the electrolyte employed in supercapacitors, greatly influences the final properties of supercapacitors. Recent years have seen a large amount of study on the development of novel electrolytes and electrode/electrolyte combinations [53, 54].

The electrolyte, one of the essential components of supercapacitors, is composed of additives, solvents, and electrolytes. The electrolyte has a big impact on how well supercapacitors perform, especially the Cs and Ed. The electrolyte acts as an ion source that produces more ions, a binding

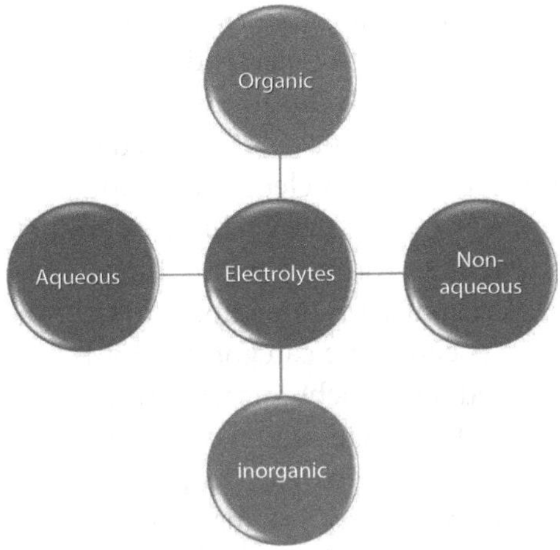


**Figure 8.2** The different types of electrolytes based on aqueous, non-aqueous, organic, and inorganic electrolytes.

agent for the electrode particles, and an accelerator of ion conduction. The electrolyte concentration must be high to prevent problems with depletion when the supercapacitor is charging. The equivalent series resistance (ESR) of the supercapacitor is primarily affected by the electrolyte's conductivity and temperature coefficient. High-conductivity electrolytes are required in supercapacitors to reduce internal resistance. The electrolyte's chemical and electrochemical stability should be good, and the use temperature range should be wide to accommodate the supercapacitor's operating circumstances. The ions' sizes in the electrolyte must match the pore size of the electrode material (for EDLCs) and it must be normal and ecologically benign. The working electrolytes used by supercapacitors include water-based electrolytes, organic electrolytes, solid electrolytes, and colloidal electrolytes [55]. Additionally, the different types of electrolytes based on aqueous, non-aqueous, organic, and inorganic electrolytes are shown in Figure 8.2.

### 8.3.1 Types of Electrolytes

**a) Aqueous Electrolyte**

The water-based solvent for the chemicals gives the aqueous electrolyte its name as an electrolyte, following its dissociation by supplying conducting ions. Alkaline solutions, specifically potassium hydroxide, are used frequently because of their strong ionic conductivity when creating aqueous electrolyte solutions for metal-air batteries, excellent air electrode performance, and respectable base metal electrode stability. When compared to organic and ionic liquid electrolytes, aqueous electrolytes (such as $H_2SO_4$, KOH, $Na_2SO_4$, and $NH_4Cl$ aqueous solution) usually have greater conductivity. However, compared to other electrolytes, aqueous electrolytes have a smaller potential window (1 V to 1.3 V), higher conductivity, smaller ionic sizes, and an abundance of protons. Despite this, aqueous electrolytes are not chosen for commercial use in supercapacitors. Acidic, alkaline, or neutral aqueous electrolytes can all be categorized [56].

To achieve increased electrochemical performance, the appropriate selection of an aqueous electrolyte, which depends on pore size and specific surface area, is essential. Aqueous electrolytes have been utilized extensively in research and development because they are easier to handle in the lab than organic electrolytes and ILs, which need purification techniques. They are electrolytes based on water. Materials with working potentials between the $O_2$ evolution potential and the $H_2$ evolution potential can be used in aqueous alkali-metal ion batteries. The pH should also be taken into account when choosing the electrode materials because it is directly related to the $O_2$ and $H_2$ growth potentials of the aqueous system. Aqueous electrolytes are divided into alkaline, neutral, and acidic solutions. The most often used aqueous electrolytes are NaOH, LiOH, $Na_2SO_4$, $H_2SO_4$, $(NH_4)_2SO_4$, $K_2SO_4$, $Li_2SO_4$, $MgSO_4$, $CaSO_4$, $BaSO_4$, KCl, NaCl, LiCl, HCl, CsCl, $CaCl_2$, $KNO_3$, $LiNO_3$, $Na(CH_3COO)$, $Li(CH_3COO)$, $Mg(CH_3COO)_2$, $Na_2HPO_4$, $NaHCO_3$, $Na_2B_4O_7$, Li-SiW, Na-SiW, and K-SiW [53]. Acid, base, and neutral electrolytes are the three different types of aqueous electrolytes.

**i) Acid Electrolytes**

When an acid dissolves in water, it divides into hydrogen ions (H+) and anions. For instance, hydrochloric acid (HCl) separates into H+ and Cl- in water. The hydrogen ions in the solution could serve as charge carriers, making it electrically conductive. The degree of dissociation and conductivity of the solution depends on the acidity and solute content. A strong acid, such as sulfuric acid ($H_2SO_4$), completely dissociates in water and produces a high concentration of hydrogen ions. In water, a weak acid, such as acetic acid ($CH_3COOH$), partially dissociates and produces a few hydrogen ions. The conductivity of an acid electrolyte can be determined using electrodes attached to a voltage source and a meter. The electrodes are submerged in the solution, and the conductivity of the solution has an inverse relationship with the circuit's current. Acid electrolytes are used in numerous scientific and technological procedures, such as batteries, fuel cells, electroplating, corrosion, pH measurement, and titration [34, 53].

**ii) Base Electrolytes**

In an electrolyte, a base act as the solute, and a polar solvent, typically water, acts as the solvent. A base is a substance that can provide an electron pair to a chemical reaction or accept a proton ($H^+$). When a base dissolves in water, it splits into cations and hydroxide ions ($OH^-$). For instance, in water, sodium hydroxide (NaOH) splits into $Na^+$ and $OH^-$. Due to the hydroxide ions' ability to act as charge carriers, the solution has the potential to become electrically conductive. The level of dissociation and conductivity of the solution depends on the strength of the base and the solute content. A strong base, such as potassium hydroxide (KOH), dissociates in water and produces a significant amount of hydroxide ions. A minor number of hydroxide ions are produced when the ammonia ($NH_3$) partially dissociates in water. A base electrolyte's conductivity can be determined using electrodes connected to a voltage source and a meter. The electrodes are immersed in the solution, and the current across the circuit is inversely proportional to the conductivity of the solution. Base electrolytes have some benefits over acid electrolytes for the usage of supercapacitors, such as greater pH stability, less corrosion, and more capacitance. A smaller voltage window, a lower energy density, and a quicker rate of self-discharge are only a few of their drawbacks. Thus, when selecting base electrolytes for supercapacitors, performance and stability trade-offs should be considered. Hence, when selecting base electrolytes for supercapacitors, performance and stability trade-offs should be considered [56].

**iii) Neutral Electrolytes**

An electrolyte that uses water as the solvent and a neutral or weak salt as the solute is known as a neutral electrolyte. A salt is a substance that can split into cations and anions in a solution. When a salt dissolves in water, it separates into its ions, which makes the resulting solution electrically conductive. For instance, when sodium sulfate ($Na_2SO_4$) dissolves in water, $Na^+$ and $SO_4^{2-}$ are produced. The cations and anions in the solution can act as charge carriers and permit the passage of an electric current. The kind and quantity of salt have an impact on the solution's conductivity and degree of dissociation. With a pH level close to 7, a neutral electrolyte is neither acidic nor basic. In comparison to acid or base electrolytes, a neutral electrolyte offers several benefits for the usage of supercapacitors, including improved safety, less corrosion, and higher voltage. Increased self-discharge, decreased capacitance, and decreased ionic conductivity are some of the drawbacks of a neutral electrolyte. Therefore, the trade-off between performance and stability should be taken into account when choosing neutral electrolytes for supercapacitors. By immersing the aqueous salt solution in a biopolymer matrix, such as xanthan gum, neutral electrolytes can work better [8].

**iv) Properties of Aqueous Electrolytes**

**1) Ionic Conductivity:** This represents the electrolyte's ability to transport ions between the electrodes. The mobility and concentration of the ions in the solution affect the ionic conductivity. In general, higher concentration causes more conductivity, but it also makes a substance more viscous and reduces the ion's ability to diffuse. Ions move more quickly and have greater kinetic energy at higher temperatures, which affects the ionic conductivity.

**2) Viscosity:** This indicates how the flow resistance of the electrolyte is measured. The diffusion and transport of ions in the solution are impacted by viscosity. Poorer ionic conductivity and a poorer supercapacitor power density are caused by higher viscosity. The concentration and size of the solute molecules, as well as the solution's temperature and pressure, all affect the viscosity.

**3) Dielectric Constant:** This measures the electrolyte's ability to decrease the electric field between the electrodes. The dielectric constant affects the supercapacitor's capacitance and energy density, resulting in a higher capacitance and a smaller voltage drop across the electrodes. The solvent and solute molecules' polarity, structure, and interactions with one another all affect the dielectric constant.

**4) Electrochemical Stability:** This measures the electrolyte's stability under electrochemical conditions. The cycling efficiency and safety of the supercapacitor are impacted by the electrochemical stability since unstable electrolytes may degrade or break down over time and result in leakage, corrosion, or explosion. The electrochemical stability is influenced by the

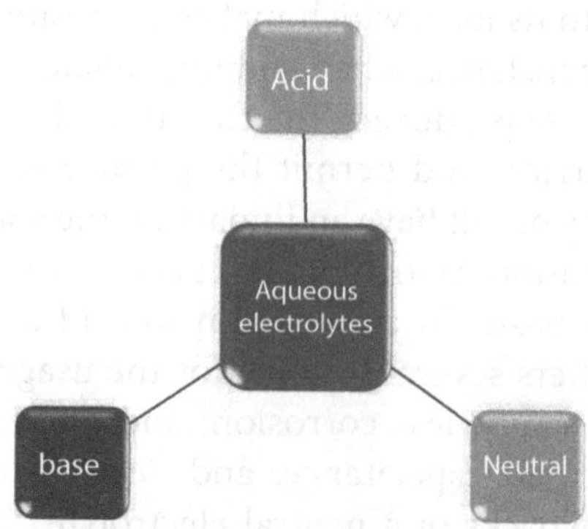


**Figure 8.3** The different types of aqueous electrolytes.

chemical composition, physical structure, and compatibility of the electrolyte with the electrode material. Additionally, the schematic for the different types of aqueous electrolytes is shown in Figure 8.3.

**b) Non-Aqueous Electrolyte**

The selection of a non-aqueous electrolyte for supercapacitors is influenced by several variables, including the electrode material, operating temperature, power, and energy requirements, as well as the supercapacitor's cost and safety considerations. Applications that require high energy density, high voltage, and high temperature can benefit from non-aqueous electrolytes. They are ineffective, however, for applications that need a high power density, a quick charge/discharge rate, a lengthy cycle life, and low cost. Another type of inorganic electrolyte is known as a non-aqueous electrolyte, which uses organic solvents instead of water to dissolve the ions. In comparison to aqueous electrolytes, non-aqueous electrolytes have various advantages, such as a higher voltage window (up to 3.5 V) and a higher energy density, which can improve the performance of supercapacitors. Non-aquatic electrolytes do, however, have several disadvantages over aqueous electrolytes, including reduced ionic conductivity, higher cost, and lower safety. Depending on the type of salt and solvent used, non-aqueous electrolytes can be classified into two groups: organic or ionic liquids. Tetraethylammonium tetrafluoroborate and lithium hexafluorophosphate are just two examples of the organic salts and solvents that are used in organic electrolytes, along with propylene carbonate and acetonitrile. A large portion of commercial supercapacitors use organic electrolytes because of their high voltage window and compatibility with a variety of electrode materials. However, organic electrolytes may be hazardous due to their high flammability, toxicity, and low thermal stability [57].

**c) Organic Electrolytes**

A particular kind of electrolyte called an organic electrolyte comprises organic components like carbon, hydrogen, and oxygen. In batteries, fuel cells, supercapacitors, and other electrochemical devices, organic electrolytes are frequently utilized. Supercapacitor organic electrolytes are made by dissolving salts with high conductivity in organic solvents. Due to its greater operating voltage, which is typically between 2.5 and 2.8 V, OEs currently predominate the market. The energy and power densities greatly increase as the cell operating voltage rises. Additionally, organic electrolytes enable the use of inexpensive packaging materials and current collectors, lowering the overall cost of the supercapacitor cell. Acetonitrile

(ACN) and propylene carbonate (PC) are the two solvents most frequently utilized with organic electrolytes. While acetonitrile has a greater ability to dissolve organic salts than other solvents, its application is restricted by its toxicity. The internal resistance of the organic electrolytes is significant. While having higher operating cell voltages, organic electrolytes have disadvantages such as expensive costs, low specific capacitance, low conductivity, and most importantly, hazards relating to flammability and toxicity [58].

**d) Inorganic Electrolytes**

A form of electrolyte called an organic electrolyte lacks organic components like carbon, hydrogen, and oxygen. Inorganic electrolytes are typically salts, acids, or bases that separate into ions in water. Supercapacitors are electrochemical devices that store energy through the generation of electric double layers or pseudocapacitance at the interface between the electrode and the electrolyte. Inorganic electrolytes are one of the essential elements of these devices. Depending on the solvent used, inorganic electrolytes can be divided into aqueous and non-aqueous categories. However, the aqueous electrolytes offers a high ionic conductivity, and it is environmentally friendly. Due to their lack of flexibility and bendability in comparison with polymer electrolytes, inorganic solid-state electrolytes have drawn less attention. $Li_2$-S-$PS_5$ is an inorganic solid-state electrolyte developed by Francisco *et al.* that functions as both an ion conductor and a separator. The high Li-ion conductivity of this electrolyte was discovered, and devices made with nanostructured electrodes showed a specific capacitance of 7.75 F $g^{-1}$. The concentration, temperature, solvent, and structural characteristics of inorganic electrolytes are just a few of the variables that affect their properties. The ionic conductivity, the electrochemical stability window, viscosity, melting point, and toxicity are a few important characteristics. The efficiency with which ions can travel through the electrolyte is determined by its ionic conductivity. The voltage range in which the electrolyte does not disintegrate or interact with the electrodes is known as the electrochemical stability window. The electrolyte's viscosity serves as a gauge for its thickness or fluidity. The temperature at which the electrolyte transitions from a solid to a liquid form is known as the melting point. Toxicology is a measure of how harmful an electrolyte is to the environment or living things [32, 58].

## 8.4 Recent Trends in the Synthesis Mechanism for Sustainable SCs

### 8.4.1 Chemical Synthesis

Chemical synthesis is necessary for the development of sustainably constructed SCs. SCs are energy storage devices that store and release energy through electrostatic charge separation at the electrode–electrolyte interface. SCs are also known as ultracapacitors or electrochemical capacitors. High power density, rapid charge and discharge speeds, and long cycle life are just a few of the ways they exceed conventional batteries. To make SCs more ecologically friendly, researchers are now focusing on the following aspects of chemical synthesis: Researchers are using chemical synthesis methods to produce novel materials including graphene, metal oxides, and carbon nanotubes. These materials can offer significant surface areas and high conductivity to enhance SC performance. The sustainable SCs, on the other hand, make an effort to reduce the damaging consequences of the materials they use on the environment [59, 60]. Green chemistry concepts and chemical synthesis pathways utilizing renewable or recyclable precursors are being researched as strategies to lower the carbon footprint of the supercapacitor production process. For supercapacitors to be widely employed, chemical synthesis procedures used to produce electrode materials and other components must be effective and scalable. Scientists are working on scalable synthesis techniques to meet the growing demand for supercapacitors in a variety of applications. Chemical synthesis approaches that facilitate better material and component recycling can improve the overall sustainability of supercapacitor technology. The energy efficiency of the chemical synthesis process itself is an important consideration. Energy-intensive production methods can negate the environmental benefits of sustainable supercapacitors. As a result, researchers are looking into low-energy synthesis techniques. The development of energy-efficient, scalable, and environmentally friendly electrode materials, electrolytes, and other supercapacitor components depends heavily on chemical synthesis. These programs aim to increase the performance and applications of supercapacitor technology across a range of industries, including electric vehicles, portable electronics, and renewable energy storage while reducing the impact on the environment [61].

**a) Hydrothermal**

The hydrothermal method has attracted the most admiration from scientists and technologists across disciplines. The word "hydrothermal" derives exclusively from geology. The term was initially used by British geologist Roderick Murchison (1792–1871) to explain how water acting under high pressure and temperature causes changes in the earth's crust that result in the formation of rocks and minerals [62].

A possible way to generate the electrode materials needed for environmentally friendly supercapacitors is the hydrothermal method. This technique makes use of high-pressure, high-temperature water or other solvents to help nanomaterials grow and assemble. It can be used in the creation of supercapacitors and offers several benefits:

1. **Controlled Nanostructure Formation:** The structure, crystallinity, and size of electrode materials can be precisely controlled using hydrothermal synthesis. The electrochemical performance of supercapacitors must be optimized by this regulation.
2. **High Surface Area:** To increase the number of active sites for the electrochemical reactions that store and release energy, the hydrothermal approach can produce nanomaterials with high surface areas, which is beneficial for supercapacitors.
3. **Tunable Properties:** By changing synthesis variables including temperature, pressure, reaction time, and precursor concentration, scientists can customize the properties of electrode materials. Due to this tunability, materials can be improved for certain supercapacitor applications.
4. **Improved Conductivity:** Some hydrothermal treatments can improve the crystallinity of electrode materials and increase their electrical conductivity, which will enhance supercapacitor performance overall.
5. **Environmentally Friendly:** In contrast to several other chemical synthesis techniques that may rely on hazardous substances, hydrothermal synthesis normally uses water as a solvent. This supports sustainability objectives.
6. **Versatile Materials:** Metal oxides, metal sulfides, conductive polymers, and carbon-based compounds are just a few of the materials that can be synthesized hydrothermally. This adaptability enables researchers to investigate diverse materials for supercapacitor electrodes.

7. **Scalability:** Supercapacitor electrode materials may be produced in large quantities using a variety of hydrothermal synthesis techniques, making them appropriate for use in commercial settings.

It is crucial to remember that even if the hydrothermal approach has numerous benefits, the selection of materials and the synthesis conditions should be carefully taken into account to suit the needs of sustainable supercapacitors. To create high-performance, environmentally friendly supercapacitor electrode materials, researchers are still investigating and refining hydrothermal synthesis methods.

In conclusion, due to its capacity to regulate nanostructures, tweak material properties, and create high-surface-area materials using environmentally benign techniques, the hydrothermal method presents a viable path for the sustainable synthesis of electrode materials for supercapacitors. As a result, numerous applications for supercapacitor technology are made more effective and environmentally benign. Additionally, the schematic for the synthesis of nanomaterial using the simple hydrothermal method is shown in Figure 8.4.

**b) Chemical Bath Deposition**

Chemical bath deposition (CBD) is a low-temperature technology that makes it possible to deposit material on large-area substrates relatively

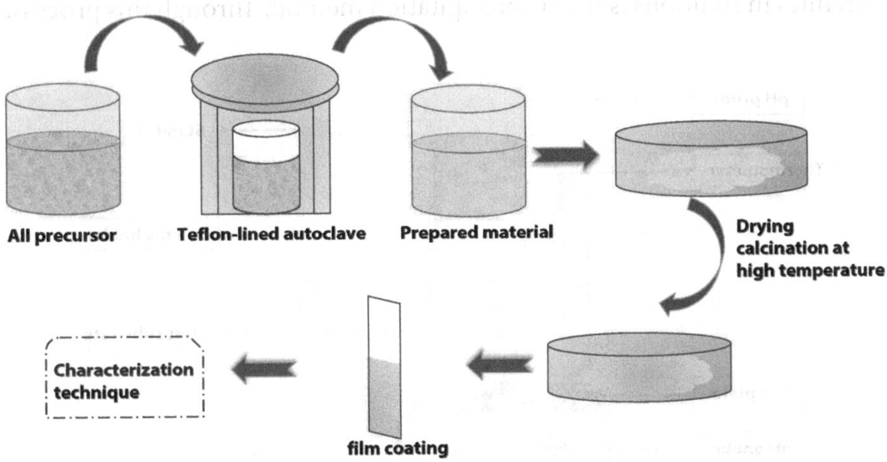


**Figure 8.4** Schematic for the synthesis of nanomaterial using the simple hydrothermal method.

affordably. Without using current or voltage, a material is directly deposited from a solution medium using the CBD method. CBD is produced utilizing a reaction between the substrate and a solution that contains various precursors, either in ionic or molecular form. These interact chemically with the substrate, causing film nucleation. Another benefit of using CBD is that it is extremely simple, especially when compared to electrodeposition, in addition to being reasonably inexpensive, having a broad substrate area and processing at moderate temperatures. The versatility of the CBD technique has limitations as a result of inadequate fundamental knowledge and control of the underlying chemistry, which is one of its downsides. Recent initiatives, though, have sought to alleviate this. The limited material yield for film creation when CBD is used in a traditional beaker setup is another disadvantage. In this configuration, the volume-to-surface ratio is high, with the majority of the solution contributing to colloid formation in the bulk of the solution while just a small percentage of the solution participates in film creation. By running the CBD process between two substrates with a tiny space between them that is filled with solutions, this issue can be solved. After that, this arrangement can be regarded as two-dimensional without area restrictions [63–65]. Additionally, the schematic for the synthesis of nanomaterial using the simple chemical bath deposition method is shown in Figure 8.5.

**c) Simple Co-precipitation Method**

Another widely used approach for producing multicomponent oxide ceramics in solutions is the co-precipitation method. Through this process,

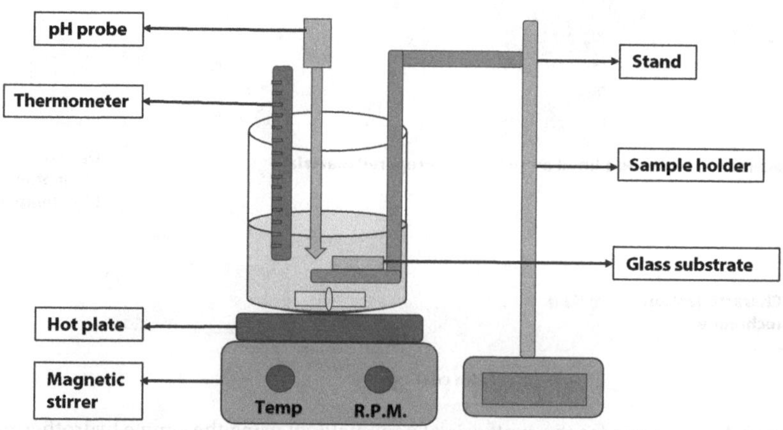


**Figure 8.5** Schematic for the synthesis of nanomaterial using the simple chemical bath deposition method.

two or more insoluble species are combined to form sediment that is simultaneously removed from the solution. The main precursors used in this method are mineral salts (such as nitrate, chloride, and sulfate), which dissolve in water or other appropriate media and produce a homogeneous solution of ionic clusters. After that, the solution is exposed to pH changes or evaporates, which act as a catalyst for the deposition of salts (hydroxides, aqueous oxides, or oxalates). Crystal growth and agglomeration are influenced by several variables, including temperature, salt concentration, the real pH of the solution, and the rate of pH change. After deposition, the solid mass is collected, cleaned, and gradually dried by heating to the boiling point. The washing and drying steps in the precipitation of hydroxide materials affect how agglomerated the final powder is. If the synthesis's objective is to produce nanoscale powders, this subject needs to be carefully considered. Further heat treatment is required to obtain products with the required crystal structure because many of the compounds generated by this technique (especially at low temperatures) are amorphous. Typically, hydroxides are converted to crystalline oxides through calcination. The majority of one-, two-, or three-component systems require a crystallization step, which is frequently carried out by calcination or hydrothermal processing in high-pressure autoclaves. Numerous studies have been reported on the manufacture of ultra-fine oxide powders, oxide–oxide composites, and biomaterials by co-precipitation techniques [66–68]. Additionally, the schematic for the synthesis of nanomaterial using the simple co-precipitation method is shown in Figure 8.6.

**d) Sol-Gel Method**

To create a network in the continuous liquid phase (gel), gelatin and colloidal suspension (sol) must first develop. Aloxysilane ions and metal alkoxides serve as the precursors for the synthesis of these colloids. Tetraethoxysilanes (TEOS), which are used to make silica gels, and tetraethoxysilane (TMOS) are the most often utilized. In water, alkoxides cannot mix. They are predecessors to many other organometallic elements, including zirconium, titanium, aluminum, silica, and many more. Alcohol is a mutual solvent. Initially, one or more chosen alkoxides are homogeneously dissolved in the sol-gel process [69–71]. These are organic precursors for silica, alumina, titania, and zirconia, among others. Sol-gel formation occurs in four stages.

1. Hydrolysis
2. Condensation
3. Growth and agglomeration of particles

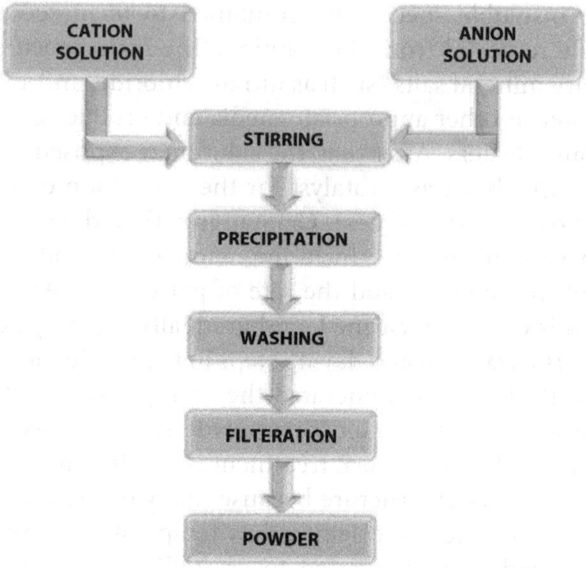


**Figure 8.6** Schematic for the synthesis of nanomaterial using the simple co-precipitation method.

1. Hydrolysis

The [OH-]group replaces the [OR] group during hydrolysis, which is when water is added. Oxygen attacks silicon atoms in silica gel, causing hydrolysis to happen. A catalyst, such as HCl and $NH_3$, can speed up the hydrolysis process. Until all alkoxy groups have been replaced by hydroxyl groups, hydrolysis will continue. Siloxane linkages (Si-O-Si), alcohol, and water were created by the silanol group's subsequent condensation (Si-OH). As the oxygen in the water attacks the silicon atom, hydrolysis takes place. Additionally, the schematic for the synthesis of nanomaterial using the sol-gel method is shown in Figure 8.7.

2. Condensation

Either an alcohol- or a water-producing condensation process leads to the polymerization that creates siloxane linkages. High-order rings, cyclic tetramers, dimers, and monomers are created as a result of condensation products. pH, reagent concentration, and the $H_2O$/Si molar ratio (in the case of silica gels) all have an impact on how quickly an object hydrolyzes. Drying and aging are other significant factors. The structure and attributes of inorganic networks produced from sol-gels can be changed by adjusting these variables.

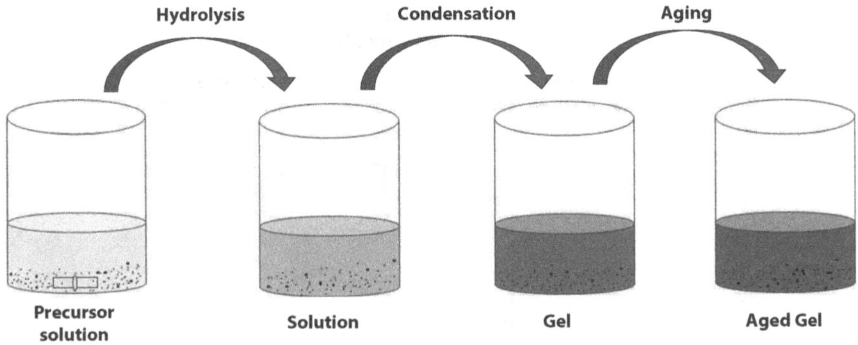


**Figure 8.7** Schematic for the synthesis of nanomaterial using the sol-gel method.

3. Growth and Agglomeration

The molecules gather in the solution as the number of siloxane linkages rises, forming a network that, when dried, solidifies as a gel. The network gets smaller as the water and alcohol are pushed out. When there is no longer a discernible difference in solubility between the smallest and largest particles, growth ceases. At greater temperatures, larger particles develop.

**e) Electrodeposition**

The electrodeposition method, which involves using an electric current to deposit a coating or film of material on a surface, is the most fundamental type of electrochemical deposition. When a substrate is submerged in a salt solution containing the material to be deposited, a negative charge is applied to the substrate, causing the material to be deposited to adhere to the substrate. The positive charge of the salt ions attracts the cathode or the negatively charged substrate. When salt ions arrive at the substrate, they take electrons from the cathode and reduce the substrate to create material. The three-electrode design, which enables precise potential adjustments, is a more complex variant of the fundamental two-electrode setup. Supercapacitor materials can be deposited using this setup. The substrate can serve as the working electrode, the counter electrode can be made of platinum foil or carbon, and the reference electrode can be made of saturated calomel or silver/silver chloride. The electrodeposition method has advantages like mass production, minimal capital investment costs, and exact control of parameters including layer thickness, homogeneity, and deposition rate [72, 73].

Electroless deposition, an autocatalytic variation of electrochemical deposition carried out without the application of an electric field, is a

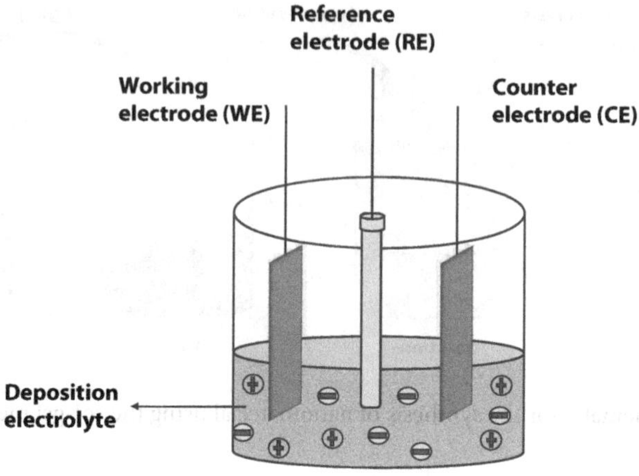


**Figure 8.8** Schematic for the synthesis of nanomaterial using the electrodeposition method.

substitute for electrolytic deposition. In an electroless process, the source of the electrons comes from a reducing agent in the solution as opposed to an external power source in an electrodeposition method. To create ruthenium oxide–carbon composites for supercapacitors, electroless methods were used. The electroless method is appropriate in some circumstances where material must be electrochemically deposited on non-conducting surfaces. The procedure known as electrophoretic deposition is another sort of electrochemical deposition. Under the influence of an electric field, charged particles suspended in a liquid medium flow toward an electrode using this technique. The coagulation of the particles at the electrode causes the material to be deposited. The synthesis and production of supercapacitor material can also be done using this technique. Additionally, the schematic for the synthesis of nanomaterial using the electrodeposition method is shown in Figure 8.8.

## 8.5 Green Synthesis of the Sustainable SCs

The production of nanoparticles usually involves common physical and chemical processes such as chemical reduction, chemical vapor deposition, hydrothermal, laser ablation, electrochemical, sonochemical, and hydrothermal. These procedures produce high-quality, industrially significant goods, but they are energy- and time-intensive, use expensive and dangerous chemicals, and produce non-renewable end products. Green techniques for manufacturing nanoparticles using plants and other microbial resources have made an impact on the developing area of nanotechnology to address these shortcomings. This environmentally friendly manufacturing process is safe for the environment because both the waste it generates and the materials it consumes are readily available plant extracts. The majority of plants may hyperaccumulate and go through intense ion reduction, or phytomining, which is significant in the synthesis of nanoparticles because plants are the most abundant resource. It was decided to use these plants because of their ability to produce nanoparticles. Plant extracts or plant components were used in the procedure as organic solvents or aqueous extracts. Only a few examples of secondary metabolites with varied functional moieties that are regularly involved in bioreduction and other postproductive activities include alkaloids, flavonoids, terpenoids, tannins, sugars, proteins, enzymes, amino acids, and peptides [74, 75].

Supercapacitors can be made with "green synthesis," a method that lessens the impact on the environment, by employing environmentally friendly techniques and materials. Electrodes, electrolytes, and synthesis phenomena are the key areas of focus for ecologically friendly supercapacitors. Choose materials for the electrodes that are plentiful and sustainable, such as graphene, activated carbon derived from renewable sources like coconut shells, or conducting polymers like polyaniline. Use eco-friendly, non-toxic electrolytes like water-based electrolytes, organic electrolytes derived from natural sources, or ionic solutions. If you need graphene, you can either use graphite that has been exfoliated using a green solvent, such as ethyl alcohol, or graphene oxide that has been created using the Hummers process. Ascorbic acid is a good choice for employing non-toxic reducing agents when reducing graphene oxide, while moderately heated pyrolysis of biomass-derived precursors like coconut shells creates the activated carbon electrodes. Green techniques, such as plasma treatment or functionalization using ecologically friendly substances, are frequently employed to modify the electrode surface and improve performance. Ionic liquids, which are non-toxic and biodegradable electrolytes, have also been employed to test the electrochemical performance of SCs. For organic electrolytes, mix natural sources with environmentally friendly solvents, such as propylene carbonate, citric acid, or other organic acids. A biodegradable substance, such as cellulose or bacterial nanocellulose, should be used to create the separator between the electrodes in the devices. When it comes to environmental concerns, recycling, life cycle analysis, and waste reduction help ensure that the supercapacitor's components can be easily separated and recycled at the end of their useful lives. Life cycle assessments help determine the supercapacitor's overall environmental impact, from material extraction to disposal, and they help reduce waste generation during the synthesis process. These techniques can be utilized to create environmentally friendly supercapacitors that help reduce the negative effects of energy storage technologies on the environment. Green manufacturing practices and materials are also included. These environmentally friendly supercapacitors can be utilized in numerous industries, including the storage of renewable energy and electric vehicles while minimizing their impact on the environment [76–78].

## 8.6 Conclusion and Future Prospects of Sustainable SCs

Here, we have addressed a thorough analysis of the state of the art in sustainable supercapacitors based on (1) recent developments in electrode materials, (2) the state of the needed electrolytes, and (3) the mechanism for synthesis of sustainable supercapacitors. Simultaneously, we critically examined the current challenges associated with various electrode materials such as carbon and metal oxide-based electrode materials, with a focus on graphene oxide, reduced graphene oxide, carbon aerogel, carbon nanotubes, and single, binary, ternary, and quaternary metal oxide-based electrodes for sustainable SCs. In addition, we offered our opinions on potential future research areas and goals, as well as current trends in the use of various electrode, electrolyte, and synthesis techniques to enhance the performance of SCs for real-world applications. In the realm of energy storage systems, carbon-based materials with exceptional performance, such as CNT, graphene, GO, and their composites, can be employed as either conductive materials or active materials. Sustainable supercapacitors have a promising future since they have the power to revolutionize many industries and provide solutions to pressing issues with energy storage and sustainability. The following are some crucial potential applications for sustainable supercapacitors:

1. Progresses in Energy Storage
It is projected that sustainable supercapacitors will play a key role in the advancement of energy storage technologies. They have the potential to supplement or perhaps take the place of conventional batteries in a variety of applications because of their fast energy storage and release. This might

lead to energy storage systems that are more effective for renewable energy sources like wind and solar power.

2. Electric Vehicles (EVs)
The performance of electric vehicles can be greatly improved using supercapacitors. The quick charging and discharging speeds made possible by their high-power density increase the effectiveness of regenerative braking and improve acceleration. In EVs of the future, hybrid energy storage systems with batteries and supercapacitors may be used to manage energy more effectively.

3. Consumer Electronics
Supercapacitors might produce consumer devices that are more durable and sustainable. Supercapacitors are batteries that charge quickly and maintain their charge for an extended length of time, decreasing the need for frequent battery replacements in devices like laptops and smartphones.

4. Sustainable Manufacturing
By utilizing environmentally friendly resources and manufacturing techniques, supercapacitor manufacture can be made more environmentally friendly. This fits with the rising need for eco-friendly and sustainable solutions.

5. Research Advancement in Industrial Growth
Supercapacitor performance, energy density, and cycle life will continue to be enhanced by research into new materials, electrode layouts, and production processes. Their uses will be expanded, and they will become more competitive with traditional energy storage technologies thanks to continued research. As energy efficiency and sustainability become more and more valued by businesses, the market for sustainable supercapacitors is anticipated to expand significantly. Supercapacitors will become more widely available and more affordable as a result of increased investments in research, development, and manufacturing.

In conclusion, energy-efficient and environmentally friendly features are greatly enhanced by the development of sustainable supercapacitors. They are an appealing technology for a variety of applications across numerous industries due to their adaptability, quick charging capabilities, and environmental advantages. In the upcoming years, research and development breakthroughs will be crucial to realizing their full potential.

## References


1. Liu, W., Song, M.S., Kong, B., Cui, Y., Flexible and Stretchable Energy Storage: Recent Advances and Future Perspectives. *Adv. Mater.*, 29, 1603436, 2017, https://doi.org/10.1002/adma.201603436.
2. Manjunatha, C., Shwetha, K.P., Athreya, Y., Kumar, S.G., Kamath, M.K.S., Perspective—Supercapacitor-Powered Flexible Wearable Strain Sensors. *ECS Sensors Plus*, 2, 017002, 2023, https://doi.org/10.1149/2754-2726/acb27a.
3. Shembade, U.V., Dhas, S.D., Gurav, S.R., Sonkawade, R.G., Wategaonkar, S.B., Ghatage, S.R., Gaikwad, M.A., Kim, J.H., Parale, V.G., Park, H.H., Moholkar, A.V., Acid substitutions for $WO_3$ nanostructures synthesis by the hydrothermal route and its effect on physio-chemical and electrochemical properties for supercapacitors. *J. Energy Storage*, 72, 108432, 2023, https://doi.org/10.1016/j.est.2023.108432.
4. Zhang, F., Yang, K., Liu, G., Chen, Y., Wang, M., Li, S., Li, R., Recent advances on graphene: Synthesis, properties and applications. *Compos. Part A Appl. Sci. Manuf.*, 160, 107051, 2022, https://doi.org/10.1016/j.compositesa.2022.107051.
5. Fleischmann, S., Mitchell, J.B., Wang, R., Zhan, C., Jiang, D.E., Presser, V., Augustyn, V., Pseudocapacitance: From Fundamental Understanding to High Power Energy Storage Materials. *Chem. Rev.*, 120, 6738–6782, 2020, https://doi.org/10.1021/acs.chemrev.0c00170.
6. Lota, K., Sierczynska, A., Acznik, I., Lota, G., Effect of aqueous electrolytes on electrochemical capacitor capacitance. *Chemik*, 67, 1138–1145, 2013.
7. Zhang, Y.Z., Wang, Y., Cheng, T., Lai, W.Y., Pang, H., Huang, W., Flexible supercapacitors based on paper substrates: a new paradigm for low-cost energy storage. *Chem. Soc. Rev.*, 44, 5181–5199, 2015, https://doi.org/10.1039/c5cs00174a.
8. Şahin, M.E., Blaabjerg, F., Sangwongwanich, A., A Comprehensive Review on Supercapacitor Applications and Developments. *Energies*, 15, 1–26, 2022, https://doi.org/10.3390/en15030674.
9. Huang, S., Zhu, X., Sarkar, S., Zhao, Y., Challenges and opportunities for supercapacitors. *APL Mater.*, 7, 100901, 2019, https://doi.org/10.1063/1.5116146.
10. Chen, P., Shen, G., Shi, Y., Chen, H., Zhou, C., Preparation and Characterization of Flexible Asymmetric Supercapacitors. *ACS Nano*, 4, 4403–4411, 2010.
11. Liang, R., Du, Y., Xiao, P., Cheng, J., Yuan, S., Chen, Y., Yuan, J., Chen, J., Transition metal oxide electrode materials for supercapacitors: A review of recent developments. *Nanomaterials*, 11, 1248, 2021, https://doi.org/10.3390/nano11051248.
12. Zhang, Y., Li, L., Su, H., Huang, W., Dong, X., Binary metal oxide: Advanced energy storage materials in supercapacitors. *J. Mater. Chem. A*, 3, 43–59, 2015, https://doi.org/10.1039/c4ta04996a.
13. Gurav, S.R., Sawant, S.A., Chodankar, G.R., Shembade, U.V., Moholkar, A.V., Sonkawade, R.G., Exploration of aqueous electrolyte on the interconnected

petal-like structure of Co-MOFs for high-performance paper-soaked supercapacitors. *Electrochim. Acta*, 467, 143027, 2023, https://doi.org/10.1016/j.electacta.2023.143027.
14. Shembade, U.V., Dhas, S.D., Magadum, M.G., Gurav, S.R., Raje, P.G., Sonkawade, R.G., Wategaonkar, S.B., Ghatage, S.R., Gaikwad, M.A., Hyeok, J., Parale, V.G., Park, H., Moholkar, A.V., Investigating the effect of electrolyte and its concentration dependence on $WO_3$ nanosheet as an efficient electrode for supercapacitors : Effect of Redox Additive. *J. Phys. Chem. Solids*, 183, 111609, 2023, https://doi.org/10.1016/j.jpcs.2023.111609.
15. Iro, Z.S., Subramani, C., Dash, S.S., A brief review on electrode materials for supercapacitor. *Int. J. Electrochem. Sci.*, 11, 10628–10643, 2016, https://doi.org/10.20964/2016.12.50.
16. Arunkumar, M. and Paul, A., Importance of Electrode Preparation Methodologies in Supercapacitor Applications. *ACS Omega*, 2, 8039–8050, 2017, https://doi.org/10.1021/acsomega.7b01275.
17. Forouzandeh, P., Kumaravel, V., Pillai, S.C., Electrode materials for supercapacitors: A review of recent advances. *Catalysts*, 10, 1–73, 2020, https://doi.org/10.3390/catal10090969.
18. Hiralal, P., Rius, G., Yoshimura, M., Amaratunga, G.A.J., Graphene-based electrochemical capacitors, in: *Graphene Sci. Handb. 5–6*, pp. 479–494, 2016, https://doi.org/10.1201/b19460-38.
19. Zhao, B., Liu, P., Jiang, Y., Pan, D., Tao, H., Song, J., Fang, T., Xu, W., Supercapacitor performances of thermally reduced graphene oxide. *J. Power Sources*, 198, 423–427, 2012, https://doi.org/10.1016/j.jpowsour.2011.09.074.
20. Hadadian, M., Goharshadi, E.K., Youssefi, A., Electrical conductivity, thermal conductivity, and rheological properties of graphene oxide-based nanofluids. *J. Nanoparticle Res.*, 16, 1–17, 2014, https://doi.org/10.1007/s11051-014-2788-1.
21. Zheng, L., Xia, K., Han, B., Zhou, C., Gao, Q., Wang, H., Pu, S., Wu, J., Codoped Porous Carbon-Coated Graphene Nanohybrid as a High-Performance Electrode for Supercapacitors. *ACS Appl. Nano Mater.*, 1, 6742–6751, 2018, https://doi.org/10.1021/acsanm.8b01552.
22. Cheng, Q., Tang, J., Ma, J., Zhang, H., Shinya, N., Qin, L.C., Graphene and nanostructured MnO2 composite electrodes for supercapacitors. *Carbon N. Y.*, 49, 2917–2925, 2011, https://doi.org/10.1016/j.carbon.2011.02.068.
23. Wimalasiri, Y., Fan, R., Zhao, X.S., Zou, L., Assembly of Ni-Al layered double hydroxide and graphene electrodes for supercapacitors. *Electrochim. Acta*, 134, 127–135, 2014, https://doi.org/10.1016/j.electacta.2014.04.129.
24. Oliveira, A.E.F., Braga, G.B., Tarley, C.R.T., Pereira, A.C., Thermally reduced graphene oxide: synthesis, studies and characterization. *J. Mater. Sci.*, 53, 12005–12015, 2018, https://doi.org/10.1007/s10853-018-2473-3.
25. Buldu-Akturk, M., Toufani, M., Tufani, A., Erdem, E., ZnO and reduced graphene oxide electrodes for all-in-one supercapacitor devices. *Nanoscale*, 14, 3269–3278, 2022, https://doi.org/10.1039/d2nr00018k.

26. Kumar, R., Sahoo, S., Tan, W.K., Kawamura, G., Matsuda, A., Kar, K.K., Microwave-assisted thin reduced graphene oxide-cobalt oxide nanoparticles as hybrids for electrode materials in supercapacitor. *J. Energy Storage*, 40, 102724, 2021, https://doi.org/10.1016/j.est.2021.102724.
27. Zheng, H., Wang, J., Jia, Y., Ma, C., *In-situ* synthetize multi-walled carbon nanotubes@MnO 2 nanoflake core-shell structured materials for supercapacitors. *J. Power Sources*, 216, 508–514, 2012, https://doi.org/10.1016/j.jpowsour.2012.06.047.
28. Wang, C.-M., Wen, C.-Y., Chen, Y.-C., Chang, J.-Y., Ho, C.-W., Kao, K.-S., Shih, W.-C., Chiu, C.-M., Shen, Y.-A., The Influence of Specific Surface Area on the Capacitance of the Carbon Electrodes Supercapacitor. *The Proceedings of the Second International Conference on Industrial Application Engineering*, 439–442, 2015, https://doi.org/10.12792/iciae2015.077.
29. Huang, K.J., Wang, L., Zhang, J.Z., Xing, K., Synthesis of molybdenum disulfide/carbon aerogel composites for supercapacitors electrode material application. *J. Electroanal. Chem.*, 752, 33–40, 2015, https://doi.org/10.1016/j.jelechem.2015.06.005.
30. Mirzaeian, M., Abbas, Q., Gibson, D., Mazur, M., Effect of nitrogen doping on the electrochemical performance of resorcinol-formaldehyde based carbon aerogels as electrode material for supercapacitor applications. *Energy*, 173, 809–819, 2019, https://doi.org/10.1016/j.energy.2019.02.108.
31. Oschatz, M., Boukhalfa, S., Nickel, W., Hofmann, J.P., Fischer, C., Yushin, G., Kaskel, S., Carbide-derived carbon aerogels with tunable pore structure as versatile electrode material in high power supercapacitors. *Carbon N. Y.*, 113, 283–291, 2017, https://doi.org/10.1016/j.carbon.2016.11.050.
32. Eftekhari, A., Supercapacitors utilising ionic liquids. *Energy Storage Mater.*, 9, 47–69, 2017, https://doi.org/10.1016/j.ensm.2017.06.009.
33. Ramachandran, R., Chen, S.M., Gnana Kumar, G.P., Gajendran, P., Devi, N.B., An overview of fabricating nanostructured electrode materials for biosensor applications. *Int. J. Electrochem. Sci.*, 10, 8607–8629, 2015, https://doi.org/10.1016/s1452-3981(23)11122-9.
34. Frackowiak, E., Metenier, K., Bertagna, V., Beguin, F., Supercapacitor electrodes from multiwalled carbon nanotubes. *Appl. Phys. Lett.*, 77, 2421–2423, 2000, https://doi.org/10.1063/1.1290146.
35. Lv, T., Liu, M., Zhu, D., Gan, L., Chen, T., Nanocarbon-Based Materials for Flexible All-Solid-State Supercapacitors. *Adv. Mater.*, 30, 1–17, 2018, https://doi.org/10.1002/adma.201705489.
36. Cheng, Q., Tang, J., Ma, J., Zhang, H., Shinya, N., Qin, L.C., Graphene and carbon nanotube composite electrodes for supercapacitors with ultra-high energy density. *Phys. Chem. Chem. Phys.*, 13, 17615–17624, 2011, https://doi.org/10.1039/c1cp21910c.
37. Li, H., Li, Z., Wu, Z., Sun, M., Han, S., Cai, C., Shen, W., Liu, X.T., Fu, Y.Q., Enhanced electrochemical performance of $CuCo_2S_4$/carbon nanotubes

composite as electrode material for supercapacitors. *J. Colloid Interface Sci.*, 549, 105–113, 2019, https://doi.org/10.1016/j.jcis.2019.04.056.

38. Lu, X., Dou, H., Gao, B., Yuan, C., Yang, S., Hao, L., Shen, L., Zhang, X., A flexible graphene/multiwalled carbon nanotube film as a high performance electrode material for supercapacitors. *Electrochim. Acta*, 56, 5115–5121, 2011, https://doi.org/10.1016/j.electacta.2011.03.066.
39. Simotwo, S.K., Delre, C., Kalra, V., Supercapacitor Electrodes Based on High-Purity Electrospun Polyaniline and Polyaniline-Carbon Nanotube Nanofibers. *ACS Appl. Mater. Interfaces*, 8, 21261–21269, 2016, https://doi.org/10.1021/acsami.6b03463.
40. Li, Q., Ni, H., Cai, Y., Cai, X., Liu, Y., Chen, G., Fan, L.Z., Wang, Y., Preparation and supercapacitor application of the single crystal nickel hydroxide and oxide nanosheets. *Mater. Res. Bull.*, 48, 3518–3526, 2013, https://doi.org/10.1016/j.materresbull.2013.05.049.
41. Bin Liu, Y., Lin, L.Y., Huang, Y.Y., Tu, C.C., Investigation of the electroactive capability for the supercapacitor electrode with cobalt oxide rhombus nano-pillar and nanobrush arrays. *J. Power Sources*, 315, 23–34, 2016, https://doi.org/10.1016/j.jpowsour.2016.03.035.
42. Liu, D., Liu, Y., Liu, X., Xu, C., Zhu, J., Chen, H., Growth of uniform $CuCo_2O_4$ porous nanosheets and nanowires for high-performance hybrid supercapacitors. *J. Energy Storage*, 52, 105048, 2022, https://doi.org/10.1016/j.est.2022.105048.
43. Sun, J., Xu, C., Chen, H., A review on the synthesis of $CuCo_2O_4$-based electrode materials and their applications in supercapacitors. *J. Mater.*, 7, 98–126, 2021, https://doi.org/10.1016/j.jmat.2020.07.013.
44. Hong, W.L. and Lin, L.Y., Design of nickel cobalt oxide and nickel cobalt oxide@nickel molybdenum oxide battery-type materials for flexible solid-state battery supercapacitor hybrids. *J. Power Sources*, 435, 226797, 2019, https://doi.org/10.1016/j.jpowsour.2019.226797.
45. Yang, D., Pulsed laser deposition of cobalt-doped manganese oxide thin films for supercapacitor applications. *J. Power Sources*, 198, 416–422, 2012, https://doi.org/10.1016/j.jpowsour.2011.10.008.
46. Ishaq, S., Moussa, M., Kanwal, F., Ayub, R., Van, T.N., Azhar, U., Losic, D., One step strategy for reduced graphene oxide/cobalt-iron oxide/polypyrrole nanocomposite preparation for high performance supercapacitor electrodes. *Electrochim. Acta*, 427, 140883, 2022, https://doi.org/10.1016/j.electacta.2022.140883.
47. Dinesh, M., Haldorai, Y., Rajendra Kumar, R.T., Mn–Ni binary metal oxide for high-performance supercapacitor and electro-catalyst for oxygen evolution reaction. *Ceram. Int.*, 46, 28006–28012, 2020, https://doi.org/10.1016/j.ceramint.2020.07.295.
48. Peng, H., Cui, S., Xie, X., Wei, G., Sun, K., Ma, G., Lei, Z., Binary tungsten-molybdenum oxides nanoneedle arrays as an advanced negative electrode

material for high performance asymmetric supercapacitor. *Electrochim. Acta*, 322, 134759, 2019, https://doi.org/10.1016/j.electacta.2019.134759.

49. Dhas, S.D., Maldar, P.S., Patil, M.D., Hubali, K.M., Shembade, U.V., Abitkar, S.B., Waikar, M.R., Sonkawade, R.G., Agawane, G.L., Moholkar, A.V., Hydrothermal synthesis of mesoporous $NiMnO_3$ nanostructures for supercapacitor application: Effect of electrolyte. *J. Energy Storage*, 35, 102277, 2021, https://doi.org/10.1016/j.est.2021.102277.
50. Godlaveeti, S.K., Maseed, H., Reddy, S.A., Nagireddy, R.R., Electrochemical performance of ternary RGO/$\beta$-$Ni(OH)_2$/$CeO_2$ composite in addition with different metal oxides for supercapacitor application. *Adv. Nat. Sci. Nanosci. Nanotechnol.*, 11, 025021, 2020, https://doi.org/10.1088/2043-6254/ab8bde.
51. Sahoo, S., Nguyen, T.T., Shim, J.J., Mesoporous Fe–Ni–Co ternary oxide nanoflake arrays on Ni foam for high-performance supercapacitor applications. *J. Ind. Eng. Chem.*, 63, 181–190, 2018, https://doi.org/10.1016/j.jiec.2018.02.014.
52. Malik, S., Gul, I.H., Baig, M.M., Hierarchical MnNiCo ternary metal oxide/graphene nanoplatelets composites as high rated electrode material for supercapacitors. *Ceram. Int.*, 47, 17008–17014, 2021, https://doi.org/10.1016/j.ceramint.2021.02.272.
53. Pal, B., Yang, S., Ramesh, S., Thangadurai, V., Jose, R., Electrolyte selection for supercapacitive devices: A critical review. *Nanoscale Adv.*, 1, 3807–3835, 2019, https://doi.org/10.1039/c9na00374f.
54. Raza, W., Ali, F., Raza, N., Luo, Y., Kim, K.H., Yang, J., Kumar, S., Mehmood, A., Kwon, E.E., Recent advancements in supercapacitor technology. *Nano Energy*, 52, 441–473, 2018, https://doi.org/10.1016/j.nanoen.2018.08.013.
55. Zhang, L., Yang, S., Chang, J., Zhao, D., Wang, J., Yang, C., Cao, B., A Review of Redox Electrolytes for Supercapacitors. *Front. Chem.*, 8, 1–7, 2020, https://doi.org/10.3389/fchem.2020.00413.
56. Iqbal, M.Z., Zakar, S., Haider, S.S., Role of aqueous electrolytes on the performance of electrochemical energy storage device. *J. Electroanal. Chem.*, 858, 113793, 2020, https://doi.org/10.1016/j.jelechem.2019.113793.
57. Zhai, S., Karahan, H.E., Wang, C., Pei, Z., Wei, L., Chen, Y., 1D Supercapacitors for Emerging Electronics: Current Status and Future Directions. *Adv. Mater.*, 32, 1–19, 2020, https://doi.org/10.1002/adma.201902387.
58. Lokhande, P.E., Inorganic Electrolytes in Supercapacitor. *Mater. Res. Found,* 61, 11–30, 2019, https://doi.org/10.21741/9781644900499-2.
59. Kulal, P.M., Dubal, D.P., Lokhande, C.D., Fulari, V.J., Chemical synthesis of $Fe_2O_3$ thin films for supercapacitor application. *J. Alloys Compd.*, 509, 2567–2571, 2011, https://doi.org/10.1016/j.jallcom.2010.11.091.
60. Vangari, M., Pryor, T., Jiang, L., Supercapacitors: Review of Materials and Fabrication Methods. *J. Energy Eng.*, 139, 72–79, 2013, https://doi.org/10.1061/(asce)ey.1943-7897.0000102.
61. Rane, A.V., Kanny, K., Abitha, V.K., Thomas, S., Thomas, S., *Methods for Synthesis of Nanoparticles and Fabrication of Nanocomposites,*

*Synthesis of inorganic nanomaterials*, 2018, https://doi.org/10.1016/B978-0-08-101975-7.00005-1.
62. Kouchachvili, L., Yaïci, W., Entchev, E., Hybrid battery/supercapacitor energy storage system for the electric vehicles. *J. Power Sources*, 374, 237–248, 2018, https://doi.org/10.1016/j.jpowsour.2017.11.040.
63. Kariper, A. and Meydaneri Tezel, F., Gökkuş, Synthesize of $WO_3$ thin film supercapacitor and its characterization. *Phys. Lett. Sect. A Gen. At. Solid State Phys.*, 388, 1–6, 2021, https://doi.org/10.1016/j.physleta.2020.127059.
64. Chodankar, N.R., Dubal, D.P., Gund, G.S., Lokhande, C.D., Flexible all-solid-state $MnO_2$ thin films based symmetric supercapacitors. *Electrochim. Acta*, 165, 338–347, 2015, https://doi.org/10.1016/j.electacta.2015.02.246.
65. Singu, B.S. and Yoon, K.R., Synthesis and characterization of $MnO_2$-decorated graphene for supercapacitors. *Electrochim. Acta*, 231, 749–758, 2017, https://doi.org/10.1016/j.electacta.2017.01.182.
66. Yadav, M.S., Singh, N., Kumar, A., Synthesis and characterization of zinc oxide nanoparticles and activated charcoal based nanocomposite for supercapacitor electrode application. *J. Mater. Sci. Mater. Electron.*, 29, 6853–6869, 2018, https://doi.org/10.1007/s10854-018-8672-5.
67. Singh Yadav, M., Sinha, A.K., Singh, M.N., Kumar, A., Electrochemical study of copper oxide and activated charcoal based nanocomposite electrode for supercapacitor. *Mater. Today Proc.*, 46, 5722–5729, 2021, https://doi.org/10.1016/j.matpr.2021.02.134.
68. Vignesh, V., Subramani, K., Sathish, M., Navamathavan, R., Electrochemical investigation of manganese ferrites prepared via a facile synthesis route for supercapacitor applications. *Colloids Surfaces A Physicochem. Eng. Asp.*, 538, 668–677, 2018, https://doi.org/10.1016/j.colsurfa.2017.11.045.
69. Pilban Jahromi, S., Pandikumar, A., Goh, B.T., Lim, Y.S., Basirun, W.J., Lim, H.N., Huang, N.M., Influence of particle size on performance of a nickel oxide nanoparticle-based supercapacitor. *RSC Adv.*, 5, 14010–14019, 2015, https://doi.org/10.1039/c4ra16776g.
70. Nallamuthu, N., Packiaraj, R., Venkatesh, K.S., Devendran, P., Bahadur, S.A., Synthesis and Characterization of Sol-Gel Derived $Bivo_4$ Nanoparticles for Electrochemical Applications. *Int. J. Eng. Adv. Technol.*, 9, 390–393, 2019, https://doi.org/10.35940/ijeat.a1189.1291s419.
71. Kumar, A., Sol gel synthesis of zinc oxide nanoparticles and their application as nano-composite electrode material for supercapacitor. *J. Mol. Struct.*, 1220, 128654, 2020, https://doi.org/10.1016/j.molstruc.2020.128654.
72. Shinde, P.A., Lokhande, V.C., Patil, A.M., Ji, T., Lokhande, C.D., Single-step hydrothermal synthesis of $WO_3$-$MnO_2$ composite as an active material for all-solid-state flexible asymmetric supercapacitor. *Int. J. Hydrog. Energy*, 43, 2869–2880, 2018, https://doi.org/10.1016/j.ijhydene.2017.12.093.
73. Mohammadi Zardkhoshoui, A., Hosseiny Davarani, S.S., Hashemi, M., Fabrication of cobalt gallium oxide with zinc iron oxide on nickel foam for a

high-performance asymmetric supercapacitor. *New J. Chem.*, 43, 4590–4598, 2019, https://doi.org/10.1039/C8NJ05854G.

74. Sudhakar, Y.N., Hemant, H., Nitinkumar, S.S., Poornesh, P., Selvakumar, M., Green synthesis and electrochemical characterization of rGO–CuO nanocomposites for supercapacitor applications. *Ionics (Kiel)*, 23, 1267–1276, 2017, https://doi.org/10.1007/s11581-016-1923-7.
75. Gopalakrishnan, A., Raju, T.D., Badhulika, S., Green synthesis of nitrogen, sulfur-co-doped worm-like hierarchical porous carbon derived from ginger for outstanding supercapacitor performance. *Carbon N. Y.*, 168, 209–219, 2020, https://doi.org/10.1016/j.carbon.2020.07.017.
76. Nayak, A.K., Das, A.K., Pradhan, D., High performance solid-state asymmetric supercapacitor using green synthesized graphene–WO3 nanowires nanocomposite. *ACS Sustain. Chem. Eng.*, 5, 11, 10128-10138, 2017, http://pubs.acs.org.
77. Aruchamy, K., Nagaraj, R., Manohara, H.M., Nidhi, M.R., Mondal, D., Ghosh, D., Nataraj, S.K., One-step green route synthesis of spinel $ZnMn_2O_4$ nanoparticles decorated on MWCNTs as a novel electrode material for supercapacitor. *Mater. Sci. Eng. B.*, 252, 114481, 2020, https://doi.org/10.1016/j.mseb.2019.114481.
78. B., N., Y., V., Abdul Razack, S., Enhanced formation of ruthenium oxide nanoparticles through green synthesis for highly efficient supercapacitor applications. *Adv. Powder Technol.*, 31, 1001–1006, 2020, https://doi.org/10.1016/j.apt.2019.12.026.

# 9

# Cyclic Stability and Capacitance Retention of MXene-Based Supercapacitors

**Muhammad Akmal Kosnan[1], Mohd Asyadi Azam[1,2*] and Akito Takasaki[2]**

[1]*Fakulti Teknologi dan Kejuruteraan Industri dan Pembuatan, Universiti Teknikal Malaysia Melaka, Hang Tuah Jaya, Durian Tunggal, Melaka, Malaysia*
[2]*Department of Engineering Science and Mechanics, College of Engineering, Shibaura Institute of Technology, Toyosu, Koto-Ku, Tokyo, Japan*

## *Abstract*
In the escalating demand for sustainable energy storage solutions, researchers have been exploring supercapacitors as a promising option for high-efficiency energy storage device. A recent progress in sustainable supercapacitors, emphasizing on the application of two-dimensional (2D) materials, is comprehensively studied. 2D materials possess exceptional traits, such as elevated surface area, excellent electrical conductivity, and eco-friendliness, providing enhanced supercapacitor performance. This paper delves into the utilization of MXenes, a type of 2D material, specifically within the realm of supercapacitor applications. The primary focus lies on examining their capacitance retention properties while addressing associated challenges and prospects. Numerous real applications of MXenes in supercapacitors are highlighted through this case study, demonstrating their versatility across various sectors, including electronics, electric vehicles, and grid-based energy storage. The study also assesses the long-term stability and performance sustainability of MXene-based supercapacitors under cyclic charge–discharge conditions. Experimental data and analysis reveal insights into the cyclic stability and the ability of MXene-based supercapacitors to retain their capacitance over extended operational lifetimes. It provides valuable insights into the promising advancements and sustainability aspects of this kind of supercapacitor emphasizing their pivotal role in the shift towards cleaner and more efficient energy storage solutions. It also may develop the potential to transform the energy storage landscape, contributing to a greener and further environmentally responsible future.

**Corresponding author*: asyadi@utem.edu.my

Basheer Ahamed and Chaudhery Mustansar Hussain (eds.) Sustainable Supercapacitors: Next-Generation of Green Energy Storage Devices, (265–298) © 2025 Scrivener Publishing LLC

***Keywords*:** MXene, supercapacitor, nanomaterials, energy storage devices, electrochemical performance

## 9.1 Introduction

Over the last 15 years, scientists have been studying two-dimensional (2D) materials ever since they discovered unique properties in single-layer graphene [1]. This inquisitiveness has brought about a wave of study into several new 2D materials as well as the discovery of many previously unknown 2D materials, such as metal dichalcogenides and boron nitride [2, 3]. While some of these materials are mainly of interest to academics, others have gained recognition because of their appealing properties, which have practical uses. One of them is MXene, which is a class of 2D nanomaterials composed of transition metal carbides, nitrides, or carbonitrides [4, 5]. The name "MXene" is derived from its chemical formula, which typically follows the pattern Mn+1XnTx, where "M" represents a transition metal element such as titanium, tantalum, or niobium, "X" represents carbon and/or nitrogen, "n" is the number of carbide or nitride layers, and "Tx" constitute surface functional groups or terminations. Typically, MXenes are produced by selectively etching the "A" layers from the parent MAX phases, which are layered materials [6]. Figure 9.1 shows the potential elements that can be included in the construction of MXene materials.

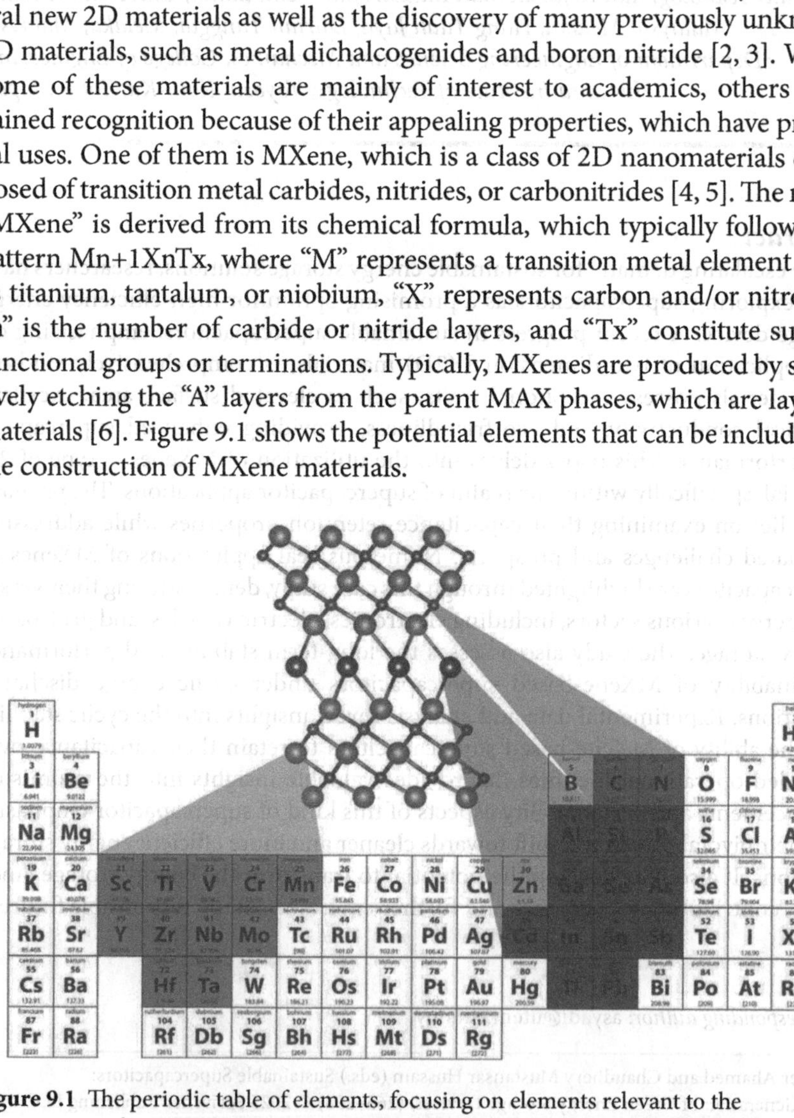


**Figure 9.1** The periodic table of elements, focusing on elements relevant to the construction of MXene materials. Reproduced with permission from Wiley. Ref. [7].

MXene has quickly received the title of the "next wonder material" since its inception in 2011, unlike its other 2DM counterparts [8]. Its remarkable features include high electrical conductivity, a significant specific surface area, flexible surface chemistry that can be operationalized, easy dispersion in various solvents such as water, and, most importantly, exceptional electrochemical performances that make it suitable for energy storage applications. MXene possesses these features due to several reasons, such as its 2D structure that facilitates rapid ion transport through 2D/3D channels, oxidoreduction that introduces a pseudocapacitive module to the electrical double-layer mode of charge storage, and balanced mechanical flexibility at the nanosheet level that can accommodate a large number of ions for lithium-ion batteries and beyond [9]. Since the finding of $Ti_3C_2$ in Drexel University, the study on 2D transition metal carbides and nitrides has grown rapidly. In recent years, a novel exfoliation process has been successfully used to generate a new member of 2D transition metal carbides, nitrides, and carbonitrides, a double MXene, which refers to a specific type of MXene material where two layers of MXene are stacked on top of each other, obtained from layered precursors [10].

This etching process upshot in the delamination of MXene layers creates 2D nanosheets with high surface areas and unique properties [11]. MXenes are highly valued in a variety of applications, including energy storage, such as supercapacitors and batteries, catalysis, sensing, and more. This is due to their superior electrical conductivity, high mechanical strength, and chemical diversity. Due to their high conductivity and vast surface area, they have recently demonstrated success in applications like Li-ion batteries, supercapacitors, and energy storage components. MXenes, which offer high energy densities and quick discharge rates, can be employed as electrode materials in flexible supercapacitors (SCs) and micro-supercapacitors (MSCs) [12]. High volumetric capacitance, a measurement of the amount of charge that can be stored per unit volume, has been shown in MXene-based supercapacitors. For instance, a recent study found that an MXene-based electrode has a volumetric capacitance of 1632 F $cm^{-3}$ [13]. They have also shown an ultrahigh rate capability, allowing for efficient charge and discharge processes. One study achieved a rate capability of 827 F $cm^{-3}$ at 5000 mV $s^{-1}$, surpassing the performance of other materials. It can also be fabricated in flexible and stretchable forms, enabling their integration into various wearable and portable devices. Moreover, MXenes have undergone annealing treatments to improve their cycling stability and electrochemical performance, making them more appropriate for long-term energy storage applications. Research is still needed to fully realize the potential of MXene-based supercapacitors and to solve issues like stability, scalability, and affordability. However, MXenes are intriguing prospects for next-generation supercapacitor applications due to their unique qualities and adaptability.

Supercapacitors are commonly evaluated for their electrochemical performance using a variety of methods, including cyclic voltammetry (CV), galvanostatic charge–discharge (GCD), and cyclic stability tests [14]. These techniques shed light on the capacitance, stability, and charge storage mechanisms of the supercapacitor electrodes. Supercapacitors and other capacitive electrochemical cells can have their behavior studied using the electrochemical technique known as CV. The voltage window, capacitance, and cycle life of the cell are only a few of the fundamental details that CV offers in the context of the creation of supercapacitors. In the field of researching supercapacitors, it is a crucial instrument. In CV, a supercapacitor electrode is subjected to a potential sweep while the resulting current is monitored. Depending on the exact MXene composition, synthesis technique, and experimental settings, the average specific capacitance of MXene materials when applying CV can change. For example, $Ti_3C_2Tx$ MXene in 3 M KOH alkaline electrolyte recorded up to 92 F $g^{-1}$ at 2 A $g^{-1}$ while bare MXene approximately recorded 246 F $g^{-1}$ in supercapacitors [15].

A common technique for assessing the performance of supercapacitors is GCD, much like CV. It entails providing the supercapacitor with a steady current while monitoring the voltage shift that results over time. It offers details about the supercapacitor's charge and discharge behavior, including the voltage rise during charge and the voltage drop during discharge. Understanding the energy storage and release processes in the device depends on having this knowledge. You may calculate the specific capacitance of a supercapacitor by dividing the charge (in coulombs), the voltage change (in volts), and the mass of the active material by the GCD curve. This merit represents the amount of charge that can be stored per unit mass of the electrode material and is a main parameter for assessing the performance of the supercapacitor.

Cyclic stability tests are essential for evaluating the long-term performance and longevity of supercapacitors. These tests involve subjecting the supercapacitor to repeat charge and discharge cycles and monitoring the changes in its capacitance over time. Cyclic stability tests can be performed using various techniques, such as GCD, CV, and constant current cycling. These methods provide insights into the changes in the supercapacitor's performance and behavior over multiple cycles. A supercapacitor's lifetime cyclic stability is sometimes expressed in terms of the number of cycles it can survive while still retaining a specific amount of capacitance.

For instance, after 10,000 cycles of capacitance retention, the typical supercapacitor may show a 91.7% [16]. Supercapacitors with increased cycle stability have recently been developed thanks to improvements in electrode materials and device design. For instance, some electrodes containing carbonaceous nanostructures have shown high cyclic stability, with 15,000 cycles and 92% capacitance retention. In addition, several variables, such as the electrode material, electrolyte composition, and device design, can affect a supercapacitor's cyclic stability. By understanding the principles and applications of cyclic stability tests, researchers can optimize the performance and durability of supercapacitors, leading to more efficient and reliable energy storage solutions.

Figure 9.2a shows the CV of two electrodes: one composed of pristine $Ti_3C_2Tx$-MXene and the other of N-$Ti_3C_2Tx$-MXene. These CV graphs encompass a potential window range of -0.8 to 0.2 V and exhibit the characteristic rectangular shape indicative of capacitive behavior [17]. However, a notable discrepancy arises when considering the individual capacitances discerned by the area under the CV curves. Notably, N-$Ti_3C_2Tx$-MXene,

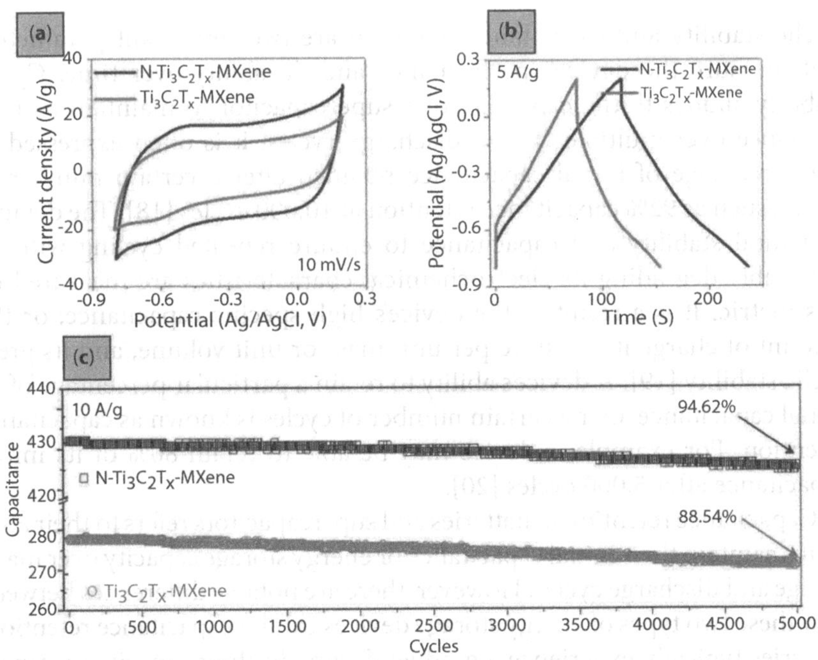


**Figure 9.2** (a) CVs of $Ti_3C_2T_x$-MXene and N-$Ti_3C_2T_x$-MXene at -0.8–0.2 V; (b) GCDs of pristine $Ti_3C_2Tx$-MXene and N-$Ti_3C_2T_x$-MXene and (c) cyclic stability. Copyright from Ref. [17].

owing to its exceptional conductivity and interconnectivity among nanosheets, presents a significantly larger CV area compared to pristine $Ti_3C_2Tx$-MXene. This implies an enhanced specific capacitance in the case of N-$Ti_3C_2Tx$-MXene. Figure 9.2b showcases the GCD of these electrodes at a current density of 5 A $g^{-1}$. Here, the N-$Ti_3C_2Tx$-MXene electrode exhibits an extended charge–discharge duration of 253.35 s, surpassing the duration observed for the pristine $Ti_3C_2Tx$-MXene electrode, which stands at 152.69 s. This observation aligns consistently with the findings from the CV analysis. To evaluate the cyclic stability of these electrodes, Figure 9.2c presents the outcomes. N-$Ti_3C_2Tx$-MXene demonstrates remarkable cyclic stability with a capacitance retention of 99.62% after 5000 cycles. In contrast, pristine $Ti_3C_2Tx$-MXene exhibits a slightly lower retention rate of 88.54% under the same cycling circumstances. The excellent cyclic performance of the N-$Ti_3C_2Tx$-MXene electrode is evident from these results.

## 9.2 Cyclic Stability and Capacitance/Capacity Retention of Supercapacitors and Batteries

Cyclic stability and capacitance retention are two important parameters that measure the device's performance and durability over time. Cyclic stability alludes to the capability of a supercapacitor to maintain its performance over multiple charge–discharge cycles. It is often expressed as the percentage of initial capacitance retained after a certain number of cycles, such as 92% capacitance retention at 10,000 cycles [18]. The device's structural stability and capacitance to endure repeated cycling without noticeably degrading its electrochemical characteristics are measured by this metric. It is a result of the device's high specific capacitance, or the amount of charge it can store per unit mass or unit volume, and its great cyclic stability [19]. A device's ability to retain a particular percentage of its initial capacitance after a certain number of cycles is known as capacitance retention. For example, a device may be able to retain 80% of its initial capacitance after 5,000 cycles [20].

Capacitance retention in batteries and supercapacitors refers to their ability to maintain their initial capacitance or energy storage capacity over many charge and discharge cycles. However, there are notable differences between how these two types of energy storage devices exhibit capacitance retention. Batteries typically experience a gradual decline in their capacity or capacitance over time [21]. This phenomenon, known as capacity fade, occurs due to various factors, which are electrode degradation, solid–electrolyte interphase (SEI) emergence, and chemical reactions within the battery. Many rechargeable batteries have relatively modest capacity fading, which enables them to retain a sizeable amount of their initial capacity across hundreds to thousands of cycles. Batteries are made to withstand more charge and discharge cycles while maintaining a sizable amount of their initial capacity. This is known as having a longer cycle life. For instance, depending on the particular chemistry and design, lithium-ion batteries frequently retain roughly 80%–90% of their capacity after 500 to 1,000 cycles. The goal of battery technology is to strike a balance between energy density and capacity retention. Even when both dimensions are being worked on, there is frequently a trade-off between them. Energy density may suffer if high-capacity retention is used, and vice versa.

Supercapacitors, however, show quick capacitance retention. Even after tens of thousands or hundreds of thousands of charge–discharge cycles, they can maintain a high degree of capacitance. This attribute makes them well-suited for applications requiring frequent and rapid energy storage and release. While supercapacitors excel in rapid charge and discharge, they typically have a shorter cycle life compared to batteries. The materials used in supercapacitor electrodes can experience degradation due to the high energy and power density cycling. This means that, over time, supercapacitors may gradually lose a larger percentage of their initial capacitance compared to batteries. Supercapacitors prioritize power density over energy density. They can deliver high power outputs, which makes them suitable for implementation such as regenerative braking in vehicles, peak power shaving in renewable energy systems, and rapid energy delivery in electronics.

Due to a number of important variations in their fundamental energy storage methods and the materials used in their fabrication, supercapacitors can tolerate longer cycle retention than batteries. Supercapacitors, for instance, store energy by separating charges electrostatically at the electrode–electrolyte interface. Ions are adsorbed onto porous electrode materials in this method, creating an electric double layer. Over numerous charge and discharge cycles, this method causes minimal wear and tear on the electrode materials since it does not require chemical reactions or phase shifts inside the materials. The materials in the positive and negative electrodes of batteries, on the other hand, depend on chemical reactions (oxidation and reduction) to store and release energy. Over time, these chemical processes may cause the structural integrity of the electrodes to deteriorate and the active materials to degrade, which will shorten the cycle life and cause capacity fading. Another issue is that the movement of ions inside the electrolyte and the creation of the double layer at the electrode–electrolyte interface are key components of the energy storage mechanism in supercapacitors. These procedures cause minimal stress on the electrode materials and enable a greater cycle life because they do not significantly alter their physical or chemical properties. While the insertion and extraction of ions or molecules into the electrode materials in batteries causes the chemical reactions that take place during charge and discharge. These procedures may subject the electrode materials to mechanical stress, expansion, and contraction, which could eventually cause wear and tear over a number of cycles.

The key difference in capacitance retention between batteries and supercapacitors lies in the rate of capacity fade and the trade-offs between energy density and cycle life. Batteries tend to offer slower capacity fade and longer cycle life but may sacrifice some power density. Supercapacitors, on the other hand, excel in rapid charge and discharge but have a shorter cycle life and are optimized for high power density applications. The choice between them depends on the specific requirements of the application [22]. Figure 9.3 demonstrates that whereas the supercapacitor voltage generally exhibits a linear and decreasing behavior from the highest value up to 50%, the battery exhibits a constant discharge voltage performance over the spectrum of usable energy. To put it another way, batteries, the most popular forms of energy storage, have large energy densities, but their power densities are low in comparison to a supercapacitor of the same size. Also, batteries have a low life cycle compared to a supercapacitor, which has an or thereabouts 1000× longer life cycle than battery also as shown in Table 9.1.

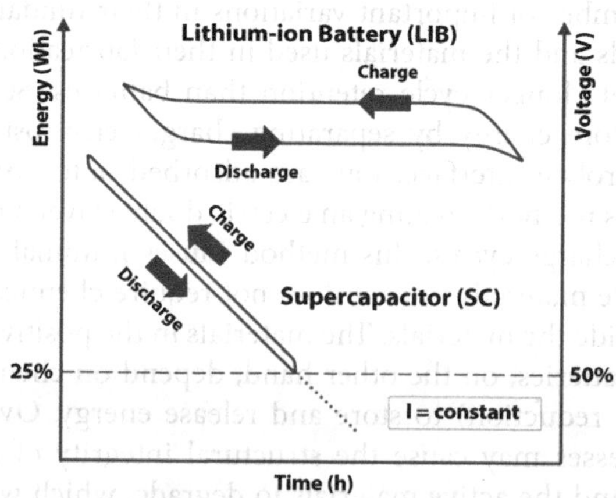


**Figure 9.3** Supercapacitor and battery charging/discharging characteristics comparison. Copyright from Ref. [22].

**Table 9.1** Performance comparison of supercapacitors and batteries. Illustrated from Ref. [22].

| Type | Battery | Supercapacitor |
|---|---|---|
| Energy density (Wh $kg^{-1}$) | 150–250 | 5–10 |
| Power density (W $kg^{-1}$) | 50–2,000 | 100,000 |
| Life cycles (cycles) | 500–2,000 | 500,000–2,500,000 |
| Voltage (V/cell) | 3.6–4.2 | 2.7–3.0 |
| Charging/discharging time | Minutes to hours | Seconds to minutes |

### 9.2.1 Cyclic Stability and Capacitance Retention of MXene-Based Supercapacitors

#### *9.2.1.1 Individual MXene-Based Supercapacitors*

A study on $Ti_3C_2Tx$ MXene-based supercapacitors demonstrated cyclic stability at 5000 cycles with high capacitance retention at a scan rate of 1000 mV $s^{-1}$ as shown in Figure 9.4a [22]. The implement showed a reduced series resistance, indicating its structural stability and ability to withstand repeated cycling without significant degradation in its electrochemical properties, while MXene-based high-temperature supercapacitors exhibited a capacitance retention of 80.3% at 5000 cycles of charging and discharging at conditions of 120°C as shown in Figure 9.4b [23].

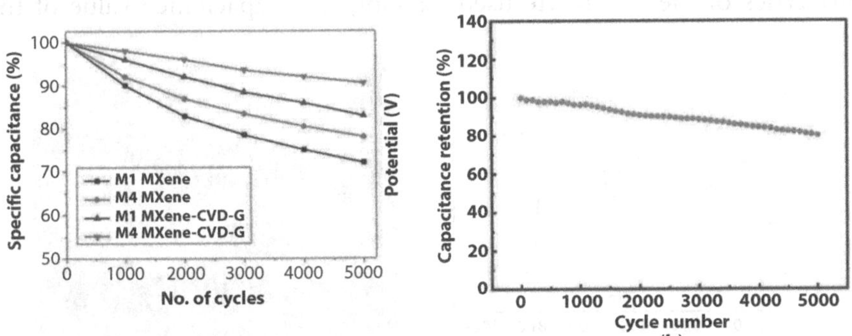


**Figure 9.4** (a) Cyclic stability of Ti3C2Tx MXene-based supercapacitors at a scan rate of 1000 mV $s^{-1}$. Copyright from Ref. [22]. (b) Capacitance retention of high-temperature MXene supercapacitors. Copyright from Ref. [23].

In one study, a hybrid material was developed by coating a $Nb_2CTx$ MXene framework with $Co_3O_4$, forming a robust 2D cross-linked structure [24]. In this innovative configuration, the self-assembly of $Co_3O_4$ within the interlayers of $Nb_2C$ MXene effectively mitigates the tendency of MXene sheets to stack upon one another. The exceptional conductivity and the abundance of active surface groups in MXenes offer numerous active sites for the unvarying dispersion of $Co_3O_4$ on the Nb2C MXene substrate. Notably, the Co-MXene electrode attains a specific capacitance of 1061 F $g^{-1}$ at a 2 A $g^{-1}$ current density, surpassing the performance of both the pure $Co_3O_4$ electrode and the $Nb_2C$ MXene electrode. Furthermore, when integrated into asymmetric supercapacitors (ASCs) as Co-MXene//AC, these devices exhibit specific energy densities of 60.3, 42.5, 31.0, and 19.7 Wh $kg^{-1}$ at specific power densities (Ps) of 0.67, 2.0, 3.4, and 6.8 kW $kg^{-1}$, respectively. Figure 9.5 shows a capacitance retention of the study and the experiment to lights up a small bulb. Impressively, the Co-MXene//AC system showcases a remarkable capacitance retention rate of 93% even after 1000 cycles at 5 A $g^{-1}$. This study gives valuable insights into the fabrication of great performing electrochemical energy storage devices using a practical and effective approach.

In accordance with the study on the performance of $Ti_3C_2Tx$ MXene nanosheets in both alkaline and neutral electrolytes, it is evident that the $Ti_3C_2Tx$ MXene material exhibits superior performance when paired with alkaline electrolytes compared to neutral ones. This distinction becomes apparent when examining the power density and correlation with energy density of $Ti_3C_2Tx$ MXene, as illustrated in Figure 9.6a within the Ragone plot. Energy and power densities are reliant on factors such as the potential window and specific capacitance, which, in turn, are influenced by the properties of the electrolyte used. Notably, the capacitance value of the

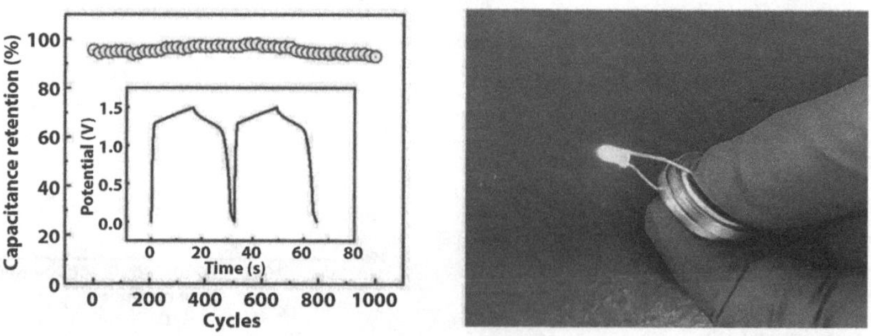


**Figure 9.5** Capacitance retention of the Co-MXene//AC ASCs at 5 A $g^{-1}$; the GCD curse of the 1000 cycles is seen in the inset. An experiment using two Co-MXene//AC ASCs lights up a small lightbulb. Copyright from Ref. [24].

electrode material displays variations between the two electrolytes under identical experimental perimeters, highlighting the impact of dampened ion size on power and energy characteristics. Specifically, the energy density of $Ti_3C_2Tx$ MXene in the alkaline electrolyte decreases from 3.22 to 0.95 Wh $kg^{-1}$, while in the neutral electrolyte, it decreases from 2.63 to 0.71 Wh $kg^{-1}$. Simultaneously, the power density increases from 500.25 to 2501.27 W $kg^{-1}$. These findings underscore the significantly higher power and energy densities achieved when employing the alkaline electrolyte in conjunction with $Ti_3C_2Tx$ MXene. To assess the cyclic stability of $Ti_3C_2Tx$ MXene electrodes, both in the alkaline and in neutral electrolytes, a cycling experiment was conducted at a current density of 10 A $g^{-1}$, spanning 10,000 cycles of GCD measurements as in Figure 9.6b. The specific capacitance exhibited a gradual decline with an increase in the number of cycles. Notably, after 10,000 cycles, $Ti_3C_2Tx$ MXene retained approximately 73% of its initial capacitance in the alkaline electrolyte, while in the neutral

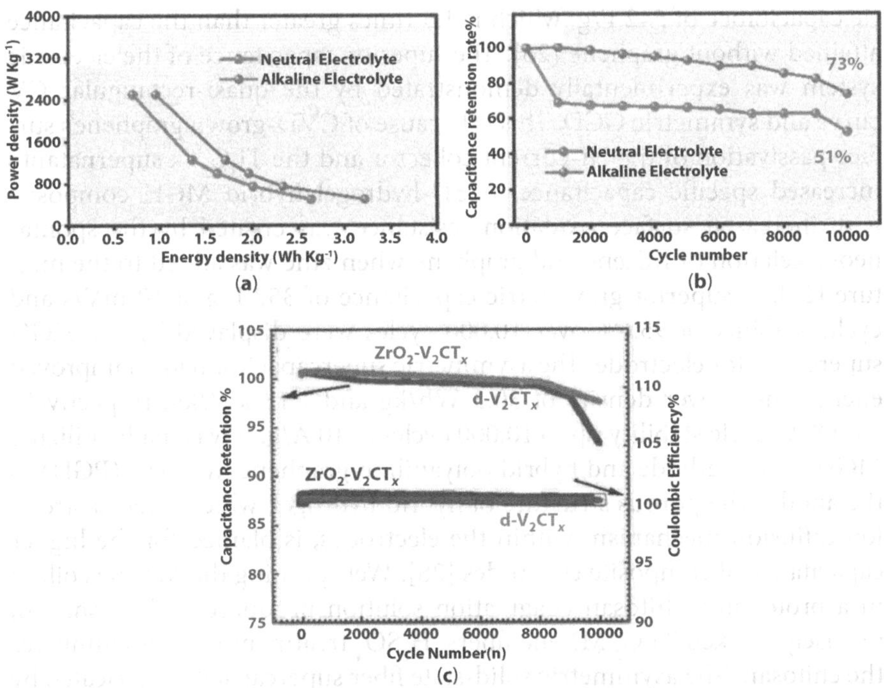


**Figure 9.6** (a) Plot shows the data for both electrolytes' power density and energy density at (2–10 $A^{-1}$) and (b) capacity retention rate for different electrolytes up to 10,000 cycles. Copyright from Ref. [24]. (c) Cyclic test of Zirconia-decorated $V_2CTx$ MXene electrodes for supercapacitors. Copyright from Ref. [25].

electrolyte, it retained approximately 51%. In contrast, other research on zirconia-decorated $V_2CTx$ MXene electrodes for application of supercapacitors demonstrated an impressive 97% capacitance retention at 10,000 cycles, showcasing exceptional cyclic stability as shown in Figure 9.6c. This nonlinear galvanometric behavior further corroborates the structural stability and long-term performance of the composite material.

### *9.2.1.2 MXene-Graphene-Based Supercapacitors*

The volumetric capacitance of the MXene-based fibers produced by wet spinning assembly method employing MXene sheets and graphene oxide liquid crystal was 372.2 F/cm$^2$ and 586.4 F/cm$^3$, respectively, with capacitance retention over 3000 cycles. The electrode for a $Ti_3C_2Tx$ MXene/graphene/Ni supercapacitor was made using CVD-grown graphene on Ni-foil as the current collector and the finest supernatant of an etched $Ti_3C_2Tx$ MXene suspension in HCl/LiF as the electrode material. At a scan rate of 5 mV/s, the manufactured supercapacitor had a higher gravimetric capacitance of 542 F/g, which is 1.5 times greater than the capacitance attained without graphene [26]. The superior capacitance of the electrode system was experimentally demonstrated by the quasi-rectangular CV curve and symmetric GCD. This is because of CVD-grown graphene's surface passivation of the Ni-current collector and the $Ti_3C_2Tx$ supernatant's increased specific capacitance. A 3D-hydrogel hybrid MGH composite with increased surface oxidation resistance was created by the spontaneous gelation of MXene and graphene when zinc was added to the mixture [27]. A superior gravimetric capacitance of 357 F/g at 10 mV/s and cyclic stability of 95.6% over 10,000 cycles were displayed by the MGH supercapacitor electrode. The asymmetric supercapacitor had an improved energy and power density of 30.3 Wh/kg and 1.13 kW/kg, respectively, with 81% cycle stability up to 10,000 cycles at 10 A/g. It was made utilizing MGH as the cathode and hybrid polyaniline-graphene hydrogel (PGH) as the anode. The porous structure of hybrid hydrogel, which offers a strong ion diffusion mechanism within the electrodes, is blamed for the higher capacitance of composite electrodes [28]. Wet spinning the MXene colloid in a protonated chitosan coagulation solution produced ribbon-shaped, precisely stacked $Ti_3C_2$ MXene fibers. $H_2SO_4$ treatment further eliminates the chitosan. The asymmetric solid-state fiber supercapacitor fabricated by using PVA/$H_2SO_4$ gel electrolyte-coated $Ti_3C_2$ MXene fiber as the cathode and rGO fiber as the anode exhibited an enhanced volumetric capacitance of 256 F/cm$^3$ at 5 mV/s, a greater volumetric power density of 7466.0 mW/cm$^3$ at 2583.0 W/kg and volumetric energy density of 58.4 mWh/cm$^3$ at

20.2 Wh/kg with 92.4% capacitance retention over 10,000 cycles at a current density of 10 A/g. They explained the higher conductivity and highly orientated stacking structure of the acid-treated $Ti_3C_2$ MXene, which creates an easier ion diffusion path and improves the access of ion electrolyte to the electrode, as the causes of the increased capacitance. The $Ti_3C_2Tx$ MXene and reduced graphene oxide (rGO) composite aerogel used to make the 3D micro supercapacitor had an outer shell made of self-healing polyurethane, and the device demonstrated a specific capacitance of 34.6 mF/cm$^2$ at a scan rate of 1 mV/s and 91% capacitance retention up to 15,000 cycles with an improved self-healing capacity [29]. The $Ti_3C_2Tx$ MXene/rGO composite demonstrates its suitability as a material for supercapacitor applications by combining the strong metallic conductivity of $Ti_3C_2Tx$ with the wide surface area of rGO. This prevents the restacking of MXene layers and inhibits the oxidation of MXene sheets. The spin coating and dip coating deposition on cotton fabric produced rGO/MXene composite solid-state supercapacitor electrodes that displayed an exceptional specific capacitance of up to 383.3 F/g [30]. Due to the highly elastic nature of the composite, the MXene/rGO composite aerogel electrode produced by multistep reduction, free-casting, and annealing technique exhibited capacitance retention of 97.2% after 1000 cycles. Between the rGO and MXene layers, one study employed a novel method by using $H_2SO_4$ electrolyte, which creates a great transport channel for ion movement [31]. At a scan rate of 10 mV/s, the symmetric supercapacitor electrode made from the RGM-$H_2SO_4$ composite exhibited a superior volumetric capacitance of 454.9F/cm$^3$, and it maintained that volumetric capacitance up to a 180° bending angle of deformation.

### 9.2.1.3 *MXene-CNT-Based Supercapacitors*

According to one study, compared to pure MXene or sheets with randomly mixed MXene/CNT particles, the sandwich-like MXene/CNT composite generated by alternately filtering MXene and CNT dispersions demonstrated enhanced volumetric capacitance of 350 F/cm$^3$ at 5 A/g [32]. The MXene/rGO composite with outstanding cycle stability was also found to have a superior volumetric capacitance of 435 F/cm$^3$, according to their findings. Another study created an asymmetric yarn supercapacitor with 100-mm-thick MXene/CNT anode and $MnO_2$/CNT cathode electrodes, PVA/LiCl gel electrolyte, and an energy density of 100 mWh/cm$^2$ [33]. The CNT intercalation between the MXene layers, which creates a simpler channel for ion diffusion, led to a higher areal capacitance of 61.38 mF/cm$^2$ at 0.5 mA/cm$^2$ in the $Ti_3C_2Tx$ MXene CNT composite electrode that was created by photolithography process and vacuum filtering [34]. In order to create an asymmetric supercapacitor, a composite cathode made of $Ti_3C_2Tx$ MXene-multiwalled CNT ($Ti_3C_2Tx$ MXene-MCNT) and an anode coated with polypyrrole were combined using $Na_2SO_4$ electrolyte. This supercapacitor showed an areal capacitance of 0.94 F/cm$^2$ over the voltage range of 0 to 1.6 V.

#### *9.2.1.4 MXene-Carbon Allotrope-Based Supercapacitors*

According to one study, the combination of MXene with acid activated carbon (AAC) increased interlamellar spacing, inhibited the aggregation of 2D MXene sheets, and considerably enhanced the electrochemical performance of electrodes [35]. The asymmetric supercapacitor produced by using MXene/AAC (2:1) as the positive electrode and AAC as the negative electrode displayed a specific capacitance of 177 F/g at 0.5 A/g with 97.4% retention of cycle life up to 10,000 cycles, according to their research. A higher potential window of up to 1.6 V was reported for the asymmetric micro supercapacitor constructed utilizing MXene as the cathode and activated carbon as the anode in neutral PVA/$Na_2SO_4$ electrolyte [36]. With a surface capacitance of 7.8 mF/cm$^2$, an energy density of 3.5 mWh/cm$^3$, and a power density of 100 mW/cm$^3$ the micro supercapacitor also demonstrated 91.4% capacitance retention after 10,000 cycles. At a current density of 10 A/g after 10,000 cycles, a $Ti_3C_2Tx$ MXene-carbon quantum dot (MXene-CQD) composite electrode demonstrated a superior capacitance of 441.3 F/g at 1 A/g [37]. The increased conductivity and larger surface area of $Ti_3C_2Tx$ MXene sheets are responsible for the increased double-layer capacitance of the MXene-CQD composite electrode, which also prevents layer restacking and stimulates surface functional groups by CQDs. The $Ti_3C_2Tx/NiCO_2S_4$@ CC composite (TNSC), which had a gravimetric capacitance of 2326 F/g at 1 A/g and better cyclic stability of 93.8% at a current density of 10 A/g, was created by spraying $Ti_3C_2Tx$ MXene onto the nickel-cobalt sulfide ($NiCO_2S_4$) @ carbon cloth (CC) substrate [38]. By adjusting the mass loading of $Ti_3C_2Tx$ MXene, which contributes to the electrical conductivity and controls the permeability of ions to nickel-cobalt sulfide, the electrochemical performance of the composite electrode can be changed. The quasi-solid-state flexible supercapacitor was created using nitrogen-doped activated carbon-coated carbon cloth (NACC) as the cathode and TNSC as the anode in a PVA/KOH gel electrolyte. It demonstrated a coulombic efficiency of 99.2% and 90.2% capacitance retention over 5000 cycles. According to a different study, the MXene-carbon fiber composite electrode heterostructure created by the single-step pyrolysis process has a porous structure that facilitates the flow of electrolyte ions into the electrode, greatly increasing capacitance [39]. A material for flexible and portable energy storage applications, the supercapacitor electrode demonstrated better capacitance and exhibited good stability after 2500 cycles of mechanical bending. By using a single-step technique of MXene ink painting and scalable chemical oxidation of N-doped carbon fiber textile (NCFT), Wang *et al.* [40] created asymmetric supercapacitor electrodes. The asymmetric supercapacitor had a high voltage window of 1.6 V and a greater areal energy density of 277.3 m Wh/cm$^2$, with 90% cycling stability after 30,000 cycles at 50 mA/cm$^2$ when it was made using MXene impinged NCFT (MNCFT) as the anode and O/N functionalized NCFT (ONCFT) as the cathode. The MNCFT//ONCFT asymmetric supercapacitor's wearable, foldable, and tailorable characteristics open the door to the creation of textile-based energy storage devices in a variety of sizes and shapes.

### 9.2.1.5 *MXene-TMD-Based Supercapacitors*

According to Rakhi *et al.* [41], the electrochemical performance of $Ti_3C_2Tx$/e-$MnO_2$ nanocomposite electrode in an aqueous electrolyte increased by nearly 3 orders of magnitude compared to pristine MXene electrodes due to the direct chemical synthesis of e$MnO_2$ whiskers on the surface of MXene sheets. A superior electrochemical capacitance was produced by the simple hydrothermal synthesis of the $Ti_3C_2Tx$ MXene/$MoS_2$ nanocomposite by combining the effects of the network topology and metallic conductivity of the $Ti_3C_2Tx$ MXene [42]. The $Ti_3C_2Tx$ MXene metallic matrix's higher electrical conductivity and improved network structure made it easier for ion electrolytes to reach the electrodes, which improved the electrochemical performance. The $Ti_3C_2Tx$ MXene/$MoS_2$ nanocomposite electrode-fabricated symmetric supercapacitor exhibited a specific capacitance of 115.2 F/g at 0.5 A/g, with a coulombic efficiency of 90%. At a scan rate of 1 mV/s, the biocompatible Ta4C3Tx MXene-tantalum oxide composite supercapacitor electrode reported volumetric and gravimetric capacitance of 447 F/cm$^3$ and 194 F/g, respectively, with a capacitance retention of 90% after 10,000 cycles [43].

### 9.2.1.6 *MXene-Metal Oxide-Based Supercapacitors*

Due to the Co-Fe oxide spacer's increased interlayer distance, the composite of Co-Fe oxide and $Ti_3C_2Tx$ MXene demonstrated an enhanced volumetric capacitance of 2467.6 F/cm$^3$ in 1 M LiCl electrolyte. When used as a symmetric supercapacitor electrode, the composite displayed an areal capacitance of 356.4 mF/cm$^2$ with a capacitance retention of 88.2% after 10,000 cycles [44]. The composite electrode maintained its electrochemical energy storage stability after 100 cycles of mechanical bending, making it an appropriate material for the creation of flexible energy storage materials. The easy hydrothermal approach was used to create a $Co_3O_4$ nanoparticles-MXene (Co-MXene) composite that prevented restacking of MXene layers and reported a superior gravimetric capacitance of 1081 F/g at 0.5 A/g, which is significantly greater than the capacitance of pure MXene electrode [45]. Co-MXene@Nickel foam was used as the anode and polyaniline-derived carbon@carbon fiber paper (PANIC@CFP) as the cathode in a 6 M KOH electrolyte with a microporous polypropylene separator between them to create an asymmetric supercapacitor. The gravimetric capacitance of the asymmetric supercapacitor was 95.71 F/g at 1 A/g, and it had a greater voltage window of 1.4 V. The zinc ion hybrid micro supercapacitor was constructed using battery-type vanadium pentoxide as the cathode and capacitor type $Ti_3C_2Tx$ MXene as the anode in $ZnSO_4$ dissolved PAM hydrogel electrolyte, and it showed an areal capacitance of 129 mF/cm$^2$ at a current density of 0.34 mA/cm$^2$ and an energy density of 48.9 mWh/cm$^2$ at 673 mW Additionally, they stated that PAM hydrogel electrolytes, as opposed to organic electrolytes, might reduce the security risk and increase the cycle stability of the supercapacitor electrodes [46].

### *9.2.1.7 MXene-Polymer-Based Supercapacitors*

Rearranging MXene layers impairs ion transport even though MXenes have demonstrated increased uses in numerous electrochemical energy storage systems. Additionally, it has been found that when electrode thickness increases, the electrochemical performance of MXene-based electrodes declines. By creating composites of MXenes with polymers, which will greatly improve the flexibility and electrochemical performance, these limitations of MXene electrodes can be eliminated [47]. Due to their higher aspect ratio than multilayered MXene sheets, single-layered MXene sheets have been found to have more accessible surface hydrophilicity and compatibility with polymers [48]. To improve the electrochemical performance of multilayered MXenes, they are delaminated before being combined to create composites. According to Yan *et al.* [49], the electrochemical deposit of conducting polypyrrole on the MXene textile electrode increased the specific capacitance up to 343.20 F/g in comparison to the pristine MXene electrode, which had a specific capacitance of 182.70 F/g. They added that the energy density of the MXene/PPy electrode-fabricated solid-state symmetrical supercapacitor was 1.30 mWh/g [49]. The volumetric capacitance of the composite supercapacitor electrode was increased to 1000 F/cm$^3$ with 92% capacitance retention after 25,000 cycles thanks to the sandwiching of $Ti_3C_2Tx$ MXene layers with PPy chain [50]. The aligned arrangement of $Ti_3C_2Tx$ MXene/PPy layers, which provides short ion diffusion paths leading to faster reversible redox reactions and greater electronic conductivity, is credited with the high capacitance. The intercalation of PPy into $Ti_3C_2$ MXene layers, as seen by Zhu *et al.*, increased the volumetric capacitance up to 406 F/cm$^3$, and the manufactured supercapacitor had a superior capacitance of 35 mF/cm$^2$ with a cycle life of 10,000 [51]. The higher capacitance of the electrodes can be attributed to the increased carrier mobility caused by the tight bonding between the PPy and MXene surfaces. In their experiment, Zhang *et al.* found that the dispersion of 1D PPy nanofibers and 2D MXene nanosheets in the 3D PVA matrix increased the composite hydrogel's tensile strength to 10.3 MPa and demonstrated a specific capacitance of 614 F/g at 1 A/g with 100% capacitance retention up to 10,000 cycles [52]. $Ti_3C_2Tx$ MXene layers were introduced with PPy particles and ionic liquid-based microemulsions (IL-mic) as a dual spacer, which greatly improved the electrochemical performance of the composite supercapacitor electrodes [53]. The ionic liquid microemulsions improved the wettability of electrodes, increasing their surface area that is accessible to ions, and the PPy particles prevented the restacking of MXene layers. With a capacitance retention rate of 91% over 2000 cycles and a coulombic efficiency of 91% between the temperature range of 4 C and 50 C, the symmetric supercapacitor made from the PPy-MXene-IL-mic composite exhibited an energy density of 31.2 Wh kg$^{-1}$ at RT. Figure 9.7 depicts the cycling stability

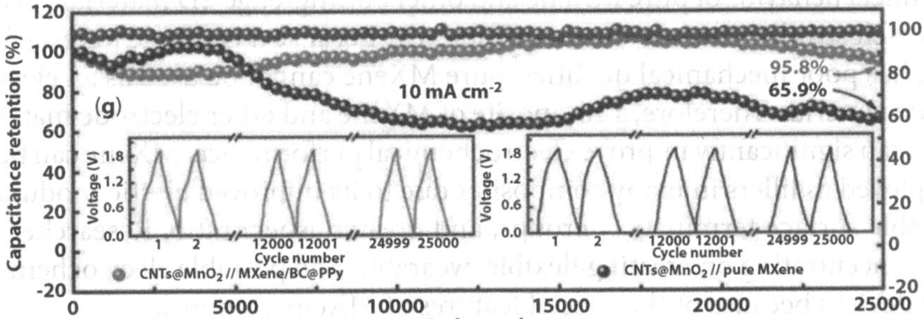


**Figure 9.7** Cycle life of ZHMSC. Copyright from Ref. [54].

of a MXene-Zinc hybrid electrode. In the 2 M $Zn(CF_3SO_3)_2$–0.1 M $MnSO_4$/PAM hydrogel, one study synthesized a zinc hybrid micro supercapacitor (ZHMSC) employing a capacitor-type MXene/bacterial cellulose (BC) @ polypyrrole (PPy) electrode and a battery-type CNTs@$MnO_2$ electrode [54]. ZHMSC reported an areal capacitance of 290 mF/cm$^2$ at 0.75 mA/cm$^2$, a higher energy density of 145.4 mWh/cm$^2$, a stronger power density of 0.36 mW/cm$^2$, and 95.8% cycle stability over 25,000 cycles. When conducting PPy polymer is added to virgin MXene-based ZHMSC, the areal capacitance, energy density, and power density all significantly rise. The retention of electrochemical performance over tensile deformation up to 400% was also found by the scientists, making it an appropriate material for the creation of flexible supercapacitors.

#### *9.2.1.8 Comparison with Other 2D Materials*

The layered structure of 2D materials ensures a higher surface-to-volume ratio, increasing the number of electrolytes exposed to the electrode's surface. Due to their poorer conductivity, weak van der Waals bonding, and hydrophobicity, these materials nevertheless displayed limits. Other 2D materials were looked at to produce better capacitance as a substitute. This led to the discovery of the 2D material MXene, which has many electrochemically active sites, a large surface-to-volume ratio, and interlayer structure for energy storage applications. In MXene-based supercapacitors, the large variety of oxidation states offered by transition metal atoms confers an effective faradic charge storage mechanism. MXenes store numerous charges and deliver them at a considerably faster rate than other pseudocapacitive materials. Instead, MXenes are an intriguing 2D material for electrochemical energy storage applications due to their outstanding conductivity and hydrophilicity. The comparison of the electrochemical behavior of pure MXene and other cutting-edge 2D materials in a three-electrode system is shown in Table 9.2. Because it is simple to restack and has poor mechanical qualities, pure MXene cannot be used as an electrode material. Therefore, a composite of MXene and other electrode materials can significantly improve electrochemical performance. MXene can be employed as fillers in many composites due to its improved elastic moduli, tunable surface termination groups, and greater aspect ratio. Researchers are concentrating on creating flexible, wearable, and portable electrochemical devices because of the special features of MXene composite.

The improvement in electrochemical behavior of MXene composite-based supercapacitors is reflected in this section's description of the

**Table 9.2** The evaluation of other cutting-edge 2D materials and pure MXene's electrochemical behavior. Copyright 2022, Elsevier from Ref. [55].

| Type | Initial capacitance ($F g^{-1}$) | Cycle life | Ref. |
|---|---|---|---|
| TMDs | 100 | 92.6%<br>2000 cycles | [56] |
| Graphene | 130 | 99%<br>4000 cycles | [57] |
| Carbon nitride | 185 | 87.2%<br>5000 cycles | [58] |
| Black phosphorene | 80 | 80%<br>15,000 cycles | [59] |
| MXene | 245 | 83%<br>10,000 cycles | [60] |

development of MXene-based supercapacitors. The electrochemical behavior of pure MXene and other cutting-edge 2D materials in a three-electrode setup is compared in Table 9.2 [55]. Because it is simple to restack and has poor mechanical qualities, pure MXene cannot be used as an electrode material. Therefore, a composite of MXene and other electrode materials can significantly improve electrochemical performance. MXene can be employed as fillers in many composites due to its improved elastic moduli, tunable surface termination groups, and greater aspect ratio. Researchers are concentrating on creating flexible, wearable, and portable electrochemical devices as a result of the special features of MXene composite.

Several factors can affect the cyclic stability and capacitance retention of supercapacitors, which is an option of electrode material that contributes to a crucial role in determining the cyclic stability and capacitance retention of supercapacitors [61]. For example, graphene-based supercapacitors exhibit good specific capacitance and great cycle stability but are prone to structural defects. MXene materials, on the other hand, have shown positive results in terms of cyclic stability and capacitance retention.

The type and composition of the electrolyte used in supercapacitors can significantly impact their cyclic stability and capacitance retention. The electrolyte should provide good ion conductivity, stability, and compatibility with the electrode materials to ensure long-term performance [62]. Aqueous electrolytes, such as sulfuric acid ($H_2SO_4$) and potassium hydroxide (KOH), are frequently used in supercapacitors due to their affordable price and high ionic conductivity. These electrolytes have been studied for their effects on cyclic stability and capacitance retention. They have a small voltage window, though, and can damage the electrodes, which causes poor cycle stability and capacitance retention. A broader voltage window is provided by organic electrolytes such acetonitrile ($CH_3CN$) and propylene carbonate (PC) compared to aqueous electrolytes. Supercapacitors can benefit from them in terms of cycle stability and capacitance retention, but they cost more and have less ionic conductivity. Ionic liquids, such as 1-ethyl-3-methylimidazolium tetrafluoroborate ($EMIBF_4$) and 1-ethyl-3-methylimidazolium bis(trifluoromethylesulfonyl) imide (EMITFSI), have been used as co-salts in electrolytes to improve the ionic conductivity and stability of supercapacitors. They can enhance the cyclic stability and capacitance retention by inhibiting electrolyte decomposition and increasing the electrochemical stability potential windows (ESPWs). Redox electrolytes, such as ferrocene/ferricenium (Fc/Fc+) and quinone/hydroquinone ($Q/QH_2$), can provide additional capacitance through faradaic reactions. However, they can also lead to electrode degradation and lower cyclic stability and capacitance retention. The potential of polymer electrolytes, such as polyvinyl alcohol (PVA) and polyacrylonitrile (PAN), to increase the cyclic stability and capacitance retention of supercapacitors has been investigated [63]. They may provide improved electrode compatibility and lessen the chance of electrode deterioration.

The structure and morphology of the electrode, particularly its particular surface area and pore size distribution, may also have an effect on the supercapacitors' cyclic stability and capacitance retention [64]. A higher active surface area for ion adsorption and desorption can be provided by well-designed electrode architectures, improving performance. The cyclic stability and capacitance retention of supercapacitors can be affected by the speed at which ions can flow inside the electrode material and electrolyte interface. Performance can be improved via faster charge/discharge rates and ion diffusion kinetics. Finally, the working circumstances, including the temperature and applied voltage range, can also affect the cyclic stability and capacitance retention of supercapacitors. To guarantee the devices' long-term performance and dependability, it is crucial to optimize these characteristics. Table 9.3 shows some recent studies on the electrochemical performance of MXene-based electrodes used in supercapacitors [65]. MXenes are potential prospects for upcoming energy storage systems due to their superior cycle stability and capacitance retention in supercapacitor applications. Researchers continue to explore various strategies to intensify

**Table 9.3** Supercapacitor MXene-based electrodes and their electrolyte used. Copyright from Ref. [36].

| Electrode | Synthesis approach | Electrolyte | Capacitance retention | Ref. |
|---|---|---|---|---|
| **Asymmetric supercapacitors** | | | | |
| HS–NCS/$Ti_3C_2Tx$ | HF-etching/ Hydrothermal | 2 M KOH | 92% for 20,000 cycles | [66] |
| $Ti_3C_2Tx$ | Etching route | 1.5 M $ZnSO_4$ | 83.58% after 5000 cycles | [67] |
| $Ti_3C_2Tx$ | Layer-by-layer assembly | CN-PC-PMMA $LiCF_3SO_3$ gel electrolyte | 90% after 200 bending cycles | [68] |
| **Symmetric supercapacitors** | | | | |
| $Nb_2C/Ti_3C_2$ | Chemical etching | 1 M PVA/$H_2SO_4$ | 98% for 50,000 cycles | [69] |
| Mo1.33 CTz | HF-etching | 19.5 M LiCl | 82.1% for 50,000 cycles | [70] |
| $Ti_3C_2Tx$ film | *In situ* etching | $H_2SO_4$ | 88.5% after 5000 cycles | [71] |

the performance and durability of MXene-based supercapacitors, paving the way for their widespread adoption in various industries.

## 9.3 Challenges, Limitations, and Future Prospects of MXene-Based Energy Storage Devices

MXene materials have garnered substantial recognition in the field of battery technology due to their unique properties, which include high conductivity, large surface area, and tunable surface chemistry. However, as with any emerging technology, there are challenges to overcome and exciting prospects to explore when considering the integration of MXene electrodes in batteries. The volume expansion that occurs during lithiation/delithiation cycles is one of the main difficulties when employing MXene electrodes in batteries. This growth may result in electrode deterioration and shortened cycle life [72]. Developing strategies to mitigate volume changes is crucial. Another challenge is the large-scale production of high-quality MXene materials. To realize the full potential of MXenes in commercial batteries, scalable and cost-effective synthesis methods need to be developed. Concerns about sustainability are developing [73]. MXene synthesis is often less hazardous to the environment than some other compounds, but it is still important to handle any environmental effects, particularly when mass production is involved.

It is crucial for consumer applications to guarantee the safety of MXene-based batteries, particularly regarding heat stability and potential toxicity. Due to their distinctive characteristics like high electrical conductivity, high specific surface area, and improved hydrophilicity, MXene-based materials have demonstrated considerable potential for energy storage applications, including batteries and supercapacitors. When compared to conventional battery materials, MXene-based electrodes frequently have lower energy densities, which restricts their usefulness in high-energy-density devices [74–76]. Future research should focus on improving the energy storage capacity of MXene materials while maintaining their other advantageous properties. During charge and discharge cycles, MXenes can experience structural changes that eventually degrade their performance. Scalability and affordability MXene materials are currently expensive and difficult to produce on a sufficient scale for use in practical energy storage systems. MXene-based materials show considerable promise for future energy storage applications despite these difficulties. To overcome these difficulties and fully realize the promise of MXene electrodes in batteries and other energy storage devices, more research and development is required.

Significant work has been made in recent years in overcoming the issues and investigating the potential uses of MXene electrodes in energy storage systems. Research is ongoing in the progression of MXene-based and hybrid electrode materials with improved performance and stability. These hybrid structures can improve total energy storage capacitance by combining the benefits of MXene materials with those of other materials, such as carbon nanotubes or metal oxides. The design of MXene structures that enable quicker ion movement and the introduction of ion-conductive additives in MXene composites are two ways that have been investigated to improve the ion diffusion and transport capabilities of MXene materials. These advancements open the door to energy storage systems with higher charge/discharge rates and overall performance. MXene-based materials have also been researched for their potential in supercapacitors and flexible batteries, two different categories of wearable energy storage gadgets. MXene are attractive candidates for the next generation of wearable electronics due to their unique characteristics and mechanical flexibility. Compared to conventional energy storage materials, MXene materials are thought to be more environmentally benign since they may be made from readily available, non-toxic precursors. This feature makes MXene-based electrodes appealing for upcoming energy storage applications, as does their excellent performance. Another possibility is MXene materials, which have good conductivity and are hence perfect for fast-charging applications. Future batteries with MXene electrodes might have quick charging periods, which is important for portable gadgets and electric cars.

Furthermore, efforts to advance green synthesis techniques for MXenes are continuing. Future prospects include refining these processes to minimize environmental impact and promote sustainability. Finally, combining MXene-based batteries with other energy storage technologies, such as supercapacitors or fuel cells, may lead to hybrid energy storage systems with improved performance and versatility. While MXene electrodes in batteries face challenges related to volume expansion, synthesis scalability, and environmental concerns, they hold tremendous promise for advancing energy storage technology. The prospects of MXene-based batteries include enhanced energy density, fast-charging capabilities, multifunctionality, flexibility, and sustainability, paving the way for innovative and eco-friendly energy solutions. To overcome obstacles and realize the full potential of MXene materials in the battery sector, ongoing research and development activities are needed.

Batteries, supercapacitors, and fuel cells are examples of energy storage technologies that play crucial roles in a variety of applications, from storing renewable energy to powering portable electronics. Nonetheless, these devices confront various challenges and limitations. Batteries, for instance, grapple with a restricted cycle life, frequently experiencing capacity fading and performance deterioration through multiple charge and discharge cycles, thus curtailing their overall lifespan. Furthermore, batteries often exhibit slow charging and discharging rates, limiting their capacity to deliver high-power outputs promptly. Moreover, environmental concerns surround certain battery chemistries, such as lead-acid and nickel-cadmium, due to their inclusion of toxic materials that can pose environmental hazards when not appropriately disposed of. Addressing these challenges is essential for the continued advancement of energy storage technologies.

Supercapacitors, while offering unique advantages in terms of rapid energy delivery, confront their own set of limitations. One prominent constraint is their comparatively low energy density, signifying that they can store less energy per unit mass or volume when compared to batteries. Additionally, supercapacitors exhibit a limited energy storage capacitance, making them better suited for applications requiring high-power bursts rather than long-term energy storage. Supercapacitors can be more expensive than conventional batteries, which may limit their broad use in some applications. Cost is another aspect to consider. Maximizing the potential of supercapacitors for diverse energy storage requirements requires striking a balance between these characteristics and overcoming these limitations.

Although they show great potential as clean energy converters, fuel cells still face a number of issues that must be resolved before they can be widely adopted. The primary issue is the high cost of making fuel cells, which can make them less commercially viable and less widely used. In addition, fuel cells require a consistent supply of fuel, such as hydrogen, which might present issues with distribution and storage. Additionally, certain fuel cell types have relatively long start-up times, which makes them unsuitable for applications that require instantaneous power output. Environmental concerns also loom large, particularly in the case of hydrogen production, a common fuel for fuel cells, which can result in greenhouse gas emissions if not sourced from renewable and sustainable production methods. Addressing these challenges is pivotal to unlocking the full potential of fuel cells as a sustainable energy solution. Energy storage devices are still necessary for many applications despite these difficulties, and continuous research and development efforts are aimed at overcoming these constraints and enhancing their performance, efficiency, and sustainability.

## 9.4 Potential Future Directions

The versatile MXene family of materials has captivated researchers and engineers alike with their exceptional properties. As we peer into the future, several exciting research directions and diverse applications of MXenes are emerging, showcasing their transformative potential. Researchers are poised to explore a wider array of MXene compositions, each designed for specific applications. For instance, tailoring MXene materials for enhanced catalytic activity could revolutionize the field of clean energy production, making water-splitting more efficient and sustainable. Besides, sustainable and environmentally friendly synthesis methods for MXenes are on the horizon. The application of green chemistry principles may lead to the development of MXene materials with reduced environmental impact. For example, MXene production from biowaste materials holds promise as a sustainable approach. MXenes have already made strides in energy storage applications. Future research aims to optimize MXene-based electrodes for batteries and supercapacitors further. For instance, graphene-coated MXene electrodes may lead to great performance lithium-ion batteries with faster charging and longer cycle life. They also are poised to become multifunctional materials. Imagine wearable health monitors featuring flexible, lightweight MXene sensors that simultaneously monitor vital signs and environmental factors, improving healthcare and safety. MXenes are set to play a pivotal role in advanced electronics. Researchers are exploring their use in next-generation transistors and nanoelectronics devices. For example, MXene-based field-effect transistors (FETs) may enable faster and more energy-efficient electronics.

MXenes are venturing into biomedical applications. The future may see MXene-based drug delivery systems that precisely target cancer cells, minimizing side effects. MXene bioelectrodes could revolutionize neuroprosthetic devices, restoring lost sensory and motor functions. Furthermore, lightweight and robust MXenes are ideal for aerospace applications. Future aircraft may incorporate MXene-based composites, enhancing structural integrity and fuel efficiency. The research and implementation of MXene materials in supercapacitors continues to be a dynamic and promising field with exciting future prospects. Future research will likely focus on fine-tuning the microstructures and morphologies of MXene-based electrodes. This includes exploring nanostructuring techniques and novel electrode architectures to enhance surface area and ion accessibility, ultimately boosting capacitance and energy density.

Sustainable and eco-friendly production methods for MXenes are gaining importance. Future research will likely focus on reducing the environmental footprint of MXene synthesis and supercapacitor operation, aligning with the global push for greener technologies. As MXene-based supercapacitors move towards practical applications, achieving scalability in both material production and device manufacturing will be essential. This includes optimizing electrode fabrication techniques for mass production. In conclusion, these future directions underscore the remarkable versatility and transformative potential of MXene materials. With ongoing research and innovative applications, MXenes are poised to reshape industries, enhance sustainability, and empower a new era of technological advancements, benefiting society and the environment alike. MXenes are not just materials of the present; they are the building blocks of a brighter future. Researchers, scientists, and engineers are actively working to prevail over existing challenges and unlock the vast probable of MXene materials for high-performance energy storage solutions. As these efforts continue, MXene-based supercapacitors are poised to make significant contributions to an extensive application, from portable electronics to sustainable energy grids, shaping the landscape of energy storage technology for years to come.

## 9.5 Conclusions

The cyclic stability and capacitance retention of MXene-based supercapacitors represent an intriguing research area with significant ramifications in the field of energy storage. It has become clear through several examinations and studies that MXene materials have extraordinary promise for resolving the major issues that face contemporary energy storage systems. The cyclic stability of MXene-based supercapacitors demonstrates their appropriateness for applications requiring long-term reliability as demonstrated by their ability to sustain high-performance features across a significant number of charge–discharge cycles. MXene-based supercapacitors are rapidly improving thanks to the combined efforts of academics and scientists around the world, despite the fact that there are still opportunities and obstacles in this field. These energy storage technologies are poised for game-changing innovations by tackling critical issues with electrode engineering, electrolyte optimization, and scalability. As we move forward, the convergence of materials science, electrochemistry, and engineering will continue to push the boundaries of cyclic stability and capacitance retention in MXene-based supercapacitors. With continued dedication and innovation, MXene-based supercapacitors are poised to participate in a pivotal role in shaping the landscape of energy storage technology, paving the way for a brighter and more sustainable tomorrow.

## Acknowledgments

The authors are very grateful to Shibaura Institute of Technology, Japan and Universiti Teknikal Malaysia Melaka, Malaysia for all technical/non-technical supports and special thanks to the Ministry of Higher Education for financially supporting this research under the Fundamental Research Grant Scheme (FRGS) research grant numbered FRGS/1/2021/TK0/UTEM/01/3/F00476.

## References

1. Armano, A. and Agnello, S., Two-Dimensional Carbon: A Review of Synthesis Methods, and Electronic, Optical, and Vibrational Properties of Single-Layer Graphene. *C — J. Carbon Res.*, 5, 4, 67, Nov. 2019, doi: https://doi.org/10.3390/c5040067.

2. Azam, M.A., Seman, R.N.A.R., Mohamed, M.A., Ani, M.H., Effect of Polytetrafluoroethylene Binder Content on Gravimetric Capacitance and Life Cycle Stability of Graphene Supercapacitor. *Int. J. Automot. Mech. Eng.*, 19, 3, 9964–9970, Sep. 2022, doi: https://doi.org/10.15282/ijame.19.3.2022.08.0768.
3. Safie, N.E. and Azam, M.A., Understanding the structural properties of feasible chemically reduced graphene. *AIMS Mater. Sci.*, 9, 4, 617–627, Jan. 2022, doi: https://doi.org/10.3934/matersci.2022037.
4. Aziz, M.F. and Azam, M.A., *MXene-Based 2D Nanomaterials for Solar Cells*, pp. 159–171, CRC Press eBooks, Apr. 2022, doi: https://doi.org/10.1201/9781003178422-10.
5. Kosnan, M.A., Azam, M.A., Safie, N.E., Munawar, R.F., Takasaki, A., Recent Progress of Electrode Architecture for MXene/MoS2 Supercapacitor: Preparation Methods and Characterizations. *Micromachines*, 13, 11, 1837, Nov. 2022, doi: https://doi.org/10.3390/mi13111837.
6. Ebrahimi, M. and Mei, C., Optoelectronic properties of Ti3C2Tx MXene transparent conductive electrodes: Microwave synthesis of parent MAX phase. *Ceram. Int.*, 46, 18, 28114–28119, Dec. 2020, doi: https://doi.org/10.1016/j.ceramint.2020.07.307.
7. Ronchi, R.M., Arantes, J.T., Santos, S.F., Synthesis, structure, properties and applications of MXenes: Current status and perspectives. *Ceram. Int.*, 45, 15, 18167–18188, Oct. 2019, doi: https://doi.org/10.1016/j.ceramint.2019.06.114.
8. Das, P. and Wu, Z., MXene for energy storage: present status and future perspectives. *J. Phys. Energy*, 2, 3, 032004–032004, Jul. 2020, doi: https://doi.org/10.1088/2515-7655/ab9b1d.
9. Wang, Y. and Wang, Y., Recent progress in MXene layers materials for supercapacitors: High-performance electrodes. *Smart Mat.*, 4, 1, e1130, Jun. 2022, doi: https://doi.org/10.1002/smm2.1130.
10. Anasori, B. and Gogotsi, Y., MXenes: trends, growth, and future directions. *Graphene 2D Mater.*, 7, 3–4, 75–79, Nov. 2022, doi: https://doi.org/10.1007/s41127-022-00053-z.
11. Jiang, L. *et al.*, 2D single- and few-layered MXenes: synthesis, applications and perspectives. *J. Mater. Chem. A*, 10, 26, 13651–13672, Jan. 2022, doi: https://doi.org/10.1039/d2ta01572b.
12. Wang, H. *et al.*, Solid solution reinforced V3CrC3Tx MXene cathodes for Zn-ion micro-supercapacitors with high areal energy density and superior flexibility. *J. Mater. Chem. A*, 10, 39, 20953–20963, Jan. 2022. doi: https://doi.org/10.1039/d2ta04747k.
13. Li, K. *et al.*, An Ultrafast Conducting Polymer@MXene Positive Electrode with High Volumetric Capacitance for Advanced Asymmetric Supercapacitors. *Small*, 16, 4, 1906851–1906851, Dec. 2019, doi: https://doi.org/10.1002/smll.201906851.
14. Gohar, R.S., Ahmad, I., Shah, A., Majeed, S., Najam-ul-Haq, M., Ashiq, M.N., Fabrication of transition-metal oxide and chalcogenide nanostructures with

enhanced electrochemical performances. *J. Energy Storage*, 31, 101621–101621, Oct. 2020. doi: https://doi.org/10.1016/j.est.2020.101621.
15. Hussain, S. *et al.*, Ultrasonically derived WSe2 nanostructure embedded MXene hybrid composites for supercapacitors and hydrogen evolution reactions. *Renew. Energy*, 185, 585–597, 2021, doi: https://doi.org/10.1016/j.renene.2021.12.065.
16. Xu, L. *et al.*, 3D frame-like architecture of N-C-incorporated mixed metal phosphide boosting ultrahigh energy density pouch-type supercapacitors. *Nano Energy*, 91, 106630–106630, Jan. 2022, doi: https://doi.org/10.1016/j.nanoen.2021.106630.
17. Mateen, A. *et al.*, *In Situ* Nitrogen Functionalization of 2D-Ti3C2Tx-MXenes for High-Performance Zn-Ion Supercapacitor. *Molecules*, 27, 21, 7446–7446, Nov. 2022, doi: https://doi.org/10.3390/molecules27217446.
18. Guo, Q., Zhao, X., Li, Z., Wang, D., Nie, G., A novel solid-state electrochromic supercapacitor with high energy storage capacity and cycle stability based on poly(5-formylindole)/WO3 honeycombed porous nanocomposites. *Chem. Eng. J.*, 384, 123370–123370, Mar. 2020, doi: https://doi.org/10.1016/j.cej.2019.123370.
19. Fan, Z. *et al.*, Modified MXene/Holey Graphene Films for Advanced Supercapacitor Electrodes with Superior Energy Storage. *Adv. Sci.*, 5, 10, 1800750–1800750, Aug. 2018, doi: https://doi.org/10.1002/advs.201800750.
20. Yang, Y., Han, Y., Jiang, W., Zhang, Y., Xu, Y., Ahmed, A., Application of the Supercapacitor for Energy Storage in China: Role and Strategy. *Appl. Sci.*, 12, 1, 354–354, Dec. 2021, doi: https://doi.org/10.3390/app12010354.
21. Benoy, S.M., Pandey, M., Bhattacharjya, D., Saikia, B.K., Recent trends in supercapacitor-battery hybrid energy storage devices based on carbon materials. *J. Energy Storage*, 52, 104938–104938, Aug. 2022, doi: https://doi.org/10.1016/j.est.2022.104938.
22. Kim, K.-H., An, J., Park, K., Roh, G., Chun, K., Analysis of a Supercapacitor/Battery Hybrid Power System for a Bulk Carrier. *Appl. Sci.*, 9, 8, 1547–1547, Apr. 2019, doi: https://doi.org/10.3390/app9081547.
23. Li, C. *et al.*, High-Temperature Supercapacitors Based on MXene with Ultrahigh Volumetric Capacitance. *ACS Mater. Lett.*, 5, 8, 2084–2095, 2023, https://pubs.acs.org/doi/10.1021/acsmaterialslett.3c00223.
24. Murugesan, R.A. and Chandar, N.K., Capacitance performance of Ti3C2Tx MXene nanosheets on alkaline and neutral electrolytes. *Mater. Res. Bull.*, 163, 112217–112217, Jul. 2023, doi: https://doi.org/10.1016/j.materresbull.2023.112217.
25. Zahra, S.A., Ceesay, E.K., Rizwan, S., Zirconia-decorated V2CT MXene electrodes for supercapacitors. *J. Energy Storage*, 55, 105721–105721, Nov. 2022, doi: https://doi.org/10.1016/j.est.2022.105721.
26. Kumar, S. *et al.*, Supercapacitors based on Ti3C2Tx MXene extracted from supernatant and current collectors passivated by CVD-graphene. *Sci. Rep.*, 11, 1, 649–649, Jan. 2021, doi: https://doi.org/10.1038/s41598-020-80799-9.

27. Sikdar, A. *et al.*, Spontaneous three-dimensional self-assembly of MXene and graphene for impressive energy and rate performance pseudocapacitors. *Electrochim. Acta*, 391, 138959–138959, Sep. 2021, doi: https://doi.org/10.1016/j.electacta.2021.138959.
28. Zhu, C. and Geng, F., Macroscopic MXene ribbon with oriented sheet stacking for high-performance flexible supercapacitors. *Carbon Energy*, 3, 1, 142–152, Jul. 2020, doi: https://doi.org/10.1002/cey2.65.
29. Yue, Y. *et al.*, Highly Self-Healable 3D Microsupercapacitor with MXene-Graphene Composite Aerogel. *ACS Nano*, 12, 5, 4224–4232, 2018, https://pubs.acs.org/doi/10.1021/acsnano.7b07528.
30. Zheng, X. *et al.*, Multifunctional RGO/Ti3C2Tx MXene fabrics for electrochemical energy storage, electromagnetic interference shielding, electrothermal and human motion detection. *Mater. Des.*, 200, 109442–109442, Feb. 2021, doi: https://doi.org/10.1016/j.matdes.2020.109442.
31. Zhang, M. *et al.*, Electrolyte-mediated dense integration of graphene-MXene films for high volumetric capacitance flexible supercapacitors. *Nano Res.*, 14, 3, 699–706, Nov. 2020, doi: https://doi.org/10.1007/s12274-020-3100-6.
32. Zhao, M. *et al.*, Flexible MXene/Carbon Nanotube Composite Paper with High Volumetric Capacitance. *Adv. Mater.*, 27, 2, 339–345, Nov. 2014, doi: https://doi.org/10.1002/adma.201404140.
33. Park, J.W. *et al.*, Highly loaded MXene/carbon nanotube yarn electrodes for improved asymmetric supercapacitor performance. *MRS Commun.*, 9, 1, 114–121, Mar. 2019, doi: https://doi.org/10.1557/mrc.2019.7.
34. Wang, R. *et al.*, MXene-carbon nanotubes layer-by-layer assembly based on-chip micro-supercapacitor with improved capacitive performance. *Electrochim. Acta*, 386, 138420–138420, Aug. 2021, doi: https://doi.org/10.1016/j.electacta.2021.138420.
35. Li, Y., Pan, C., Kamdem, P., Jin, X.J., Binder-Free Two-Dimensional MXene/Acid Activated Carbon for High-Performance Supercapacitors and Methylene Blue Adsorption. *Energy Fuels*, 34, 8, 10120–10130, 2020, https://pubs.acs.org/doi/10.1021/acs.energyfuels.0c01352.
36. Xie, Y. *et al.*, High-voltage asymmetric MXene-based on-chip micro-supercapacitors. *Nano Energy*, 74, 104928–104928, Aug. 2020. doi: https://doi.org/10.1016/j.nanoen.2020.104928.
37. Wang, Y. *et al.*, Stimulation of surface terminating group by carbon quantum dots for improving pseudocapacitance of Ti3C2Tx MXene based electrode. *Carbon*, 180, 118–126, Aug. 2021, doi: https://doi.org/10.1016/j.carbon.2021.04.089.
38. Li, L. *et al.*, Hierarchically layered nanocomposite electrodes formed by spray-injected MXene nanosheets for ultrahigh-performance flexible supercapacitors. *Appl. Surf. Sci.*, 549, 149226–149226, May 2021, doi: https://doi.org/10.1016/j.apsusc.2021.149226.
39. Sun, L., Fu, Q., Pan, C., Hierarchical porous 'skin/skeleton'-like MXene/biomass derived carbon fibers heterostructure for self-supporting, flexible

all solid-state supercapacitors. *J. Hazard. Mater.*, 410, 124565–124565, May 2021, doi: https://doi.org/10.1016/j.jhazmat.2020.124565.

40. Wang, Y. *et al.*, A High-Performance, Tailorable, Wearable, and Foldable Solid-State Supercapacitor Enabled by Arranging Pseudocapacitive Groups and MXene Flakes on Textile Electrode Surface. *Adv. Funct. Mater.*, 31, 7, 2008185–2008185, Nov. 2020, doi: https://doi.org/10.1002/adfm.202008185.
41. Rakhi, R.B., Ahmed, B., Anjum, D., Alshareef, H.N., Direct Chemical Synthesis of MnO2 Nanowhiskers on Transition-Metal Carbide Surfaces for Supercapacitor Applications. *ACS Publications*, 8, 29, 18806–18814, 2016, https://pubs.acs.org/doi/10.1021/acsami.6b04481.
42. Hou, W. *et al.*, Mixed-dimensional heterostructure of few-layer MXene based vertical aligned MoS2 nanosheets for enhanced supercapacitor performance. *J. Alloys Compd.*, 859, 157797–157797, Apr. 2021, doi: https://doi.org/10.1016/j.jallcom.2020.157797.
43. Rafieerad, A. *et al.*, Development of Fluorine-Free Tantalum Carbide MXene Hybrid Structure as a Biocompatible Material for Supercapacitor Electrodes. *Adv. Funct. Mater.*, 31, 30, 2100015–2100015, May 2021, doi: https://doi.org/10.1002/adfm.202100015.
44. Xie, W. *et al.*, MOF-derived CoFe2O4 nanorods anchored in MXene nanosheets for all pseudocapacitive flexible supercapacitors with superior energy storage. *Appl. Surf. Sci.*, 534, 147584–147584, Dec. 2020, doi: https://doi.org/10.1016/j.apsusc.2020.147584.
45. Zhang, Y., Cao, J., Yuan, Z., Zhao, L., Wang, L., Han, W., Assembling Co3O4 Nanoparticles into MXene with Enhanced electrochemical performance for advanced asymmetric supercapacitors. *J. Colloid Interface Sci.*, 599, 109–118, Oct. 2021, doi: https://doi.org/10.1016/j.jcis.2021.04.089.
46. Li, X. *et al.*, A flexible Zn-ion hybrid micro-supercapacitor based on MXene anode and V2O5 cathode with high capacitance. *Chem. Eng. J.*, 428, 130965–130965, Jan. 2022, doi: https://doi.org/10.1016/j.cej.2021.130965.
47. Gao, L. *et al.*, MXene/Polymer Membranes: Synthesis, Properties, and Emerging Applications. *Chem. Mater.*, 32, 5, 1703–1747, 2020, https://pubs.acs.org/doi/10.1021/acs.chemmater.9b04408.
48. Zhan, X., Chen, S., Zhou, J., Sun, Z., MXene and MXene-based composites: synthesis, properties and environment-related applications. *Nanoscale Horiz.*, 5, 2, 235–258, Jan. 2020, doi: https://doi.org/10.1039/c9nh00571d.
49. Yan, J. *et al.*, Polypyrrole–MXene coated textile-based flexible energy storage device. *RSC Adv.*, 8, 69, 39742–39748, Jan. 2018, doi: https://doi.org/10.1039/c8ra08403c.
50. Boota, M., Anasori, B., Voigt, C., Zhao, M.-Q., Barsoum, M.W., Gogotsi, Y., Pseudocapacitive Electrodes Produced by Oxidant-Free Polymerization of Pyrrole between the Layers of 2D Titanium Carbide (MXene). *Adv. Mater.*, 28, 7, 1517–1522, Feb. 2016, doi: https://doi.org/10.1002/adma.201504705.
51. Zhu, M. *et al.*, Highly Flexible, Freestanding Supercapacitor Electrode with Enhanced Performance Obtained by Hybridizing Polypyrrole Chains with

MXene. *Adv. Energy Mater.*, 6, 21, 1600969–1600969, Aug. 2016, doi: https://doi.org/10.1002/aenm.201600969.
52. Zhang, W. *et al.*, A multidimensional nanostructural design towards electrochemically stable and mechanically strong hydrogel electrodes. *Nanoscale*, 12, 12, 6637–6643, Jan. 2020, doi: https://doi.org/10.1039/d0nr01414a.
53. Fan, Q.H. *et al.*, Ti3C2-MXene composite films functionalized with polypyrrole and ionic liquid-based microemulsion particles for supercapacitor applications. *Chem. Eng. J.*, 428, 131107–131107, Jan. 2022, doi: https://doi.org/10.1016/j.cej.2021.131107.
54. Cheng, W., Fu, J., Hu, H., Ho, D., Interlayer Structure Engineering of MXene-Based Capacitor-Type Electrode for Hybrid Micro-Supercapacitor toward Battery-Level Energy Density. *Adv. Sci.*, 8, 16, 2100775–2100775, Jun. 2021, doi: https://doi.org/10.1002/advs.202100775.
55. Panda, S., Deshmukh, K., Pasha, K., Theerthagiri, J., Manickam, S., Choi, M.Y., MXene based emerging materials for supercapacitor applications: Recent advances, challenges, and future perspectives. *Coord. Chem. Rev.*, 462, 214518–214518, Jul. 2022, doi: https://doi.org/10.1016/j.ccr.2022.214518.
56. Xing, J., Zhu, Y., Zhou, Q., Zheng, X., Jiao, Q., Fabrication and shape evolution of CoS2 octahedrons for application in supercapacitors. *Electrochim. Acta*, 136, 550–556, Aug. 2014, doi: https://doi.org/10.1016/j.electacta.2014.05.118.
57. Lyu, J., Mayyas, M., Salim, O., Zhu, H., Chu, D., Joshi, R., Electrochemical performance of hydrothermally synthesized rGO based electrodes. *Mater. Today Energy*, 13, 277–284, Sep. 2019, doi: https://doi.org/10.1016/j.mtener.2019.06.006.
58. Shen, C. *et al.*, Rational design of activated carbon nitride materials for symmetric supercapacitor applications. *Appl. Surface Sci.*, 455, 841–848, Oct. 2018, doi: https://doi.org/10.1016/j.apsusc.2018.06.065.
59. Wen, M. *et al.*, Synthesis of high-quality black phosphorus sponges for all-solid-state supercapacitors. *Mater. Horiz.*, 6, 1, 176–181, Jan. 2019, doi: https://doi.org/10.1039/c8mh00708j.
60. Qin, S. *et al.*, High and Stable Ionic Conductivity in 2D Nanofluidic Ion Channels between Boron Nitride Layers. *J. Am. Chem. Soc*, 139, 18, 6314–6320, 2017, doi: https://pubs.acs.org/doi/10.1021/jacs.6b11100.
61. Afzal, A., Abuilaiwi, F.A., Habib, A., Awais, M., Waje, S.B., Atieh, M.A., Polypyrrole/carbon nanotube supercapacitors: Technological advances and challenges. *J. Power Sources*, 352, 174–186, Jun. 2017, doi: https://doi.org/10.1016/j.jpowsour.2017.03.128.
62. Abdel, A. *et al.*, Advanced materials and technologies for supercapacitors used in energy conversion and storage: a review. *Environ. Chem. Lett.*, 19, 1, 375–439, Aug. 2020, doi: https://doi.org/10.1007/s10311-020-01075-w.
63. Altın, Y. and Bedeloğlu, A. Ç., Polyacrylonitrile/polyvinyl alcohol-based porous carbon nanofiber electrodes for supercapacitor applications. *Int. J. Energy Res.*, 45, 11, 16497–16510, May 2021, doi: https://doi.org/10.1002/er.6896.

64. Wang, H. *et al.*, Micro-meso porous structured carbon nanofibers with ultra-high surface area and large supercapacitor electrode capacitance. *J. Power Sources*, 482, 228986–228986, Jan. 2021, doi: https://doi.org/10.1016/j.jpowsour.2020.228986.
65. Ampong, D.N. *et al.*, MXene: fundamental to applications in electrochemical energy storage. *Discov. Nano*, 18, 1, 3–3, Feb. 2023, doi: https://doi.org/10.1186/s11671-023-03786-9.
66. Javed, M.S. *et al.*, Heterostructured bimetallic–sulfide@layered Ti3C2T – MXene as a synergistic electrode to realize high-energy-density aqueous hybrid-supercapacitor. *Nano Energy*, 101, 107624–107624, Oct. 2022, doi: https://doi.org/10.1016/j.nanoen.2022.107624.
67. Shi, B., Li, L., Chen, A., Jen, T., Liu, X., Shen, G., Continuous Fabrication of Ti3C2Tx MXene-Based Braided Coaxial Zinc-Ion Hybrid Supercapacitors with Improved Performance. *Nano Micro Lett.*, 14, 1, 34–34, Dec. 2021, doi: https://doi.org/10.1007/s40820-021-00757-6.
68. Yun, J. *et al.*, Layer-by-Layer Assembly of Reduced Graphene Oxide and MXene Nanosheets for Wire-Shaped Flexible Supercapacitors. *ACS Appl. Mater. Interfaces*, 13, 12, 14068–14076, 2021, doi: https://pubs.acs.org/doi/10.1021/acsami.0c19619.
69. Nasrin, K., Sudharshan, V., Arunkumar, M., Sathish, M., 2D/2D Nanoarchitectured Nb2C/Ti3C2 MXene Heterointerface for High-Energy Supercapacitors with Sustainable Life Cycle. *ACS Appl. Mater. Interfaces*, 14, 18, 21038–21049, 2022, doi: https://pubs.acs.org/doi/10.1021/acsami.2c02871.
70. Zheng, W. *et al.*, MXene//MnO 2 Asymmetric Supercapacitors with High Voltages and High Energy Densities. *Batteries Supercaps*, 5, 10, e202200151, Jun. 2022, doi: https://doi.org/10.1002/batt.202200151.
71. Fan, Z. *et al.*, A Compact MXene Film with Folded Structure for Advanced Supercapacitor Electrode Material. *ACS Appl. Energy Mater.*, 3, 2, 1811–1820, 2020, doi: https://pubs.acs.org/doi/10.1021/acsaem.9b02259.
72. Szuplewska, A. *et al.*, The 10th anniversary of MXenes: Challenges and prospects for their surface modification toward future biotechnological applications. *Adv. Drug Deliv. Rev.*, 182, 114099–114099, Mar. 2022, doi: https://doi.org/10.1016/j.addr.2021.114099.
73. Nahirniak, S., Ray, A., Saruhan, B., Challenges and Future Prospects of the MXene-Based Materials for Energy Storage Applications. *Batteries*, 9, 2, 126–126, Feb. 2023, doi: https://doi.org/10.3390/batteries9020126.
74. Lamiel, C., Hussain, I., Ogunsakin, O.R., Zhang, K., MXene in core–shell structures: research progress and future prospects. *J. Mater. Chem. A*, 10, 27, 14247–14272, Jan. 2022, doi: https://doi.org/10.1039/d2ta02255a.

75. Deshmukh, K., Muzaffar, A., Kovářík, T., Ahamed, M.B., Pasha, K., *Introduction to 2D MXenes: fundamental aspects, MAX phases and MXene derivatives, current challenges, and future prospects*, pp. 1–47, Elsevier eBooks, Amsterdam, Jan. 2022, doi: https://doi.org/10.1016/b978-0-12-823361-0.00009-5.
76. Ain, *et al.*, MXene-based aptasensors: Advances, challenges, and prospects. *Prog. Mater Sci.*, 129, 100967–100967, Aug. 2022, doi: https://doi.org/10.1016/j.pmatsci.2022.100967.

# 10

# Current Status of Sustainable Supercapacitors

**Priya R. and S. Sonia***

*Department of Physics, Holy Cross College (Autonomous), Nagercoil, Tamil Nadu, India*

***Abstract***

An essential tool for tackling the problem of power restriction in renewable power sources and storage of electricity is the supercapacitor. It is essential to develop novel materials for electrodes and prepare them for usage in energy storage applications. The electrode material must have a significant area of surface-to-volume ratio in order to attain considerable energy storage densities. The material used for electrodes must also be extremely conductive for successful charge transfer. Several novel materials that can be used to boost the capacitance of supercapacitors have been developed in recent years. This overview concentrates on the three types of supercapacitors, most recent advancements in energy storage systems, necessity of supercapacitors, current challenges, and the most current developments in the field of material preparation. It also suggests potential answers to problems that arose throughout the fabrications of supercapacitors and predicts the next path that supercapacitor researchers will pursue in the future.

***Keywords*:** Eco-friendly, green synthesis method, green materials, energy storage, renewable energy, supercapacitor, novel electrode, MOF

## 10.1 Introduction

Since more people are using electrical equipment, there is a growing need for electricity. On the one hand, the transformation of electrical energy via energy sources that are renewable is inconsistent. A temporary energy storage system is needed to ensure a consistent supply. Researchers are

**Corresponding author*: sonia.s@holycrossngl.edu.in

Basheer Ahamed and Chaudhery Mustansar Hussain (eds.) Sustainable Supercapacitors: Next-Generation of Green Energy Storage Devices, (299–322) © 2025 Scrivener Publishing LLC

looking for electrochemical systems for storing energy that have significant energy as well as power densities, optimized life span, economic viability, and environmental friendliness. Supercapacitors are becoming more and more crucial in a variety of high-power and high-reliability applications, including energy storage for electric buses, regenerative braking energy capture in passenger cars, backup power systems for emergency aircraft, defibrillators, and wind turbine pitch control [1–6]. Supercapacitors have higher power densities and cycle lifetimes compared to conventional battery technologies, and they also possess higher specific energy than ordinary electrolytic capacitors.

In the foreseeable future, supercapacitors are considered to be the most intriguing options for energy storage according to the researches in the field of energy storage technologies. The batteries work better in applications requiring more energy density, like those in which a device must run constantly for a long time. In contrast, compared to batteries, supercapacitors offer a far greater energy density. Therefore, they are perfect for demanding duties such as operating an electric vehicle. Additionally, supercapacitors last a lot longer than batteries do. An ultracapacitor can frequently withstand more than 1,000,000 charging and discharging cycles, as compared to 2000–3000 cycles for a typical battery. It might lead to significant material and cost reductions. Supercapacitors can be recharged quickly. Supercapacitors do not degrade over time.

## 10.2 Supercapacitors

Supercapacitors, also referred to as ultracapacitors and electrochemical capacitors, create capacitances that can be several orders of magnitude higher compared to normal capacitors by using thin electrolytic dielectric materials and vast surface areas of materials for electrodes [7–11]. This allows supercapacitors to maintain the standard higher-power densities of conventional capacitors while at the same time attaining larger energy densities. Supercapacitors are the most advanced electrochemical energy storage technology, covering the gap created by conventional capacitors and battery while assisting in a variety of power and energy-related applications.

Because supercapacitors can form two independent layers of separated charges between their plates, which are often made from porous, carbon-based materials, they can store more energy than standard capacitors. Supercapacitors store energy through the adsorption of ions at the interface in between the electrode and the electrolyte, therefore referred to as electrical double-layer capacitors (EDLC). As a result, it is capable of storing and releasing energy in a couple of seconds and has a long life span. Electrostatic double-layer electrodes are commonly made of activated carbon, carbon fiber cloth, aerogel, graphite, graphene, and carbon nanotubes in various forms of carbon [12–14]. Because not all materials are appropriate to be utilized as an electrode in pseudo capacitors, a pseudocapacitive material employs electron conductive polymers that have low electron spin resonance, high capacitance, and cycles [15]. The hybrid-type SCs combine two electrodes made of carbon-based materials with high pseudo capacitance and double-layer capacitance. Rechargeable battery electrodes have been influenced by the invention of hybrid-type supercapacitor electrodes.

Supercapacitors can be used to bridge the energy storage gap among capacitors and batteries in a variety of applications. Rather than being employed in applications that require a long-term compact storage of energy, such as smartphones and consumer-grade devices, they tend to be utilized in applications that required rapid charging and discharging cycles, such as regenerative brakes in cars. Small supercapacitor units provide additional power for static random-access memory (SRAM). However, supercapacitors serve a purpose for certain consumer-grade applications that require features such as rapid recharging or a long-life cycle. MP3 players and camera flashes are examples of professional-grade equipment. Supercapacitors are also commonly utilized in wind turbines. Here, large supercapacitors help to balance the unsteady wind energy supply. Figure 10.1 showcases the supercapacitor's application in a real-world electronic setup, emphasizing its practical utility.

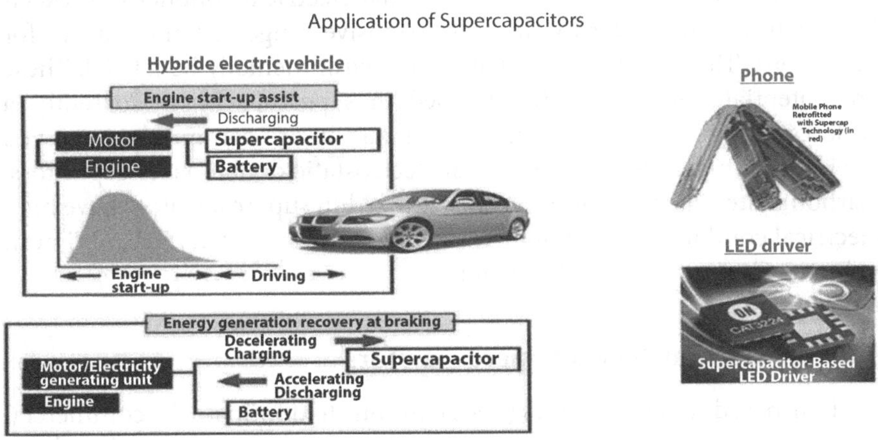


**Figure 10.1** Applications of supercapacitors.

## 10.3 Necessity of Supercapacitors

The supercapacitor is one of the greatest powerful systems to store energy. Supercapacitors, in contrast to batteries, are frequently utilized in devices and systems that require a high-power source due to their numerous advantages. Because of their exceptional performance and commercial potential, supercapacitors are now employed in a variety of business sectors, such as commercial and industrial use, energy generation, transport, electronics for consumers, national defense, communication, health care gadgets, hybrid and electric vehicles, and so on [16–18]. One of the primary applications of supercapacitors is energy storage. The characteristics of supercapacitors, such as charging discharging, life span, size, weight, and environmental friendliness, make them suitable for a variety of applications. Supercapacitors are being utilized more frequently in situations where it is required to store and release huge amounts of energy quickly. Today's market applications include the automotive sector, hybrid modes of transportation across the globe, utilized in automobiles, plus rail-system powered concepts [19]. Furthermore, the minimal maintenance requirements of supercapacitors, as well as their ability to survive harsh conditions, make them appropriate for renewable energy-related applications.

## 10.4 Electrostatic Double-Layer Capacitors

At the moment, the major goals of EDLC research are to increase their operating temperature range and energy efficiency. Carbon materials have a large specific area of surface, exceptional electrical conductivity, excellent chemical stability, as well as an extensive range of temperatures for operations. They are also abundant and environmentally beneficial. These are potential electrode materials used in supercapacitors. Currently, a wide variety of compounds made from carbon are commonly employed as electrode materials in commercial electrostatic double-layer capacitors. Carbon materials used for electrodes utilized in supercapacitors have high electrical conductivity, wettability, effective large area of surface, optimum volume of pores, and size dispersion.

### 10.4.1 Carbon-Based Supercapacitors

Carbon-based supercapacitors (CSs) continue to dominate the commercial industry, offering intriguing qualities such as fast energy pulses, hundreds of millions of charging/discharging cycles, a broad temperature range of operation, and excellent Coulombic efficiencies. Historically, activated carbons with a high degree of micro porosity were the most commonly used supercapacitor electrode materials. To construct spherical carbon nano architectures with a small surface-to-volume ratio as well as close-packing nature, a variety of synthesis procedures such as emulsion polymerization, modified Stöber synthesis, self-assembly, and hydrothermal carbonization can be applied. Carbon-based materials are often used in supercapacitors due to their excellent physical, chemical, and electrochemical properties, which include their large surface area, ability to regulate porous structures, excellent electrical conductivities, chemical inertness, stability in structure, control in modification, as well as adaptability for creating a variety of composites [20, 21]. Graphene, carbon nanotubes, and activated carbon are the three most common carbon-based materials utilized as materials for electrodes in supercapacitors. Nanomaterials used as electrodes in supercapacitors include carbon nano fibers, carbon nano onions, diamond-like carbons, carbide-derived carbons, and carbon nano horns, but their use is currently limited to testing in laboratories due to high production costs, limited scalability, and potential environmental effects.

A carbon-based supercapacitor with comb-like planar electrodes was created by Ukrit Jitropas *et al.* (2023). Additionally, three different fringe widths were patterned onto planar electrode samples. With a maximum 86.7% boost, all planar electrode samples can offer a greater specific capacitance than the typical sandwich configuration [22].

In order to create symmetric electrodes, Aziz Ahmad *et al.* (2023). physically activated carbon from date seed biomass, and PVA/KOH were employed as gel polymer electrolytes for all-solid-state supercapacitors. The date seed biomass was first carbonized at 600°C, and it was later utilized to create physically activated carbon by activating $CO_2$ at 850°C. The porous, flaky, and layered morphologies of C-850 were clearly visible in the SEM and TEM images. The finest electrochemical performances in supercapacitors were displayed by the C-850-fabricated electrodes with PVA/KOH electrolytes [23].

$MoS_2$ nanobelts and a hybrid carbon nanostructure were created by Lina Khandare *et al.* (2023) utilizing a straightforward hydrothermal process. The Lemon grass-derived Carbon (LG-C) interlayers were sprinkled with $MoS_2$ nanobelts to offer active sites and prevent the sheets from being restacked. At a current density of 0.5 A $g^{-1}$, the specific capacitance of $MoS_2$/LG-C and LG-C exhibits 77.5 F $g^{-1}$ and 30.1 F $g^{-1}$, respectively. The maximal power density and energy density of the $MoS_2$/LG-C-based supercapacitor were 273.2 W $kg^{-1}$ and 2.1 Wh $kg^{-1}$, respectively. Additionally, charging–discharging was used to test $MoS_2$/LG-C's cyclic stability up to 3000 times, with the results showing that only 71.6% of the capacitance was retained at a current density of 3 A $g^{-1}$. The findings demonstrated that $MoS_2$/LG-C is a superb, reasonably priced electrode material with excellent electrochemical performance [24].

### 10.4.2 Graphene-Based Supercapacitors

The unique and desirable features of graphene drive its application in supercapacitors. Graphene, one of numerous carbon allotropes, has a vast area of surface, exceptional chemical stability, and a high mechanical stiffness. In addition, it makes a fantastic platform for the decoration of oxides of transition metals, conducting polymers, and elements like boron (electron withdrawal) and nitrogen (electron contributing) that may attract or add to the pi electronic charge density of the graphene plane. By offering the optimum combination of enormous electrochemical surface area, porous nature, tunable electron concentrations, and other properties, graphene is important in charge storage techniques. Graphene composite materials can have a variety of charge storage processes, even though nanostructured graphene and its pores can store charge through the double-layer process. The immense versatility of graphene in the construction of composites and its capacity to operate via a variety of charge storage mechanisms account for its prominence in the field of supercapacitors. Graphene sheets' appealing morphological flexibility, conductivity, plus mechanical robustness makes them ideal for creating graphene/polymer blends for potential uses in electrical devices. Graphene, oxide of graphene, modified graphene oxide, as well as its composites are being extensively used as electrodes in supercapacitors for both academic and industrial applications. Graphene and its derived compounds and hybrids are ideal active materials in electrochemical energy storage and the conversion utilization in general, as well as supercapacitors in particular, due to their vast surface area, outstanding mechanical and chemical stabilities, chemical inertness, and extensive surface chemistry.

Several energy storage research fields have resulted in the creation of electrode materials with excellent electrochemical performance to date. Wu *et al.*, for example, reported the synthesis of hydrous $RuO_2$/graphene sheets with varying ruthenium loadings that had a specific capacitance of 570 F $g^{-1}$. Furthermore, utilizing $RuO_2$/CNT nanocomposites, Bi *et al.* improved the capacitive performance of $RuO_2$ and reached a specific capacitance of

935 F $g^{-1}$. However, because of its exorbitant price, practical applications of $RuO_2$ have been limited. As a result, inexpensive metal oxides such as manganese dioxide, nickel oxide, tin dioxide, and others have been increasingly explored in order to find an ideal replacement for the costly ruthenium oxide [25].

Tsinghua University researchers revealed a flexible graphene supercapacitor that retained over 99% of its efficiency after 10,000 cycles and a charge/discharge voltage window of 3 V in late 2022. This supercapacitor operated numerous small electrical devices, including an LED and a calculator, but only for a few seconds at a time. In another 2022 work, Imperial College London researchers created a woven graphene supercapacitor. It had a quick response time of 0.6 seconds when used as a pressure sensor, but its capacitance degraded to around 90% after only 10,000 cycles. Two publications from a study at Rice University and Queensland University of Technology were published in the *Journal of Nanotechnology and Power Sources*. Two graphene layers with an electrolyte layer sandwiched between them would make up the proposed solution, according to researchers at these universities. The resulting film is robust, thin, and capable of quickly releasing significant amounts of energy. The researchers propose that in next electric vehicles, the smaller, slimmer supercapacitors might take the place of larger batteries.

Despite the aforesaid problems, supercapacitors have found real-world uses; however, those based on graphene are still in the early stages. The transportation sector has seen some of the most high-profile applications:

- Bus fleets powered by supercapacitors have been introduced in China and Serbia. One of these fleets has a reported 25-kilometer range and charges in six to seven minutes. To compensate for the low energy density of supercapacitors and consequently the limited range per charge, buses are recharged at the depot or bus stops.
- Skeleton Technologies manufactures a graphene-based supercapacitor for use in trains that can recover up to 30% of braking energy. This technology has been chosen for use in new trains for Spain's Granada metro system, which are set to start service in the summer of 2024 [26].
- Samsung promotes a lithium-ion hybrid for automotive applications that is ideal for low-voltage systems that require rapid charging and discharging.

## 10.5 Hybrid-Based Supercapacitors

Because of its exceptional electrochemical qualities, safeguarding, economic feasibility, as well as ecological reliability, supercapacitor battery hybrid (SBH) storage of energy devices have attracted the curiosity of modern researchers and engineers. Carbon derivative from 0 to 3 dimensions, such as activated carbon, graphene, porous carbon, and others, are being utilized as a promising electrode material because of their availability in reserves, economic feasibility, variable pore size, and broad range of operational potential. Hybrid supercapacitors are finding new applications in electricity, storing energy, transport, manufacturing, and hybrid energy vehicles [27].

At the Auto Expo 2020, the Automotive Research Association of India (ARAI) displayed its unique hybrid energy storage system developed in partnership with the Indian Space Research Organisation (ISRO) on the Tata Ace platform. The hybrid energy storage system combines battery and supercapacitor devices. They are now collaborating on battery and supercapacitor technologies for two reasons: to lower battery size or to exchange energy between systems; as the battery size decreases, the weight of the supercapacitor decreases as well. As the weight decreases, the efficiency increases, and the cost decreases as well. Those battery-electric vehicles has a range of 120 km. We can reach the same range with this technique, but they used a smaller supercapacitor, which produces 40% of the total energy required by the car. This will result in weight loss, money savings, and an overall improvement. The vehicle was developed before; however, in 2019, they replaced the power pack with a supercapacitor [28]. Figure 10.2 delineates the intricately engineered hybrid vehicle model, a collaborative

**Figure 10.2** Hybrid vehicle [28].

development spearheaded by the Automotive Research Association of India in conjunction with the Indian Space Research Organisation.

A solid-state ultracapacitor that combines significant capacitance with battery-like discharging properties has been developed at NASA's Marshall Space Flight Center. When combined with having the ability to provide sustained power like a battery, a new type of ultracapacitor with high capacitance may open the door to an entirely new class of energy storage technologies. Significant capacitance and an odd discharge pattern more akin to a battery have been seen in test devices. Records show that these testing devices quickly drain until they reach a certain voltage, at which point they slowly discharge like batteries. In light of this, the technology is known as a hybrid ultracapacitor. In an attempt to substitute range safety batteries, which are needed to power the devices that destroy space launch rockets, this technology was developed. The main distinction for this particular technique is the integration of co-dopants to alter the dielectric behavior of the perovskite materials based on barium titanate. Co-dopants include hydroxyl ions, lanthanum, and other rare earth elements. Both nanoscale heat treatments and meticulously crafted heat treatments are applied to the materials. It has been discovered that adding hydroxyl ions increases the dielectric material's capacitance by several orders of magnitude. Additionally, the comparatively low voltages are used in today's consumer and commercial electronics to achieve this high capacitance levels. The capacitors that have been evaluated so far are straightforward, single-layer devices. Efforts will be undertaken in the future to develop manufacturing and packaging procedures in order to boost device energy density, including multi-layer devices and packages [29].

El Paso, Texas (July 18, 2023) researchers detail how they were able to obtain the greatest amount of energy storage, referred to as capacitance, in a supercapacitor ever accomplished in a new groundbreaking chemistry research. Marta Plonska-Brzezinska, Ph.D., of the Medical University of Bialystok in Poland, and Luis Echegoyen, Ph.D., emeritus professor at The University of Texas at El Paso, conducted the research. It was recently published in the journal *Scientific Reports*, which is released by Nature Portfolios, a recognized research author. However, the energy that current supercapacitors can store is restricted, which restricts the range of uses for them. The ability to construct supercapacitors to store additional energy would make them significantly lighter and charges much more quickly than batteries, which would have a significant influence on the market. The newly developed supercapacitor created by Echegoyen and Plonska-Brzezinska used a material with a carbon "nano-onion" core framework that has multiple porosity that enable higher energy storage to achieve a record level of storage or capacitance [30].

A team of researchers from Korea and India (2023) created a portable supercapacitor that can be charged by light. This innovation has the potential to replace traditional lithium-ion batteries in portable and wearable devices such as smartphones, tablets, laptop computers, and smartwatches [31].

In the present work, published in *ACS Energy Letters*, researchers at IISc's Department of Instrumentation and Applied Physics (IAP) created their supercapacitor employing "Field Effect Transistors" as charge collectors, rather than the metallic electrodes used in normal capacitors. As a result, Misra and her coworkers decided to develop the hybrid field-effect transistors by switching between layers of molybdenum disulfide and graphene that are only a few atoms' thickness. These layers are then connected to gold contacts to boost electron mobility. To build a solid-state supercapacitor, a solid gel electrolyte is sandwiched between a pair of field effect transistor electrodes. The material silicon dioxide/silicon makes up the whole structure. The supercapacitor was created, and the researchers then used different voltages to assess its electrochemical capacitance, or capacity to hold a charge. Researchers discovered that capacitance is increased by 3000% in specific circumstances. A capacitor with only $MoS_2$ and no graphene shows merely an 18% increase in capacitance under the similar circumstances. In the future, the researchers plan to look into whether using materials other than $MoS_2$ could potentially increase the capacitance of their supercapacitor to a greater extent [32]. Figure 10.3 illustrates the intricate schematic of state-of-the-art supercapacitors with the capacity for photonic charging, ingeniously engineered by the researchers at the Indian Institute of Science's Department of Instrumentation and Applied Physics (IAP).

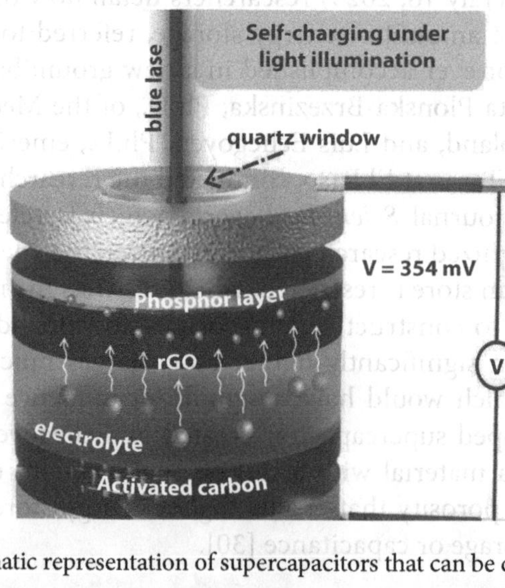


**Figure 10.3** Schematic representation of supercapacitors that can be charged by light [31].

## 10.6 Pseudo Capacitors

The promising electrode materials known as pseudocapacitive materials have two distinct types of capacitances. One is the faradaic redox pseudo capacitance, which only develops on the material's surface and never allows electrolyte ions to enter the electrode's interior. The ions in the bulk material occupy tunnels or vacant sites in integration pseudocapacitance, which is the other type. Pseudocapacitive materials typically have better rate capabilities than battery materials along with their intercalation behavior. As evidenced by the numerous publications over the past few decades, pseudocapacitive materials typically have better chemical stability, acceptable electronic conductivity, and excellent electrochemical performance, which has led to their widespread and growing interest for use in supercapacitors.

### 10.6.1 Polymer-Based Supercapacitors

Polymers, with their wide variety, exceptional flexibility, and ease of processing, can effectively improve the functionality of supercapacitors and broaden their practical applications. Polymer materials used in supercapacitors are currently non-biodegradable, causing environmental risks.

Recently, a research team lead by Prof. Wu Zhong Shuai of the Chinese Academy of Sciences, Dalian Institute of Chemical Physics explained the significance of biodegradable polymers, a component of future green supercapacitors. "A green supercapacitor requires good biodegradability of different components in order to achieve harmless disposal in various environments," Prof. Wu says. "To meet these requirements, device preparation must rely heavily on biodegradable polymers and innovative design strategies. Biodegradable polymers are the ideal raw material, and appropriate research strategies and preparation methods are powerful means to realise environmentally friendly supercapacitor treatment," Prof. Wu stated [33].

Scholars from Imperial College London and University College London have created an electrode material for supercapacitors that is more affordable, eco-friendly, and energy dense. This development may help these high-power, quick-charging electric vehicle technology gain more commercial interest. The researchers used lignin, a bio-based by-product of the paper industry, to develop freestanding electrodes with enhanced energy storage capability. By utilizing lignin rather than expensive graphene-based carbon, the team was able to produce a freestanding structure that is smaller and lighter than current models without reducing energy storage

**Figure 10.4** Eco-friendly vehicles [34].

capability. They can charge in the time it takes for people to exit and enter short distance electric vehicles like buses, cabs, and trains; hence, they are appropriate for usage in these types of vehicles. By producing stable multi-functional materials using waste biomass streams, for example, lignin, we can lessen our dependency on essential minerals like lithium and create a sustainable and cost-effective supply network for energy resources in the future. They are presently developing a brand-new supercapacitor that can be applied commercially and has an electrolyte that is less expensive and non-corrosive [34]. Figure 10.4 delineates the conceptual model of an environmentally sustainable vehicle, meticulously developed by scholars from Imperial College London and University College London.

### 10.6.2 Metal–Organic Framework-Based Supercapacitors

Porous coordination polymers, which is another term for MOFs, are created from organic linkers and metal-containing nodes, which were first described by Yaghi and Li in 1995. More than 20,000 categories of MOFs have been reported in prior literature as of this writing. Additionally, they have sparked a lot of curiosity in a variety of sectors, including sensors, medication delivery, gas adsorption, and catalysts. Metal–organic frameworks have recently been used directly as active components with enormous potential for supercapacitors. Due to several special properties of metal–organic frameworks, their application in supercapacitors has recently gained a lot of interest. The large surface and pore structure, for example, enable the high storage capacity, and the metallic clusters therein are able to be used to build electrode materials with built-in redox active sites that interact between the electrode and the electrolyte.

In the study of Hoa Bui Thi *et al.* (2023), Cu@Fe-MOF along with Fe-MOF thin films were created on NF substrates employing a drop-cast approach. In comparison to the pure electrode Fe-MOF/NF, the Cu@Fe-MOF/NF electrode demonstrated relatively superior electrochemical performance when utilized in electrochemical supercapacitor applications. Interestingly, at 3 A $g^{-1}$, the Cu@Fe-MOF/NF film electrode produced a maximum specific capacitance of 562.0 F $g^{-1}$, which is two times greater than the value of the pristine Fe-MOF/NF electrode. Amazingly, the Cu@Fe-MOF/NF electrode showed 89.5% capacitive retention even after 5000 charging/discharging cycles, demonstrating improved cyclic stability [35].

Kimia Zarean Mousaabadi *et al.*'s study includes the direct synthesis of a di metal–organic framework (Ni/Co-Hemin MOF), upon the phosphorus-doped reduced graphene oxide (PrGO) for utilization as an active component for improving the performance of the asymmetrical supercapacitors. The nanocomposite was used as an active component in supercapacitors and demonstrated an impressive specific capacitance of 963 C $g^{-1}$ at 1.0 A $g^{-1}$, and a high-rate capability of 68.3% when the current density was increased by 20 times, and enhanced cyclic stability (2023) [36].

With capacitances of 1518.8 F $g^{-1}$ at 1 A $g^{-1}$ and 244 mF $cm^{-2}$ at 0.5 mA $cm^{-2}$, respectively, a new layered structural nickel metal–organic framework nanosheet has been successfully created and employed for the first time as an excellent electrode for a supercapacitor and a flexible all solid-state asymmetric supercapacitor (Chengyu Yang *et al.*-2020).

One-pot solvothermal synthesis of Fe (BPDC) nanosheet array on nickel foam has been achieved by Jicheng Zhao *et al.* An improved Fe (BPDC)-3 electrode with a very high specific capacitance of up to 17.54 F $cm^{-2}$ at 1 mV/s and outstanding cyclic stability with 129% retention even after 10,000 cycles was created by controlling the solvothermal duration. Furthermore, Fe (BPDC)-3 along with activated carbon (AC) were used as a cathode and anode materials in the construction of a hybrid supercapacitor, respectively. The supercapacitor as-fabricated has an excellent specific capacity of 26.45 F $cm^{-2}$, a maximum density of energy of 45.64 Wh $kg^{-1}$, and a capacity retention of 86.41% after 10,000 cycles. Such favorable outcomes indicate the Fe (BPDC) electrodes' potential as an ideal cathode for energy storage, and their high flexibility to the environment gives a novel concept for general use as a new energy source [37].

## 10.7 Green Supercapacitors

As a result of decreasing fossil fuel resources and greater environmental awareness, there has been an increase in demand for green and renewable energy generation and storage technologies. Energy storage devices must consider other issues in the upcoming years, such as sustainability through the use of green materials and the possibility of providing both affordable materials used in batteries right now, which account for as much as seventy percent of total cell costs, and powerful devices to alleviate energy poverty or meet the increasing requirement for electric vehicles. Green synthesis is popular because it is a green technique that does not involve the use of hazardous chemicals during the experimental procedure. Because of their renewability and sustainability, bio-sourced materials have gained in appeal.

Because of their renewability and sustainability, bio-sourced materials have become more important in the production of energy devices such as batteries and solar cells. Many battery types, including lithium-ion and sodium-ion, have demonstrated promising performance when biosourced materials were used as electrolytes and electrode materials. Environmentally friendly materials, such as biopolymers derived from chlorophyll, porous carbon based on chitosan, fallen leaves, apricot shell, or even used face masks made of polypropylene, have been examined for use in green supercapacitors. Gel electrolytes could increase specific capacitance and ionic transport capabilities while avoiding downsides like higher weight or electrolyte leakages. Some extensive research has been carried out and demonstrates the benefits of environmentally friendly gel electrolytes.

A "supercapacitor" built by MIT engineers can store a lot of energy and is made of abundant, ancient materials. The device, which is constructed entirely of water, cement, and carbon black that resembles powdered charcoal, could serve as a base for low-cost systems to store intermittent renewable energy sources such as wind or solar energy. Researchers discovered that by combining the two components with water, they could make a supercapacitor, a battery-free alternative that can store electrical energy. The scientists from MIT who developed the system, for example, believe that their supercapacitor could one day be embedded in a house's concrete foundation, where it could store an entire day's worth of energy while costing very little and still providing the foundation with the necessary strength it requires. The scientists also envision a concrete road that might provide electric cars with contactless charging as they drive over it [38].

Coconut husk was employed by Ruixia Chu *et al.* as a raw material, and it was carbonized using high-temperature pyrolysis and a common hydrothermal pre-treatment. Three-electrode systems were used to test the materials that were produced under various reaction circumstances. The electrochemical studies suggested that when the pre-treatment time was 24 hours and the activator was KOH, the carbonized coconut husk had optimum energy storage properties. This work investigated a carbon electrode for supercapacitor energy storage that was produced from a novel raw biomass material [39].

In order to create high mass-loading activated carbon electrodes with micro- and meso-pores for high voltage super capacitive swing adsorption of carbon dioxide, Muhammad Bilal *et al.* reported a straightforward synthesis and manufacturing technique. With an adsorption capacity of

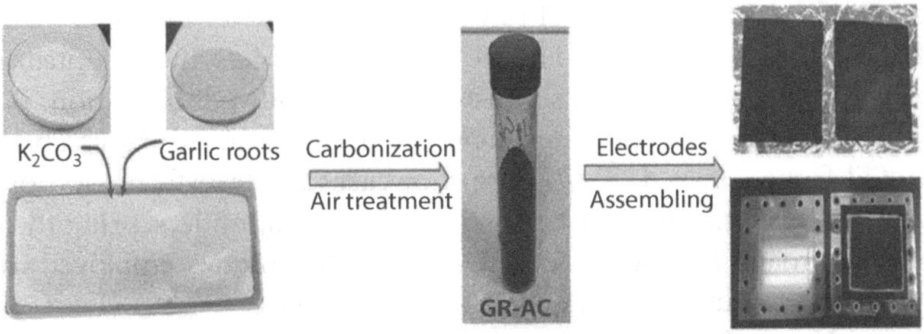


**Figure 10.5** Activated carbon generated from garlic roots and electrode assembly [40].

**Figure 10.6** Paper-based supercapacitors made from seaweed [41].

312 mmol kg$^{-1}$, the minimal energy consumption of 70 kJ mol$^{-1}$ is accomplished. By lowering the potential window to 1.4 V, the adsorptive performance of SSA can be improved to 524 mmol kg$^{-1}$ at an energy consumption of 130 kJ mol$^{-1}$. The electrodes exhibit capacitive behavior with stable SSA performance across numerous cycles up to 1.4 V. These findings suggest that the creation of novel electrodes offers a viable means of enhancing the SSA's sorption kinetics, adsorption capacity, and energy efficiency [40].

With the creation of the thinnest, lightest, and biodegradable paper-based supercapacitor, researchers at Gujarat Energy Research and Management Institute (GERMI) have made a significant advancement in energy storage technology. This novel technology, made from seaweed, has a number of benefits in terms of its traits and possible uses. The marine macroalgae known as seaweed was used by the researchers to make the paper-based supercapacitor. They created a material with great tensile strength, performance, and cost-effectiveness by extracting cellulose nanofibers from seaweed and reducing them into graphene oxide and zinc oxide [41]. Figure 10.5 illustrates the process of generating activated carbon from garlic roots and provides a detailed depiction of the subsequent electrode assembly. This figure highlights the transformation of raw garlic roots into a functional material, emphasizing the steps involved in the activation process and the integration of the resulting carbon into a sophisticated electrode configuration. Figure 10.6 presents the innovative development of paper-based supercapacitors fabricated from seaweed. This figure showcases the transformation of seaweed into a sustainable, high-performance material, underscoring the eco-friendly approach and advanced engineering techniques employed in the creation of these cutting-edge energy storage devices.

Sahana Nayak *et al.* proposed an environmentally friendly synthesis technique for producing nickel cobaltite ($NiCo_2O_4$), a nanomaterial for supercapacitor electrodes in their study. To create nanoparticles, *Moringa oleifera* (drumstick) leaf extracts were employed. At various current densities, the manufactured nanomaterial showed comparatively good specific capacitance values. Additionally, a supercapacitor made from artificial nanomaterials and activated carbon demonstrated an improved energy density of 19.8 Wh $kg^{-1}$ at a power density of 748.81 W $kg^{-1}$ and an excellent retention capacity of almost 85% after 8000 cycles at a current density of 5 A $g^{-1}$. Green synthesis is likely to be an intriguing technique in the manufacture of electrode materials for supercapacitors with sufficient energy storage capacity in the future due to its inherent benefits over traditional chemical methods [42].

Riya Thomas *et al.* proposed a cutting-edge strategy to ensure the highly valuable utilization of materials derived from petroleum and coal in supercapacitor utilization in this study, enabling an integrated brown-to-green transition for energy, ecology, and the environment. Fuel coke was utilized to create porous graphene sheets co-doped with nitrogen and phosphorus (a-NPGO), which exhibit remarkable electrochemical properties as a supercapacitor electrode material. The a-NPGO has a high specific capacitance (322 F $g^{-1}$ at 1 A $g^{-1}$), which is about 11 times that of the undoped graphene derivative based on coke. After 5000 charging and discharging cycles, this symmetrical supercapacitor constructed with a-NPGO electrodes delivered an outstanding power density around 812 W $kg^{-1}$ with an energy density of 14 Wh $kg^{-1}$ including exceptional capacitance retention of 90%. The use of coke-derived materials in supercapacitors with excellent performance has the potential to widen the scope of the electrochemical storage of energy idea and allow for the environmentally acceptable use of fossil fuels [43].

Celery leaves, a common biomass vegetable, are employed by Jiaqi Wu *et al.* in their study to make activated carbon for high voltage and high-capacity supercapacitors using a straightforward approach. Hierarchically porous carbon materials can concurrently achieve high capacity and extended cycling durability because of their higher surface area and abundance of nitrogen and sulfur heteroatoms. According to data, specific capacitance may achieve 753.0 F $g^{-1}$ at 0.5 A $g^{-1}$, and capacity retention can reach 96.7% after 5000 cycles. A high energy density of 17.36 Wh $kg^{-1}$ at 250 W $kg^{-1}$ was also provided by the built symmetric cell. Systematically investigated were the impacts of yellow and green colors, as well as various activation temperatures, on the characteristics of electrode materials. The findings demonstrated good electrochemical performance in biomass carbon electrode materials, which would aid in the creation of an eco-friendly supercapacitor and a high-quality energy system [44].

Walnut green peel from agricultural waste has the potential to be an outstanding activated carbon electrode material, and its use as a resource would significantly reduce the issue of environmental pollution. Here, utilizing melamine as the nitrogen source, synthetic N-doped activated carbon (AC) materials (N/ACs) created using walnut green peel were described by Na Tian *et al.* N/ACs possessed a high specific surface area, a low resistance, and a fast transfer of charge rate. The improved N/AC-2 had the best capacitive performance of all, with a specific capacitance of 304.5 F $g^{-1}$ at 0.5 A $g^{-1}$. N/AC-2 displayed high power density (22.5 kW $kg^{-1}$) and energy density (40.05 W h $kg^{-1}$) as the supercapacitor electrode. Furthermore, after 3000 cycles at 10 A $g^{-1}$, the capacitance retention rate of N/AC-2 was 95.2%. Yue Liu *et al.* demonstrated that the generated N/ACs are an ideal prospective material for electrodes in supercapacitors [45].

In their work, José Tiago Carvalho *et al.* developed 1D fiber-shaped supercapacitors (FSCs) using stretch-broken carbon fiber yarns (SBCFYs) as

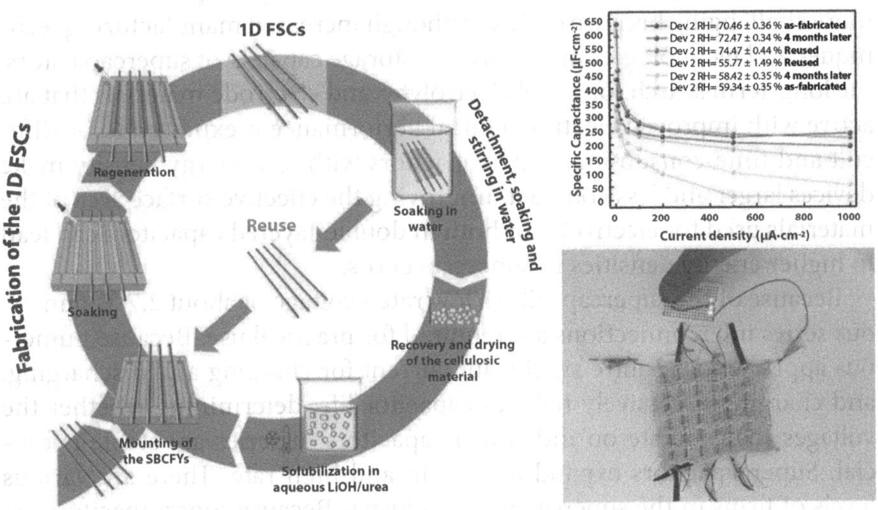


**Figure 10.7** Carbon yarn-based supercapacitors [46].

current collectors and an *in situ* regenerating cellulose-based ionic hydrogel (RCIH) to serve as an electrolyte. The areal specific capacitance at 5 A $cm^{-2}$ is 433.02 F $cm^{-2}$, and its specific energy density is $1.73 \times 10^{-2}$ Wh $cm^{-2}$. The highest specified power density is $5.33 \times 10^{-1}$ mW $cm^{-2}$ at 1 mA $cm^{-2}$. The 1D FSCs outperformed equivalent ones made using the reference PVA/$H_3PO_4$ gel electrolyte in terms of longer life cycle and capacitance retention after 10,000 successive voltammetry cycles. Furthermore, the viability and repeatability of the manufactured devices was demonstrated by connecting three devices in parallel and series configurations, with a slight change in the density of current in the flat and bend orientations. In order to ensure a circular economy path, an environmentally friendly strategy was used to recover the active ingredients from the 1-dimension fiber-shaped supercapacitors and reuse or recycle them without sacrificing the electrochemical performance [46]. Figure 10.7 illustrates the development and structure of carbon yarn-based supercapacitors. This figure provides an in-depth look at how carbon yarn is utilized to create highly efficient supercapacitors, emphasizing the advanced material properties and innovative design strategies that enhance their performance and energy storage capabilities.

## 10.8 Current Challenges of Supercapacitors

Supercapacitors have relatively low energy densities. There are still chances for advancement in the energy density of supercapacitors and batteries at the moment [47–49]. The scientific investigation of supercapacitors becomes more challenging because of this. Although increased manufacturing techniques and technologies may boost the storage capacity of supercapacitors, the long-term search for novel electrolytes and electrode materials that are active with improved electrochemical performance is expected to be difficult and time-consuming. Supercapacitors with low energy density make devices larger and less compact. Improving the effective surface area of the materials used for electrodes or both in double-layered capacitors can lead to higher energy densities in supercapacitors.

Because of the supercapacitor's low rated voltage of about 2.7 V, numerous series interconnections are required for practical use. Because numerous applications require significant current for charging and discharging, and charging excessively reduces capacitor life, determining whether the voltages that operate on individual capacitors in series are stable is crucial. Supercapacitors expand at a rapid and high rate. There are various levels of firms in the supercapacitor industry. Because supercapacitors are an innovative kind of energy storage device, their healthy development is inextricably linked to market and industry regulation, which seeks to set appropriate regional, national, and even international standards. For supercapacitors, a system of technical regulations should be established, which includes terms, a model for classification naming method, an electrical performance assessment method, safeguarding requirements, general norms, material for electrode specifications, electrolyte specifications, charging specification series, manufacturing scientific requirements, transport demands, reconstruction needs, and devastation demands.

For instance, the general specifications and the management and storage needs for supercapacitor monomers and modules in scrap disposal, including disassembling and recycling of scrap monomer and supercapacitor modules, electrolyte capacitor shell processing, plate processing, diaphragm processing, and other aspects, seek to guide and standardize the supracapacitor industry towards achieving the objective of low-cost, environmentally friendly reusing disposal [50]. It is an essential tool for boosting the industry's thriving development.

## 10.9 Future Scope of Supercapacitors

The scope is very bright. Supercapacitors are creating solutions for electrical systems, wearable electronics, and medical implants that were previously unimaginable. They will harness energy from body heat and movements to power medical implants including pacemakers, knee implants, and others. This prevents the need for replacement over the long term. Due to their ability to charge quickly, supercapacitors will be used by robots and UAV. In the future, supercapacitors applications can be used in a variety of ways. Buses, trams, and trains have already started using ultracapacitors for vehicle power. Kinetic Energy Recovery Systems (KERS) are also used by trains, which saves a significant amount of energy. Cars may begin to be powered by them when energy density rises. Currently, numerous tasks are being taken on by more people. The supercapacitor is also being used in smart grids, micro and small grids, and renewable energy systems.

Supercapacitor energy storage devices are anticipated to have a bright future, especially in the fields of renewable energy and transportation. They outperform conventional batteries in a number of ways, including quicker charging and discharging times, longer lifetimes, and greater resistance to temperature extremes. Supercapacitors are being studied and developed for use in hybrid and electric vehicles in the transportation sector. They can be used to store regenerative braking energy, which can then be used to power the car while it accelerates. This may lead to increased energy efficiency and longer electric-only driving ranges. Supercapacitors are being researched as a technique of storing energy produced by sporadic sources like solar and wind energy in the renewable energy sector. They can be utilized to store energy produced during high production periods, which can subsequently be released during low production periods. This can lessen the requirement for fossil fuel-based power generation and smooth out the energy distribution to the grid. Supercapacitor energy storage system research and development are continuing, and it is expected that new and better technologies will continue to appear in the future.

## 10.10 Conclusions

Supercapacitor energy storage devices are anticipated to have a bright future. While the market for supercapacitors expands, so does the demand for enhanced fabricating procedures and material for electrodes. As the world explores for solutions to energy problems, supercapacitors are anticipated to expand in popularity in the upcoming years. Yet, the supercapacitor implementations that have been discovered are a thrilling realization of a traditional technology that is only improving over time. The ultimate goal in both research and industry for supercapacitors is to attain a higher power density while maintaining higher energy density at the same time. In addition to the clever design of capacitors, the creation of electrodes, electrolytes, and separator materials can help to complete this challenging task. The current state of the market also predicts exponential expansion of supercapacitors in the future.

## References

1. Tie, D., Huang, S., Wang, J., Zhao, Y., Ma, J., Zhang, J., Hybrid energy storage devices: Advanced electrode materials and matching principles. *Energy Storage Mater.*, 21, 22–40, 2018.
2. Zuo, W., Li, R., Zhou, C., Li, Y., Xia, J., Liu, J., Battery- Supercapacitor Hybrid Devices: Recent progress and future prospects. *Adv. Sci.*, 4, 7, 1600539, 2017.
3. Wang, Y., Song, Y., Xia, Y. J. C. S. R., Electrochemical Capacitors: mechanism, materials, systems, characterization and applications. *Chem. Soc. Rev.*, 45, 21, 5925–5950, 2016.
4. Díaz-González, F., Sumper, A., Gomis-Bellmunt, O., Villafáfila-Robles, R., A review of Energy storage technologies for wind power applications. *Renew. Sustain. Energy Rev.*, 16, 4, 2154–2171, 2012.

5. Zou, Y., Hu, X., Ma, H., Li, S.E., Combined state of charge and state of Health estimation over lithium-ion battery cell cycle life span for electric vehicles. *J. Power Sources*, 273, 793–803, 2015.
6. Zhang, L., Hu, X., Wang, Z., Sun, F., Dorrell, D.G., A review of Supercapacitor modeling, estimation, and applications: A control/management perspective. *Renew. Sustain. Energy Rev.*, 81, 1868–1878, 2018.
7. Conway, B.E., *Electrochemical Supercapacitors: Scientific Fundamentals and Technological Applications*, New York, Kluwer-Plenum, 1999.
8. Burke, A., Ultracapacitors: why, how, and where is the technology. *J. Power Sources*, 91, 1, 37–50, 2000.
9. Kotz, R. and Carlen, M., Principles and applications of electrochemical capacitors. *Electrochim. Acta*, 45, 15-16, 2483–2498, 2000.
10. Aricò, A.S., Bruce, P. *et al.*, Nanostructured materials for advanced energy conversion and storage devices. *Nat. Mat.*, 4, 366–377, 2005.
11. Chu, A. and Braatz, P., Comparison of commercial supercapacitors and high-power lithium-ion batteries for power-assist applications in hybrid electric vehicles I. Initial characterization. *J. Power Sources*, 112, 1, 236–246, 2002.
12. Halper, M.S. and Ellenbogen, J.C., Supercapacitors: A Brief Overview, in: *MITRE*, Nano systems Group, McLean, VA, USA, 2006.
13. Pandolfo, A.G. and Hollenkamp, A.F., Carbon properties and their role in supercapacitors. *J. Power Sources*, 157, 11–27, 2006.
14. Kinoshita, K., *Electrochemical Oxygen Technology*, Wiley, Hoboken, NJ, USA, 1992, ISBN 978-0-471-57043-1.
15. Li, X. and Wei, B., Facile synthesis and super capacitive behaviour of SWNT/MnO2 hybrid films. *Nano Energy*, 1, 479–487, 2012.
16. Tie, D., Huang, S., Wang, J., Zhao, Y., Ma, J., Zhang, J., Hybrid energy storage devices: Advanced electrode materials and matching principles. *Energy Storage Mater.*, 21, 22–40, 2018.
17. Zuo, W., Li, R., Zhou, C., Li, Y., Xia, J., Liu, J., Battery-Supercapacitor Hybrid Devices: Recent Progress and Future Prospects. *Adv. Sci.*, 4, 7, 1600539, 2017.
18. Shifei, H., Zhu, X., Sarkar, S., Zhao, Y., Challenges and opportunities for supercapacitors. *APL Mater.*, 7, 100901, 2019.
19. Afif, A., Rahman, S.M.H., Azad, A.T., Zaini, J., Islan, M.A., Azad, A.K., Advanced materials and technologies for hybrid supercapacitors for energy storage – A review. *J. Energy Storage*, 25, 100852, 2019.
20. Wang, R., Li, X., Nie, Z., Zhao, Y., Wang, H., Metal/metal oxide nanoparticles-composited porous carbon for high-performance supercapacitors. *J. Energy Storage*, 38, 102479, 2021.
21. Ruiz-Montoya, J.G., Quispe-Garrido, L.V., Gómez, J.C., Baena-Moncada, A.M., Gonçalves, J.M., Recent progress in and prospects for supercapacitor materials based on metal oxide or hydroxide/biomass-derived carbon composites. *Sustain. Energy Fuels*, 5, 5332–5365, 2021.

22. Jitropas, U., Saengsui, W., Meevasana, W., Experimental study of a carbon-based planar supercapacitor in an aqueous electrolyte. *J. Energy Storage*, 66, 107399, 2023.
23. Ahmad, A., Gondal, M.A., Hassan, M., Iqbal, R., Ullah, S., Alzahrani, A.S., Memon, W.A., Mabood, F., Melhi, S., Preparation and Characterization of Physically Activated Carbon and Its Energetic Application for All-Solid-State Supercapacitors. *ACS Omega*, 8, 21653–21663, 2023.
24. Khandare, L.N., Late, D.J., Chaure, N.B., $MoS_2$ nanobelts-carbon hybrid material for supercapacitor applications. *Front. Chem.*, 11, 1166544, 2023.
25. Wu, Z.-S., Wang, D.-W., Ren, W., Zhao, J., Zhou, G., Li, F., Cheng, H.-M., Anchoring Hydrous $RuO_2$ on Graphene Sheets for High-Performance Electrochemical Capacitors. *Adv. Funct. Mater.*, 20, 3595–3602, 2020.
26. https://www.graphene.info.com/update-skeleton-technologies-graphene-based supercapacitors.
27. Molahalli, V., Chaithrashree, K., Singh, M.K., Agrawal, M., Krishnan, S.G., Hegde, G., Past decade of supercapacitor research – Lessons learned for future innovations. *J. Energy Storage*, 70, 108062, 2023.
28. ARAI showcases Tata Magic Hybrid vehicle developed with ISRO.
29. Novel, Solid-State Hybrid Ultracapacitor Battery.
30. https://www.utep.edu/newsfeed/2023/researchers-achieve-historic-milestone-in-energy-capacity-of-supercapacitors.html.
31. https://www.hindu.com/sci-tech/science/indian-korean-researches-develop-light-powered-supercapacitors/article 66950924.ece.
32. https://www.hindustantimes.com/science/iisc-researches-design-for-storing-enormous-electric charge-101680328874559.html,2023.
33. https://english.dicp.cas.cn/news/in/202302/t20230220-327450.html.
34. https://www.imperial.ac.uk/news/223353/new-generation-supercapacitors-electrify-green transportation.
35. Thi, H.B., Patil, S.A., Katkar, P.K., Kaseem, M., Nazir, G., Lu, S.-W., Patil, H., Kim, H., Magotra, V.K., Im, H., Shrestha, N.K., Cu@Fe-Redox Capacitive-Based Metal–Organic Framework Film for a High-Performance Supercapacitor Electrode. *Nanomaterials*, 13, 10, 1587, 2023.
36. Mousaabadi, K.Z., Ensafi, A.A., Naghsh, E., Hu, J.-S., Rezaei, B., Dual Ni/Co-hemin metal–organic framework-PrGO for high-performance asymmetric hybrid supercapacitor. *Sci. Rep.*, 13, 12422, 2023.
37. Zhao, J., Yang, L., Li, R., Zhou, Y., One–Step Synthesis of Fe–Based Metal–Organic Framework (MOF) Nanosheet Array as Efficient Cathode for Hybrid Supercapacitors. *Inorganics*, 11, 4, 169, 2023.
38. https://news.mit.edu/2023/mit-engineers-create-supercapacitor-ancient-materials-0731.
39. Chu, R., Tan, S., Qiao, Y., Fu, W., Zhang, K., Zhu, P., Liu, D., Huang, W., A carbonized coconut husk for supercapacitor electrode. *J. Phys. Conf. Ser.*, 2563, 012031, 2023.

40. Bilal, M., Li, J., Guo, H., Landskron, K., High-Voltage Supercapacitive Swing Adsorption of Carbon Dioxide. *National Library of Medicine*, 19, 24, 2207834, 2023.
41. https://www.gktoday.in/Paper-basedSupercapacitor-made-from-seaweed/2023.
42. Nayak, S., Kittur, A.A., Nayak, S., Green Synthesis of Nickel Cobaltite Using Moringa oleifera Plant Extract for Electrode Materials in Sustainable Supercapacitor Energy Storage. *J. Electron. Mater.*, 52, 2, 1437–1447, 2022.
43. Thomas, R. and Balachandran, M., Fuel coke derived nitrogen and phosphorus co-doped porous graphene structures for high-performance supercapacitors: The trail towards a brown-to-green transition. *J. Energy Storage*, 12, E, 108799, 2023.
44. Wu, J., Ma, Z., Wang, G., Chen, T., Biomass Porous Carbon Derived from Celery Leaves with High Capacitance for Supercapacitor. *Chem. Select*, 8, 18, e202204616, 2023.
45. Tian, N., Gao, M., Liu, X.-H., Liu, X., Yang, T., Wu, J., Activated carbon derived from walnut green peel as an electrode material for high-performance supercapacitors. *Biomass Conversion Biorefinery*, 13, 16781–16789, 2022.
46. Carvalho, J.T., Cunha, I., Coelho, J., Fortunato, E., Martins, R., Pereira, L., Carbon-Yarn-Based Supercapacitors with *In Situ* Regenerated Cellulose Hydrogel for Sustainable Wearable Electronics. *Appl. Energy Mater.*, 5, 10, 11987–11996, 2022.
47. Tie, D., Huang, S., Wang, J., Zhao, Y., Ma, J., Zhang, Hybrid energy storage devices: Advanced electrode materials and matching principles. *Energy Storage Mater.*, 21, 22–40, 2018.
48. Zhao, Y., Ran, W., He, J., Song, Y., Zhang, C., Xiong, D.-B., Gao, F., Wu, J., Xia, Y., Oxygen-rich hierarchical porous carbon derived from artemia cyst shells with superior electrochemical performance. *ACS Appl. Mater. Interfaces*, 7, 2, 1132–1139, 2015.
49. Zhao, Y., Zhang, Z., Ren, Y., Ran, W., Chen, X., Wu, J., Gao, F., Multiscale Construction of Bifunctional Electro Catalysts for long- lifespan Rechargeable Zinc- Air Batteries. *J. Power Sources*, 286, 1–9, 2015.
50. aip.scitation.org.

# 11

# Future Perspective of Sustainable Supercapacitors

**Umesh V. Shembade[1], Rishikesh A. Moholkar[1], Rohan S. Khot[2], Tushar T. Bhosale[1], Nishigandha B. Chougale[1], Mayuri G. Magadum[1], Sandeep B. Wategaonkar[2,3] and Annasaheb V. Moholkar[1*]**

[1]*Thin Film Nanomaterials Laboratory, Department of Physics, Shivaji University, Kolhapur, (M.S), India*
[2]*Department of Chemistry, D. Y. Patil Agriculture and Technical University, Talsande, (M.S), India*
[3]*Department of Chemistry, Kisanveer Mahavidyalaya, Wai, (M.S), India*

***Abstract***
Presently, the global shift towards an economy that does not rely on fossil fuels is driven to meet environmental and climate goals while promoting sustainable development. To attain this objective, it is crucial to optimize the utilization of resource materials, which can extend the lifespan and economic value of these resources. There remains a significant demand for the development of innovative energy storage solutions that possess remarkable energy and power density, along with extremely long cycling lifetimes. Recent progress in the creation of versatile energy-storage systems, capable of maintaining their electrochemical properties even when subjected to deformation, has been spurred by the increasing market for flexible and wearable electronics. Today, research efforts in sustainable supercapacitors (SCs) and lithium-ion batteries are on the rise as novel materials are sought after to craft high-energy storage devices, which have become indispensable components of modern life. Sustainable SCs exhibit immense promise for premium storage device applications due to their unique attributes such as high energy density, reduced device weight, exceptional cycling stability, and remarkable pseudocapacitive characteristics. Moreover, sustainable SCs have displayed considerable potential for integration into upcoming wearable and implantable electronic technologies. Additionally, these sustainable SCs offer enhanced safety, extended cycling lifetimes, and greater power density when compared to metal-

**Corresponding author*: avmoholkar@gmail.com

Basheer Ahamed and Chaudhery Mustansar Hussain (eds.) Sustainable Supercapacitors: Next-Generation of Green Energy Storage Devices, (323–352) © 2025 Scrivener Publishing LLC

ion batteries. This chapter delves into the chemical principles underpinning the storage mechanisms within sustainable SCs, aiming to surmount obstacles and unlock superior performance for future opportunities. Within this chapter, we explore the most recent breakthroughs and prospective directions in the burgeoning field of materials science and engineering, particularly in the realm of flexible and comfortable devices.

***Keywords*:** Sustainable SCs, storage mechanism, high energy density, symmetric and asymmetric devices, future aspects

## 11.1 Introduction

One of the significant challenges our society confronts is the quest for sustainable energy conversion and storage machinery that are reliable, cost-effective, and environmentally friendly. The new strategies aim to solve the endlessly growing demand for energy and the dwindling reserves of fossil fuels. In the past two to three decades, we have witnessed remarkable progress in the advancement of electrochemical energy-storage systems (ES), such as lithium-ion batteries (LIBs) and supercapacitors (SCs), designed for a wide range of applications, including portable devices, electric vehicles, and large-scale energy-storage systems for the grid. Lately, the drive to develop multifunctional devices and sustain their electrochemical capabilities under deformation has been fueled by the increasing demand in the burgeoning market of elastic and wearable semiconductor technology. Contemporary investigations in the field of SCs and LIBs are motivated by the necessity to employ innovative materials in the creation of high-energy storage devices, which have become indispensable components of modern-day living. Sustainable SCs show promise as top-tier energy storage device (ESD) candidates due to their attractive features, including high energy density (Ed), reduced device weight, exceptional cycling stability, an impressive pseudocapacitive nature, enhanced safety, prolonged cycling lifespan, and higher power density (Pd) compared to metal-ion batteries [1–6].

The swift progress in industries such as chemical, metallurgical, pharmaceutical, and others, coupled with modernization and expanding populations, is causing harm to the environment by releasing substantial quantities of harmful and unwanted compounds. Consequently, sustainable SCs are the reliable, eco-friendly, cost-effective, and non-hazardous procedures for nanomaterial synthesis that represent a burgeoning potential in nanotechnology for addressing this challenge. In contrast to secondary ion batteries, SCs consistently generate energy, making them well-suited for high-power devices. The nanostructured materials offer an improved reaction mechanism, chemical reactivity, quicker ionic diffusion pathways with amplified surface zones, and the presence of a large number of active sites for electrochemical processes [7, 8]. These nanostructured materials reveal different applications in the span of a wide array of fields, including control systems, power generation, transportation, smart devices, consumer electronics, national defense, communications, and medical equipment like defibrillators and pulsed lasers [9]. The combination of outstanding features such as higher Pd, Ed, and exceptional cycling stability makes SCs strong contenders in energy storage device systems (ESDs). These attributes render SCs effective not only in hybrid vehicles but also in various portable electronic devices [10].

According to their methods of energy storage, SCs can generally be divided into three groups: redox capacitors, which use both faradaic and non-faradaic processes to store charge, pseudo-capacitors (PSCs), which use faradic reactions, and electric-double layer capacitors (EDLC), which use the non-faradaic process. For charge storage, the EDLC materials depend on surface area, porosity, and the availability of various carbon-based resources like activated carbon (AC), carbon nanotubes (CNT), carbon aerogels (CA), graphene (GO), and related derivatives. On the other hand, PSCs are made of substances like conducting polymers, metal oxides, and transition metal oxides (TMOs). By promoting the transfer of electrons from an electrode to an electrolyte contact via redox reactions, these materials produce capacitance [11–13].

Nevertheless, a significant portion of the ESD market still requires adjustments to align with the intended objectives. Additionally, the endeavor to create flexible and sustainable ESDs with superior Ed and Pd demands more concerted efforts. In the context of evolving electronic devices, weight holds increasing importance, as thinner, lighter, and more cost-effective designs are sought after, and the utilization of energy work for supporting weight has gained traction in recent years [14]. There is a pressing need for a new paradigm in ESDs to cater to the demands of portable and adaptable electronics. We envision that those flexible and sustainable electronics will open up a multitude of innovative possibilities, ranging from bendable mobile phones to implantable cardiac sensors and flexible organic light-emitting diode (OLED) displays. Consequently, the power sources for flexible and sustainable electronics must deliver substantial Pd and Ed, respectively [15]. Flexible and sustainable SCs have garnered significant attention from both research and industry circles due to their exceptional power capabilities and extended lifespan when compared to traditional batteries. Nonetheless, crafting affordable, lightweight, and environmentally friendly sustainable SCs remains a formidable challenge. The future progress of sustainable SCs primarily revolves around augmenting the ED of the device through the progress of novel electrode resources, different electrolytes, and innovative network architectures for the devices. Further, it is imperative to refine and adapt sustainable SCs based on current collector technologies, allowing them to be seamlessly integrated with energy-generating equipment or power sources [16, 17].

In this book chapter, we put forth our specific perspectives concerning the challenges, possibilities, and future outlook for sustainable SCs within the domain of energy storage. The present chapter provides a comprehensive elucidation on the sustainable SCs, encompassing the appropriate selection of materials for electrodes, electrolytes, separators, and current collectors, respectively. Additionally, we have reported various carbon-based materials, including electrodes derived from activated carbon, carbon nanotubes (CNTs), and graphene, as well as metal oxides and transition metal oxide/hydroxide-based electrodes, which are briefly discussed as potential materials for SC applications. Additionally, this section delves into the role of various electrolytes in the field of sustainable SCs to address the scientific research. Furthermore, this chapter includes different aspects to overcome the drawbacks the sustainable SCs, which are beneficial for the energy storage phenomenon.

## 11.2 Research Motivation and Objectives of the Sustainable SCs

In the realm of sustainable SCs, extensive research has been dedicated to exploring various electrode materials for energy storage purposes. However, improving the device efficiency stands as a critical objective, and further, the devices have been designed in a such way as to overcome the limitations of the device in energy storage systems. According to the energy storage mechanism, the SCs are classified into three groups, which are shown in Figure 11.1. Furthermore, the utilization of oxides, sulfides, and phosphates as electrode materials gained traction with the introduction of pseudocapacitance with excellent electrochemical performance. Now, a carbon-based electrode has remained a dominant choice in SCs owing to their expansive surface area, a characteristic that has demonstrated substantial progress over time. Additionally, the motivation for developing electrode materials for energy storage was initially drawn from the field of

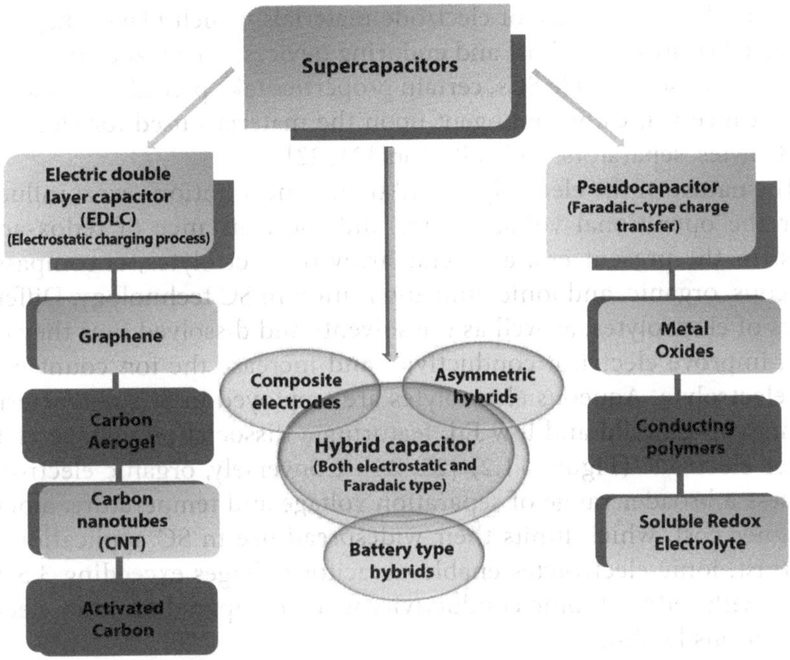


**Figure 11.1** Classification of the sustainable SCs based on the charge storage mechanism with the types of electrode materials.

batteries and the limited advancements and success in battery technologies hindered their application in SCs. The evolution of sustainable SCs gained momentum through the creation of numerous composite nanomaterials, incorporating diverse working electrode materials to enhance charge storage mechanisms [18–20]. These materials encompass conducting polymers like polyaniline, polypyrrole, and polythiophene, as well as carbon nanomaterials such as carbon nanotubes and graphene, porous carbon materials, and carbon paper. Besides, considerable efforts have been invested in manufacturing high-performance flexible SCs, employing a range of electrode materials.

In contemporary times, numerous researchers have dedicated their efforts in the charge storage field to the involvement of diverse electrode materials to meet the requirements of sustainable SCs within the domain of energy storage. The extensive research conducted in the SC field has underscored the critical role played by the choice of both electrolyte and electrode materials in shaping the properties of SCs. Recent reports emphasize that the SC performance and availability will be predominantly influenced by the choice of electrode materials, which plays a key factor in the fabrication of robust and enduring devices. To advance progress in the realm of sustainable SCs, certain properties take precedence, and their significance is greatly contingent upon the materials used for electrodes, electrolytes, separators, and collectors [21, 22].

The nature of the electrolytes and their concentrations exert influence over the operational voltage range and the resistance of redox-active sites. In the present era, a diverse array of electrolytes, encompassing aqueous, organic, and ionic, find application in SC technology. Different types of electrolytes, as well as the solvents and dissolved ions they contain, improve electrical conductivity and increase the ion count inside the electrolyte. Aqueous electrolytes are employed in SCs characterized by high specific Pd and low Ed, featuring a dissociation voltage of 1.15 V per electrode (Figure 11.2) [23, 24]. Conversely, organic electrolytes possess a broader range of separation voltage and temperature, albeit at a higher cost, which limits their widespread use in SC applications. In contrast, ionic electrolytes enable capacitor voltages exceeding 3.5 V s, albeit with reduced ionic conductivity when compared to other electrolyte options [7, 25].

Despite significant advancements in the application of SCs achieved through the use of improved electrode materials and their performance, most energy storage devices necessitate collectors to establish connections between capacitor electrodes and enhance the performance of SCs. This is primarily due to the limited conductivity of the active material, and

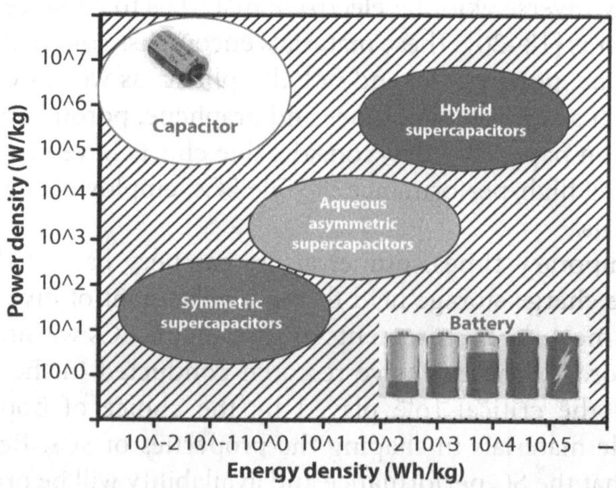


**Figure 11.2** Ragone plot of the different energy storage devices.

collectors serve as the primary conduit for charge and discharge currents. Additionally, separators play a pivotal role in shaping the performance of SCs, particularly when forming symmetric and asymmetric hybrid devices. However, to mitigate any detrimental effects, separators must possess both porosity and extreme thinness. In the current market landscape, polymer-based separators with advanced development take the lead, offering high flexibility, cost-effectiveness, and porosity.

**Research objective 1:** Establish substantial ideas that are fundamental to symmetric solid-state hybrid device (SSHD) practice and research by the electrodes that are crucial to the operation of SCs.

**Research objective 2:** Prepare a research agenda for asymmetrical solid-state hybrid devices (ASSHD) by accumulating suggestions from the literature, and finally executing the results.

## 11.3 The Challenges for Sustainable SCs

Sustainable SCs are extensively employed in a variety of industries, including transport, storage mechanisms, the detection sector, and consumer

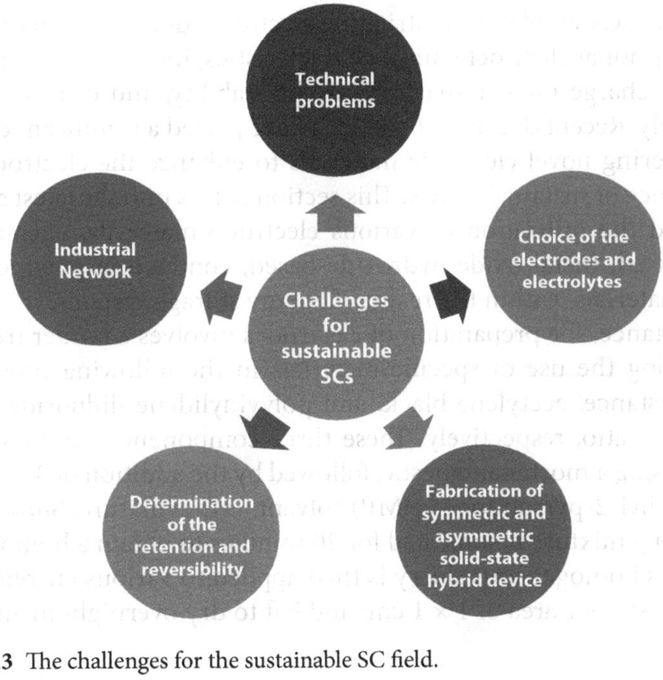


**Figure 11.3** The challenges for the sustainable SC field.

electronics as well as in the domestic field, due to their exceptional qualities such as high Ed, Pd, and excellent cycling stability over thousands of cycles. However, there are some difficulties for the above devices and the following aspects are given to highlight the present issues and how they can be resolved (Figure 11.3).

## 11.3.1 Technical Problems

The main flaw of SCs is their low Ed, which presents significant obstacles to their use in practical applications as a power storage device. Due to the low Ed, the devices show lower performance and enhance the ability or the Ed of the sustainable SCs using the effective surface area of the electrode materials, working potential window, nature of the electrodes and electrolytes, or the incorporation of the redox additives into the electrolytes. It has been also confronted with technical issues, modifying the new models, then its accuracy measurement, and the strategies for developing standards for the energy storage sector [26].

## 11.3.2 Choice of Electrodes and Electrolytes

### *11.3.2.1 Role of the Electrode Material in the Field of Sustainable SCs*

Sustainable SCs stand out as intriguing electronic devices, primarily due to their exceptional electrochemical characteristics, including high Ed, rapid charge–discharge rates, remarkable cyclic stability, and capacitance (Cs) respectively. Recent decades of research have placed a significant emphasis on uncovering novel electrode materials to enhance the electrochemical performance of sustainable SCs. This section delves into the latest advancements and the utilization of various electrode materials, encompassing carbon-based, metal oxide/hydroxide-based, conducting polymer-based, and 2D materials, within the realm of energy storage systems.

For instance, the preparation of electrodes involves a binder treatment, necessitating the use of specific materials in the following proportions: active substance, acetylene black, and polyvinylidene difluoride (PVDF) in an 8:1:1 ratio, respectively. These three components are meticulously blended using a mortar and pestle, followed by the addition of 3–4 drops of the N-methyl-2-pyrrolidone (NMP) solvent to the mixture. Subsequently, the resulting mixture is sonicated for 20 minutes to obtain a homogeneous slurry. This homogeneous slurry is then applied to various current collectors with a surface area of $1 \times 1$ cm$^2$ and left to dry overnight in an oven at 80 °C.

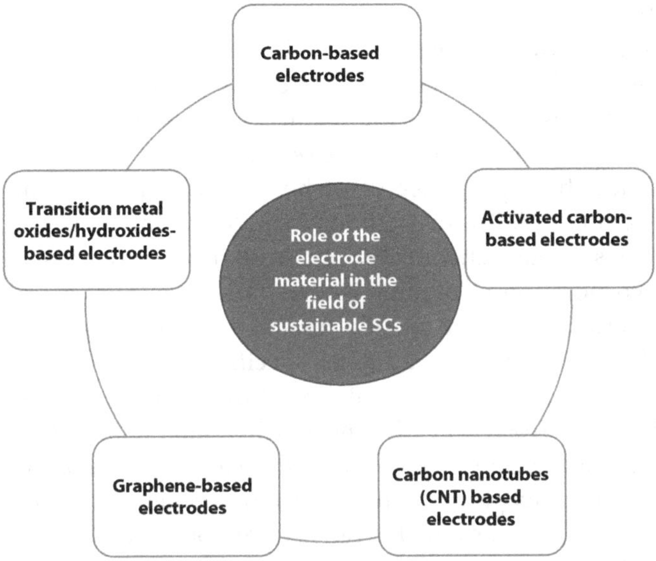


**Figure 11.4** Role of the electrode materials in the field of sustainable SCs.

The different active material-based electrodes are used for the examination of the electrochemical performance of the prepared material and the resultant classification is reported below and graphically shown in Figure 11.4.

**a) Carbon-Based Electrodes**

Carbon compounds are the most widely favored materials for constructing SC electrodes and are present in various forms. This preference arises from several key factors, including their expansive surface area, affordability, accessibility, and well-established manufacturing processes for electrodes using carbon-based substances. These carbon-based materials operate by forming an electrochemical double layer at the junction between the electrode and electrolyte, serving as a storage system. Consequently, the capacitance primarily hinges on the available surface area for interaction with electrolyte ions.

Numerous factors play a crucial role in influencing the electrochemical performance of these materials, including the specific surface area, pore structure, and arrangement, distribution of pore sizes, surface modifications, and electrical conductivity of the active substance and the nature of the working electrodes. The formation of an exceedingly thin film on the electrode material's surface, known as the Helmholtz layer, is a pivotal aspect of the charge storage mechanism for carbon-based electrodes. This layer interfaces with electrolyte ions, both non-dissolved and weakly-dissolved ions. The presence of these dissolved ions, driven by strong electrostatic forces, results in the formation of an expanded outermost Helmholtz layer where charge storage occurs.

To enhance the capacitance of carbon-based materials, various techniques are employed. These methods include the introduction of redox-active moieties through chemical doping with redox-active small molecules and the incorporation of organic isomers adsorbed onto a carbon material framework to generate pseudocapacitive carbon-based materials. In the case of carbon materials, a higher specific surface area translates to a greater capacity for charge accumulation at the interface between the electrode and the electrolyte. Furthermore, when augmenting the Cs of carbon materials, it is essential to consider surface functionalization. Further, materials such as activated carbon, carbon aerogels, carbon nanotubes, graphene, and other carbon derivatives are commonly utilized as electrodes [27–30].

**b) Activated Carbon-Based Electrodes**

The favored electrode material for SCs is activated carbon (AC) due to its substantial surface area, robust electrical properties, and cost-effectiveness. The AC can be produced through the physical or chemical activation of various carbonaceous materials, including wood, coal, nutshells, and others. The physical activation involves subjecting carbon precursors to high temperatures (typically ranging from 700 to 1200°C) in the presence of oxidizing gases such as steam, $CO_2$, or air. The resulting activated carbon exhibits a broad range of physicochemical properties, including surface area, porous structure, and pore size distribution, which depend on the carbon precursors and activation processes used for the fabrication. Additionally, the researchers have extensively explored the relationship between Cs and the specific surface area (SSA) of activated carbon in the field of energy storage. However, there exists some disagreement among researchers regarding this relationship and some studies have indicated that despite a high SSA of around 3000 $m^2/g$, relatively low Cs values were achieved [31]. This discrepancy suggests that not all pores in AC effectively contribute to charge accumulation. Consequently, while SSA remains a critical parameter for electrochemical performance in EDLCs, other factors such as pore size distribution also significantly impact performance.

For activated carbon electrodes in aqueous electrolytes, which have larger ions, the Cs values range between 100 and 300 F/g and tend to be higher compared to organic solutions (approximately 150 F/g). Research by Wang *et al.* [32] revealed that increasing the SSA of AC from 621 $m^2/g$ to 2685 $m^2/g$ resulted in enhanced Cs of electrodes, rising from 17.68 F/g to 171.2 F/g, respectively. However, excessive activation can lead to a high pore volume, which can have adverse effects, including reduced conductivity and material density. These consequences ultimately result in lower Ed and diminished power capacity.

**c) Carbon Nanotube (CNT)-Based Electrodes**

The discovery of carbon nanotubes (CNT) has significantly advanced the science and engineering of carbon materials. The power density of a supercapacitor is primarily determined by the total resistance of its components. CNT has garnered significant attention as a potential electrode material for supercapacitors due to its unique characteristics, including its exceptional mechanical and thermal stability, remarkable electrical properties, and distinctive pore structure.

CNT electrodes have been manufactured to take advantage of their special attributes, such as their high surface area (SSA) accessible exterior structure, the presence of mesopores within their interior network, as well as their chemical and thermal stability, outstanding electrical conductivity, and low mass densities. Furthermore, CNTs exhibit superior connectivity and the ability to establish a continuous distribution, which enhances their widely utilized SSA. This is achieved thanks to the mechanical flexibility of CNTs and their open network [33, 34].

**d) Graphene-Based Electrodes**

Graphene, a two-dimensional structure composed of $sp^2$ covalently bonded carbon atoms arranged in a honeycomb-like crystal lattice, is merely one atom thick. This unique manifestation of carbon possesses a plethora of outstanding attributes. These include exceptional mechanical properties, remarkable morphological characteristics, high electrical conductivity, substantial carrier mobility, a significant surface area, and remarkable chemical stability. Moreover, the two-dimensional nature of graphene results in a thinner electrode, which, in turn, yields an electrode with an enlarged potential window, enhanced flexibility, chemical stability, and thermal stability. Due to these distinctive qualities, graphene stands as a formidable candidate for applications in energy storage [35].

In light of this, graphene may be used as electrode material in high-performance SC applications since it differs from other carbon-based materials like AC and CNTs in that it does not rely on how pores are distributed in the solid state. The fact that the electrolyte can contact both of the graphene sheet's external surfaces is another benefit of using it as an electrode material. To maximize the inherent surface capacitance and specific surface area of single-layer graphene, precautions have been taken to prevent the rearrangement of graphene sheets during all stages of graphene development and subsequent electrode fabrication processes. One effective method involves creating curved graphene sheets, which eliminates the need for face-to-face redesign [36].

**e) Transition Metal Oxide/Hydroxide-Based Electrodes**

Metal oxides present a choice of materials for crafting supercapacitor electrodes, thanks to their high specific capacitance and low resistance. This characteristic simplifies the construction of supercapacitors with substantial energy storage and rapid discharge capabilities. Among the commonly employed metal oxides are nickel oxide (NiO), ruthenium dioxide ($RuO_2$), manganese oxide ($MnO_2$), and iridium oxide ($IrO_2$), respectively. These materials offer an attractive option due to their cost-effectiveness in production and compatibility with milder electrolytes. The most prevalent substances exhibiting redox activity in SCs, particularly hybrid supercapacitors (HSCs), are transition metal-based compounds, primarily in the forms of oxides and hydroxides. Transition metals like nickel (Ni), iron (Fe), cobalt (Co), titanium (Ti), and others play a pivotal role in the creation of transition metal oxides (TMOs) and transition metal hydroxides [37].

One of the alternate materials for SC electrodes with high energy and power is TMOs, which have larger Cs than carbon material, a lower resistance than carbon, as well as a high conductivity. TMOs and CPs are the most common redox pseudocapacitive materials because they both display rapid reversible redox reactions, which is one of their properties. The high redox reaction rates of these materials enable these long working cycles and large specific capacitances. There are various TMOs with charge storage mechanisms based on the faradic behavior, including $RuO_2$, $Fe_3O_4$, $MnO_2$, NiO, $Co_3O_4$, $IrO_2$, as well as $V_2O_5$ and CuO as specified by B.E. Conway. [38]. In addition, layered transition metal hydroxides (TM(OH)s) with a large space between their layers, including Ni $(OH)_2$, FeOOH, and Co (OH), offer sufficient redox reactions, raising the capacitance value of the electrodes in SCs. The latter could provide greater charge storage than the former, though, as their associated oxides have greater electronic conductivity than the TM(OH)s. The structural stability of TM(OH)s is less stable than that of equivalent oxides, which has resulted in a reduction in rate capability or cycle performance. One of many approaches to get over this restriction and improve the cycling stability of the material is layering the TM(OH)s through the anion and cation doping procedure. For example, bimetallic complexes can be created by partially substituting additional TM cations like $Co^{2+}$, $Ni^{3+}$, etc. for the $Ni(OH)^2$ to increase its cycle life [39].

### 11.3.3 Role of the Electrolytes in the Sustainable SCs

In the realm of electrochemical supercapacitors, the term "electrolyte" encompasses both the electrolyte salt and the solvent it is dissolved in. This component plays a pivotal role in shaping the formation of the double layer and the accessibility of pores to electrolyte ions. It also directly wields influence over the operational voltage range and the overall resistance of the cell. The ionic resistivity, inversely proportional to a cell's power generation capability, is a key consideration, while the energy density is directly related to the square of the voltage range. Consequently, the meticulous selection of the appropriate electrolyte stands as a critical factor in the construction of high-performance supercapacitors. Safety considerations for energy storage systems (ESs) dictate that electrolyte materials should exhibit low volatility, minimal flammability, and corrosion potential, along with broader potential windows. These features contribute to enhanced chemical and electrochemical stability without compromising performance. The nature of the electrolyte encompasses several aspects, including the size and composition of ions, their concentrations, the types of solvents they interact with, the nature of ion–solvent interactions, the interaction between electrode materials and the electrolyte, and the range of allowable voltage [23]. The resultant classification of the electrolytes that are useful in the field of sustainable SCs is presented in Figure 11.5.

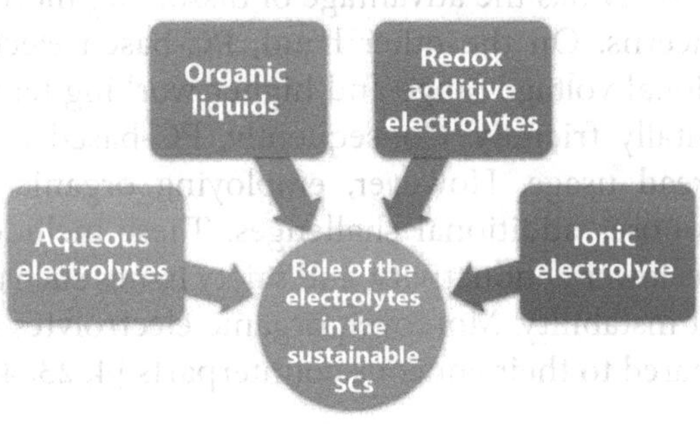


**Figure 11.5** Role of the electrolytes in the field of sustainable SCs.

**a) Aqueous Electrolytes**

In laboratory settings, aqueous electrolytes offer a more straightforward option compared to ionic liquids and organic electrolytes, which necessitate purification procedures before they can be used in the context of SCs. Aqueous electrolytes have been extensively utilized in research and development due to their widespread availability. Additionally, they exhibit superior conductivity when compared to organic and ionic electrolytes, a factor that aids in reducing equivalent series resistance and enhancing the power output of SCs. The aqueous electrolyte such as alkaline, neutral, and acidic solutions shows excellent electrochemical performance based on the reaction mechanism with electrode materials. The most common aqueous electrolytes are KOH, NaOH, LiOH, $Na_2SO_4$, $H_2SO_4$, $(NH_4)_2SO_4$, $K_2SO_4$, $Li_2SO_4$, $MgSO_4$, $CaSO_4$, $BaSO_4$, KCl, NaCl, LiCl, HCl, CsCl, $CaCl_2$, $KNO_3$, $LiNO_3$, $Na(CH_3COO)$, $Li(CH_3COO)$, $Mg(CH_3COO)_2$, $Na_2HPO_4$, etc. The drawbacks of aqueous electrolytes include worries about corrosion and a narrow stability window (V), which reduce cell performance and stability. Aqueous electrolytes frequently limit their cell voltage to 1 V due to water breakdown occurring at 1.23 V.

The maximum amount of power and energy that can be concentrated in an ES is significantly constrained by the low voltage stability of aqueous electrolytes [40, 41].

**b) Organic Electrolytes**

Typically, organic electrolytes, which consist of organic solvents with embedded conducting salts, are employed in supercapacitors (SCs). Currently, these organic electrolyte-based SCs dominate the commercial sector of electrochemical supercapacitors due to their operational range, typically falling between 2.2 and 2.7 V. The two most commonly used organic electrolyte solvents are acetonitrile (ACN) and propylene carbonate (PC). While ACN has the advantage of dissolving more salts, it comes with safety concerns. On the other hand, PC-based electrolytes offer a broader operational voltage range and higher working temperatures, and are environmentally friendly. Consequently, PC-based electrolytes have gained widespread usage. However, employing organic electrolytes in supercapacitors poses additional challenges. These include reduced specific capacitance, lower conductivity, and safety issues such as combustibility, toxicity, and instability. Moreover, organic electrolytes are often more expensive compared to their aqueous counterparts [4, 23, 42].

**c) Ionic Liquids (ILs)**

Concerns about the safety of organic solvents are starting to be addressed, and ionic liquids are improving key characteristics for use in electrochemical supercapacitors. The ILs are also known as molten salts at room temperature or below; they are defined as salts that are fully made up of ions and have a melting point of under 100 °C. The room-temperature ILs are attractive to SC application because they are non-volatile, insufficiently flammable, and heat-resistant, three characteristics that are unique and impossible to obtain with ordinary solvents. The largest operational voltage window for ILs is between 0 and 3 V, and it can withstand temperatures of up to 300°C without vaporizing. Further, ILs have attracted a lot of interest as substitute electrolytes due to their distinct structures and properties. In the class of ILs, the following electrolytes such as sulfonium anions, ammonium solvents, pyrrolidinium, imidazolium, phosphonium cations, bis(trifluoromethanesulfonyl)imide, and dicyanamide anion are present. To improve the conductivity and decrease the viscosity of ILs for that purpose, the electrolyte can be prepared using the organic solvents and IL ions, and it has been investigated as electrolytes for supercapacitive devices [43].

**d) Redox Additive Electrolyte**

The performance of supercapacitors has seen a significant boost through the utilization of redox additive electrolytes, primarily because they augment the number of redox-active sites on both the electrode surface and within the electrolyte. This contrast with aqueous electrolytes has led to notable improvements in supercapacitor performance. These redox additive electrolytes offer an extended voltage window, ultimately enhancing the energy density of supercapacitors, making them suitable for a wide range of applications. An increase in the capacitance (Cs) of SCs can be achieved by encouraging the participation of redox-active electrolytes in a pseudocapacitive manner. Moreover, the pseudocapacitance is not limited to just the electrode materials; the electrolytes, specifically the redox mediator within the electrolyte, can also contribute to pseudocapacitance. In the quest for pseudocapacitive sources, researchers have explored organic redox mediators like methylene blue, hydroquinone, p-phenylenediamine, indigo carmine, m-phenylenediamine, lignosulfonates, humic acids, and sulfonated polyaniline, further expanding the possibilities in enhancing supercapacitor performance [44, 45].

### 11.3.4 Fabrication of Symmetric and Asymmetric Solid-State Hybrid Device

The construction method employed for electrodes within an integrated device, along with the design of the electrodes themselves, plays a pivotal role in establishing the fundamental characteristics of supercapacitors (SCs), including energy storage, power delivery, and lifespan [42]. In practical applications, the rated voltage for energy storage devices (ESDs) typically remains below 2.7 V, a significant consideration for series connections of resultant devices. It is of utmost importance to ascertain the individual capacitor voltages to ensure that the charging and discharging processes remain within the operational lifespan of the manufactured devices. While the devices created resemble the ideal model often utilized in military applications, especially for rapidly supplying power to satellites and spacecraft, SCs possess the capacity for immediate high-throughput and substantial energy storage, owing to their unique properties and ample energy reserves. Consequently, a robust solid-state design is essential for widespread utilization across various sectors, offering significant benefits for system stability [46, 47]. Additionally, the schematic for the charge storage mechanism in the symmetric and asymmetric hybrid devices is reported in Figure 11.6.

**a) Spiral Wound Configuration**

SCs typically involve the utilization of sizeable active electrodes combined with solid electrolytes within a cylindrical structure when designed in a spiral wound format. This particular configuration enhances the utilization of electrode materials by maximizing the available surface area while keeping the required volume to a minimum. They can generate a substantial

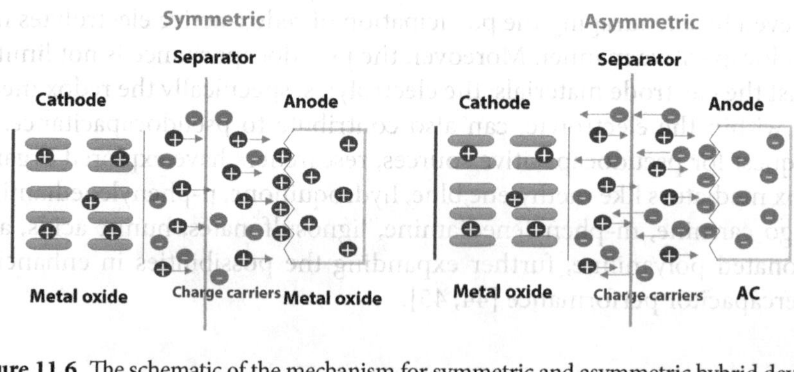


**Figure 11.6** The schematic of the mechanism for symmetric and asymmetric hybrid devices.

amount of energy when these flexible electrode-based devices are assembled in a spiral wound configuration.

**b) Symmetric Device**

Considering that the Ed is a function influenced by both capacitance and the square of the operating voltage, it becomes evident that the interplay between the electrode and electrolyte is pivotal in the pursuit of high-performance SCs. When examining device configurations, the prepared electrodes or fabricated coin cells can be organized in different ways, including series, parallel, or combinations thereof. These configurations serve to enhance the capacitive current and extend the operational voltage range. In particular, the series configuration of SCs can elevate the working voltage while concurrently diminishing the internal resistance, a factor contingent on the number of devices within the cells. This underscores the significance of devising a flexible and sustainable SC design with symmetric options, which effectively caters to energy and power requirements. Such an approach is made feasible through the expansion of the voltage window and capacitive current, thus achieving the desired performance balance [48].

**c) Asymmetric Device**

An asymmetric configuration stands out as a significant design choice for enhancing the maximum operating potential of a device, achieved through the incorporation of diverse active materials at the positive and negative electrodes. Furthermore, these asymmetric SC configurations can be categorized based on the presence of electrode materials, with one utilizing the electrical double-layer capacitor (EDLC) principle and the other relying on pseudocapacitive materials. This asymmetric device design presents advantages in terms of broader voltage ranges, elevated Ed, and overall stability. These benefits are realized through the judicious selection of electrode combinations with varying operational windows. In many instances, carbon-based electrodes are preferred due to their high surface area and superior Ed characteristics, making them suitable for constructing asymmetric configurations that encompass both pseudocapacitive and battery-like electrode components [49].

The balance of active material thickness and mass is a significant design consideration for asymmetric SCs. To increase the overall capacitance, the stored charge (Q) of each electrode must be as close as possible. The following equation determines the ideal mass ratio of the positive electrode to the negative electrode (m+/m) [50, 51].

$$Q = \frac{M_+}{M_-} = \frac{C_- V_-}{C_+ V_-} \tag{11.1}$$

The mass of carbon-based electrodes often surpasses that of metal oxides and conducting polymers, especially when considering the appropriate mass loading for configuring asymmetric setups. Initially, activated carbon served as the negative electrode in asymmetric SC devices. However, carbon-based electrodes have emerged as a viable alternative, offering prolonged cycle stability and rapid power delivery. Enhancing the relatively low energy density of capacitor electrodes is achievable by combining them with battery electrodes. Conversely, augmenting the low power output and cyclic stability of battery electrodes can be accomplished through integration with capacitor electrodes. This amalgamation of capacitor and battery components forms a hybrid technology, effectively addressing the individual limitations of each and expanding their overall capabilities [52].

### 11.3.5 Determination of the Retention and Reversibility

SCs have been acknowledged as a crucial component of future power systems due to the rapid growth of sustainable and renewable energy sources and their high levels of safety and reliability, extended lifespans, and high-power characteristics. The incorporation of sustainable SCs in current technologies has several difficulties, including high energy density, quick charge–discharge capabilities, high flexibility, and sustained cycle life.

**a) Industrial Network**

The progress of supercapacitor technology, seen as an innovative energy storage solution, is inherently intertwined with regulatory frameworks set by markets and industries. These regulations aim to establish viable standards at regional, national, and even global levels. These standards encompass an array of technical terminologies and categorization models within the technical standard system. They encompass safety specifications, general requirements, specifications for electrode materials, electrolyte specifications, charger specifications, production technical standards, transportation criteria, and recycling guidelines. Of particular significance are the comprehensive specifications that address the management and storage requirements for supercapacitor devices and modules. These guidelines span recycling procedures, the processing of electrolyte capacitor casings, plate manufacturing, and device processing. The overarching objective is to offer industry-wide guidance and standardization for modules, with the ultimate aim of achieving cost-efficiency and environmentally friendly recycling practices in the sector.

Several nations, including Australia, France, Japan, South Korea, and Russia, have been actively pursuing sustainable SCs technologies. The relatively high cost of SCs initially hindered their widespread adoption, but this has gradually stabilized within the emerging energy storage market. As technology advances and manufacturers intensify their efforts, it is expected that costs will decrease, and technical capabilities will expand, enabling a broader array of applications. A novel ecological energy system, known as the Internet of Energy (IOE), revolves around electricity and prioritizes renewable energy sources. This system is characterized by its utilization of diverse energy resources, the careful balance of energy import and export, industrial integration, and significant community involvement. It is worth highlighting that the future trajectory of intelligent urban energy development and efficient energy utilization is closely intertwined with energy storage [49].

In recent decades, the sustainable SCs have emerged as a significant energy storage device, particularly for large-scale electricity storage, and it offers a distinct advantage in energy storage when compared to batteries or fuel cells. One ongoing challenge in this field is the management of electric power sources (EPS) load scheduling, which has been the subject of long-term and continuous research efforts. Furthermore, the persistent proliferation of portable and flexible devices, coupled with the inevitable expansion of power plant equipment, characterizes EPS systems. In recent years, a notable potential has emerged for the development of energy storage sources based on supercapacitors, offering capacitive-storage capabilities. Currently, ESDs and an experimental methodology have been devised, building upon conventional capacitors as a foundation [53].

## 11.4 Technical Aspect

### 11.4.1 Flexible and Sustainable Hybrid Device

The growing interest in flexible and sustainable ESDs underscores their pivotal role in the evolving energy storage sector, driven by the demand for portable, adaptable, and wearable electronic devices. Among the array of alternatives, sustainable SCs emerge as promising energy storage solutions due to their remarkable attributes, including high power output, rapid charge/discharge rates, extended cycle life, and cost-effectiveness. The presence of oxygen vacancies assumes a critical role in shaping the ionic diffusion and conductivity of electrode materials. Elevated oxygen vacancies contribute to reduced diffusion resistance. Consequently, porous networks facilitate a greater specific surface area (SSA) and an increased abundance of oxygen vacancies, thereby augmenting specific capacity and capacitance. Moreover, the thin nanosheet framework alleviates strain resulting from volume fluctuations during the charge–discharge process, fostering enhanced interaction between the electrode and the electrolyte interface. The electrode materials designed with these structures exhibit remarkable attributes, including high specific capacity/capacitance and enduring stability over time [49, 54, 55]. These flexible devices exhibit an almost constant capacitance, showcasing no discernible performance degradation. Notably, when the device is restored to its initial flat form, its performance swiftly reverts to its original state, exemplifying exceptional resistance to bending stresses.

### 11.4.2 Composite or Doping

Furthermore, advanced electrode materials must possess specific attributes tailored to the particular application, ensuring the creation of high-power and high-energy density supercapacitor (SC) devices. To enhance the electrochemical properties of pseudocapacitive materials like $MnO_2$ and $RuO_2$, various approaches can be employed, including the development of superior pseudocapacitive materials like transition metal oxides (TMOs) and metal–organic frameworks (MOFs). Efforts have been made to augment the electrochemical performance of TMOs by functionalizing them with diverse carbon-based supporting materials. Multiple research studies have explored this avenue to increase active sites and promote efficient charge transfer. TMOs have garnered significant attention due to their exceptional physicochemical characteristics, including a high surface area, semiconducting capabilities, cost-effectiveness, chemical stability, mechanical robustness, and ease of synthesis through various techniques like mechanical exfoliation, direct sonication in solvents, and chemical vapor deposition (CVD) [45].

As an illustrative example, an all-solid-state supercapacitor (ASC) device was developed using a negative electrode composed of flower-shaped $MoS_2$ integrated with graphene nanosheets (GNSs), and a positive electrode comprising $MnO_2$ also combined with GNSs. This ASC device demonstrated impressive electrochemical performance, showcasing a notably high Cs.

These achievements underscore the remarkable potential of all-solid-state ASCs. However, it is important to note that TMOs possess inherent challenges, such as high surface energy and interlayer interactions, which can lead to restacking and consequently reduce the number of active sites available for electrochemical processes. To overcome these limitations and enhance the activity of TMOs for advanced applications, one effective strategy involves creating novel functional hybrid materials by combining TMOs with highly conductive supporting materials [56, 57].

### 11.4.3 New Approach and Transparency

A significant focus of research in the energy storage field has been addressing the pressing need for intelligent and programmable electronic devices. These electronic devices can be programmed to perform a wide range of functions, catering to various requirements in our daily lives. This points toward the emergence of a generation of adaptable devices that can facilitate user-friendly and personalized interactions between wearable and bio-integrated electronics. Achieving this relies on exploring new materials, conceptualizing innovative architectures, and leveraging computational modeling and machine learning. Notably, there is a growing interest in flexible multifunctional energy storage electrodes and devices as part of upcoming technologies. Future optoelectronic devices are likely to prioritize transparency as a desirable feature.

In the quest for sustainable development, the next generation of robust and versatile electronic devices should be designed and manufactured.

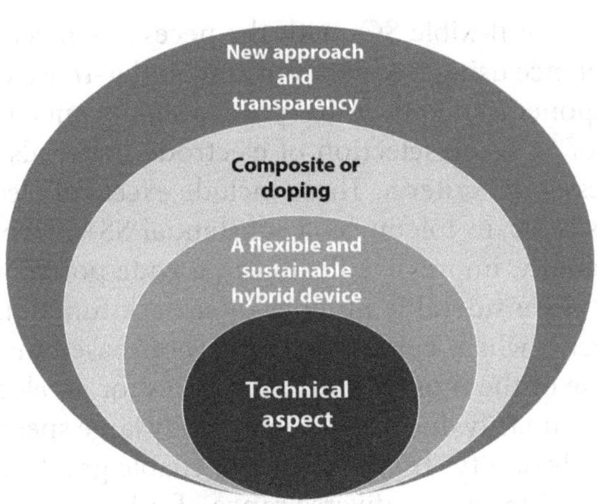


**Figure 11.7** The various technical aspects to overcome the drawbacks of sustainable SCs.

These devices are expected to fulfill various roles, including those related to supercapacitors with electrochromic features, which can have implications for applications such as hard drives, memory devices, and potentially self-repair capabilities. Additionally, the various technical aspects to overcome the drawbacks of sustainable SCs are reported in Figure 11.7.

## 11.5 Application-Level Aspects for Sustainable SCs

The utilization of cutting-edge materials in the development of asymmetrical flexible hybrid devices (ASHD) opens up new avenues for research and development in the realm of multiple electrode materials. The ASHD structures offer a significant advantage with their expansive operating potential window, a feature attributed to the distinct potential ranges of cathodes and anodes. These hybrid supercapacitors (HSCs) capitalize on the strengths of both battery and supercapacitor electrodes, yielding high energy density and high Pd, respectively. Nevertheless, the production of such systems still encounters various challenges, including the quest for scalable, cost-effective manufacturing methods and the development of efficient processes that enable better control over surface characteristics like porosity and morphology. Sustainable SCs have garnered considerable attention within the field of energy storage research. They stand out due to their exceptional capacitive behavior, elevated Pd, superior rate capability, enhanced cyclic stability compared to traditional batteries, and increased energy density when contrasted with standard capacitors. These characteristics render them highly versatile and suitable for a wide range of applications [58, 59].

The fabrication of flexible SCs with the necessary mechanical strength and high capacitance using freestanding and binder-free electrodes, which are pivotal components of FSCs, often poses a significant challenge. Similar to other types of SCs, the selection of electrode materials for FSCs must satisfy several essential criteria. These include excellent electrical conductivity, high tensile strain tolerance, a substantial SSA, favorable chemical and thermal stability, impressive flexibility, a wide potential window, and the presence of a considerable number of surface functional groups. For example, graphene, which belongs to the carbon-based material category, stands out as one of the most promising choices for an electrode material in FSCs. This is primarily due to its exceptionally large specific surface area and remarkable flexibility. These properties enable graphene to be bent at extreme angles and provide a diverse range of microstructures, making it

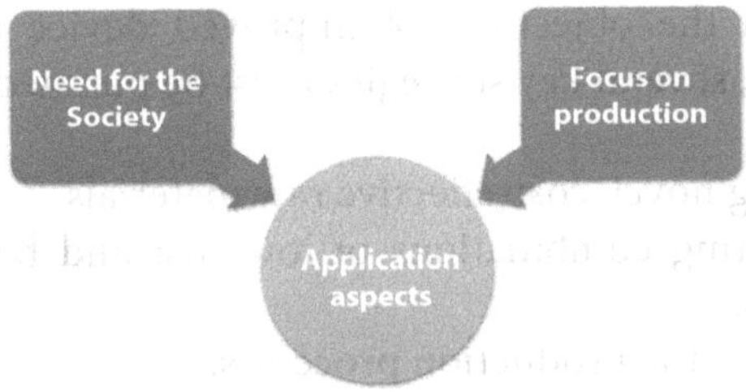


**Figure 11.8** The several application aspects for sustainable SCs.

highly suitable for FSC applications [55, 60, 61]. The utilization of electrodes, materials, and electrolytes in the field of SCs plays a pivotal role in addressing societal needs and advancing industrial development. The demand for high-capacity, portable power sources has become increasingly significant in modern technology. On one hand, there is a growing global interest in environmental preservation and sustainable energy practices. The people worldwide are actively seeking solutions and are eager to incorporate more renewable energy sources. Nowadays, society experience various difficulties in tackling the energy storage issue and building them in the upcoming techniques and the resultant aspects are shown in Figure 11.8.

The emergence of SCs stems from the urgent need for high-capacity, portable backup power for various electronic devices, driven by the rapid growth of the electronics industry. These societal demands have, to a certain extent, accelerated the development of the SC sector, presenting a substantial market opportunity. To meet the requirements of societal progress, research efforts have been elevated, leading to widespread applications in new energy passenger vehicles, rail transportation, smart metering, wind turbines, power grid infrastructure, heavy machinery, and national defense initiatives. The scope of SC applications is poised for continuous expansion, with collaboration and cooperation among upstream and downstream businesses in the fields of communication and technology playing a crucial role in this growth trajectory.

The foremost and paramount consideration across all industries revolves around enhancing product performance and reducing production costs. Research efforts in this domain primarily focus on advancing SC technology itself, along with refining manufacturing processes and techniques. This entails the development of stable and efficient electrode and electrolyte materials to boost performance while concurrently driving down costs.

To accomplish the objectives of improved device performance and cost-effectiveness, several key strategies must be pursued:

1. Exploring novel, cost-effective raw materials.
2. Investigating combinations of low-cost and high-cost raw materials.
3. Enhancing the production processes.
4. Addressing the industrialization prospects of materials and cost-related issues about the application of new electrode materials.
5. Conducting compatibility research on existing electrode materials, including their compatibility with electrolytes.
6. Focusing on the overall scalability in the research of group modules.

In summary, the growth of SCs is intrinsically tied to advancements in science and technology, as well as the increasing demand for applications. It is anticipated that the development of SCs for use in new energy vehicles and intelligent wearable technology will progress rapidly and gain wider acceptance.

## 11.6 Future Perspectives and Challenges for the Sustainable SCs

To assist future researchers in their endeavors, it is essential to emphasize several crucial points. One of the foremost considerations during that period was the selection of synthesis methods and raw materials that could yield an economical, environmentally friendly, non-toxic, and sustainable solution to fulfill the future energy requirements of everyday human life. The following advanced materials displayed the most promising potential for supercapacitor applications: metal–organic frameworks (MOFs), metal nitrides, phosphorene, and MXenes respectively. To meet evolving demands, there is a pressing need to advance manufacturing processes, enabling the creation of highly porous materials characterized by hollow structures, controlled diffusion pathways, and remarkable structural density. Moreover, it is imperative to incorporate organic and redox additive electrolytes alongside aqueous electrolytes to enhance the electrochemical performance of electrode materials. This augmentation will result in elevated potential thresholds, expanded surface areas, increased Ed, enhanced Cs, and improved cyclic stability. Furthermore, the development of asymmetric SCs warrants heightened attention due to their capability to deliver superior power and energy densities.

## 11.7 Conclusion

SCs have emerged as a promising and environmentally friendly energy storage solution with significant potential across various industries and applications. The expansive market ahead offers numerous opportunities for SC development. However, despite their merits, these advanced energy storage systems face certain challenges that need to be addressed. To foster wider adoption and meet the demands of alternative fuel vehicles and smart clothing, it is imperative to enhance supercapacitors' utilization, reduce production costs, and elevate Ed. Future research will continue to prioritize the refinement of electrode materials, a key factor limiting supercapacitor performance and cost-effectiveness. Unquestionably, both LIBs and SCs have profoundly impacted various practical applications, reshaping industries and lifestyles worldwide. This chapter has highlighted recent advancements in the realm of stretchable and flexible energy storage, encompassing material exploration, structural designs, manufacturing processes, and integrated assembly. The chapter thoroughly elucidates the creation of supercapacitors, encompassing the careful selection of electrodes and electrolytes. Below, we briefly touch upon several carbon-based materials with potential applications as supercapacitor electrode materials. These materials encompass activated carbon-based electrodes, carbon nanotubes (CNT)-based electrodes, graphene-based electrodes, metal oxides, and transition metal oxide/hydroxide-based electrodes. Recent breakthroughs and achievements underscore the intriguing prospects of flexible and sustainable energy storage systems for real-world applications such as wearable electronics. Nevertheless, numerous challenges remain to be addressed before practical implementation can be realized. Ongoing research efforts are essential to gain a comprehensive understanding of electrochemical performance and mechanical properties. The development of robust materials exhibiting superior electrochemical and mechanical attributes, improved safety for flexible and stretchable electrodes, and the exploration of large-scale industrial production methods to reduce fabrication costs are critical objectives.

## References

1. Sun, J., Luo, B., Li, H., A Review on the conventional capacitors, supercapacitors, and emerging hybrid ion capacitors: past, present, and future. *Adv. Energy Sustain. Res.*, 3, 2100191, 2022, https://doi.org/10.1002/aesr.202100191.
2. Wu, L., Shi, X., Wu, Z.S., Recent advancements and perspectives of biodegradable polymers for supercapacitors. *Adv. Funct. Mater.*, 33, 2211454, 2023, https://doi.org/10.1002/adfm.202211454.
3. Liu, W., Song, M.S., Kong, B., Cui, Y., Flexible and stretchable energy storage: recent advances and future perspectives. *Adv. Mater.*, 29, 1603436, 2017, https://doi.org/10.1002/adma.201603436.
4. Lv, T., Liu, M., Zhu, D., Gan, L., Chen, T., Nanocarbon-based materials for flexible all-solid-state supercapacitors. *Adv. Mater.*, 30, 1–17, 2018, https://doi.org/10.1002/adma.201705489.
5. Zhai, S., Karahan, H.E., Wang, C., Pei, Z., Wei, L., Chen, Y., 1D Supercapacitors for emerging electronics: current status and future directions. *Adv. Mater.*, 32, 1–19, 2020, https://doi.org/10.1002/adma.201902387.
6. Biswas, S. and Chowdhury, A., Organic supercapacitors as the next generation energy storage device: emergence, opportunity, and challenges. *Chem. Phys. Chem.*, 24, e202200567, 2023, https://doi.org/10.1002/cphc.202200567.
7. Huang, S., Zhu, X., Sarkar, S., Zhao, Y., Challenges and opportunities for supercapacitors. *APL Mater.*, 7, 100901, 2019, https://doi.org/10.1063/1.5116146.
8. Afif, A., Rahman, S.M., Tasfiah Azad, A., Zaini, J., Islan, M.A., Azad, A.K., Advanced materials and technologies for hybrid supercapacitors for energy storage – A review. *J. Energy Storage*, 25, 100852, 2019, https://doi.org/10.1016/j.est.2019.100852.
9. Wang, S., Ma, J., Shi, X., Zhu, Y., Wu, Z.S., Recent status and future perspectives of ultracompact and customizable micro-supercapacitors. *Nano Res. Energy*, 1, 500–526, 2022, https://doi.org/10.26599/NRE.2022.9120018.
10. Shembade, U.V., Dhas, S.D., Gurav, S.R., Sonkawade, R.G., Wategaonkar, S.B., Ghatage, S.R., Gaikwad, M.A., Kim, J.H., Parale, V.G., Park, H.H., Moholkar, A.V., Acid substitutions for $WO_3$ nanostructures synthesis by the hydrothermal route and its effect on physio-chemical and electrochemical properties for supercapacitors. *J. Energy Storage*, 72, 108432, 2023, https://doi.org/10.1016/j.est.2023.108432.
11. Zhao, B., Liu, P., Jiang, Y., Pan, D., Tao, H., Song, J., Fang, T., Xu, W., Supercapacitor performances of thermally reduced graphene oxide. *J. Power Sources*, 198, 423–427, 2012, https://doi.org/10.1016/j.jpowsour.2011.09.074.
12. Chen, J., Sheng, K., Luo, P., Li, C., Shi, G., Graphene hydrogels deposited in nickel foams for high-rate electrochemical capacitors. *Adv. Mater.*, 24, 4569–4573, 2012, https://doi.org/10.1002/adma.201201978.
13. Hiralal, P., Rius, G., Yoshimura, M., Amaratunga, G.A.J., Graphene-based electrochemical capacitors, in: *Graphene Sci. Handb., 5–6*, pp. 479–494, 2016, https://doi.org/10.1201/b19460-38.

14. Dhas, S.D., Maldar, P.S., Patil, M.D., Hubali, K.M., Shembade, U.V., Abitkar, S.B., Waikar, M.R., Sonkawade, R.G., Agawane, G.L., Moholkar, A.V., Hydrothermal synthesis of mesoporous $NiMnO_3$ nanostructures for supercapacitor application: Effect of electrolyte. *J. Energy Storage*, 35, 102277, 2021, https://doi.org/10.1016/j.est.2021.102277.
15. Troschke, E., Oschatz, M., Ilic, I.K., Schiff-bases for sustainable battery and supercapacitor electrodes. *Exploration*, 1, 20210128, 2021, https://doi.org/10.1002/exp.20210128.
16. Zhang, Y.Z., Wang, Y., Cheng, T., Lai, W.Y., Pang, H., Huang, W., Flexible supercapacitors based on paper substrates: a new paradigm for low-cost energy storage. *Chem. Soc. Rev.*, 44, 5181–5199, 2015, https://doi.org/10.1039/c5cs00174a.
17. Pérez-Madrigal, M.M., Edo, M.G., Alemán, C., Powering the future: Application of cellulose-based materials for supercapacitors. *Green Chem.*, 18, 5930–5956, 2016, https://doi.org/10.1039/c6gc02086k.
18. Muzaffar, A., Ahamed, M.B., Deshmukh, K., Thirumalai, J., A review on recent advances in hybrid supercapacitors: Design, fabrication and applications. *Renew. Sustain. Energy Rev.*, 101, 123–145, 2019, https://doi.org/10.1016/j.rser.2018.10.026.
19. Augustyn, V., Simon, P., Dunn, B., Pseudocapacitive oxide materials for high-rate electrochemical energy storage. *Energy Environ. Sci.*, 7, 1597–1614, 2014, https://doi.org/10.1039/c3ee44164d.
20. Gogotsi, Y. and Penner, R.M., Energy storage in nanomaterials-capacitive, pseudocapacitive, or battery-like? *ACS Nano*, 12, 2081–2083, 2018, https://doi.org/10.1021/acsnano.8b01914.
21. Goodenough, J.B., Lee, H.Y., Manivannan, V., Supercapacitors and batteries. *Mater. Res. Soc. Symp. - Proc.*, 548, 656–665, 1999, https://doi.org/10.1557/proc-548-655.
22. Joseph, A. and Thomas, T., Recent advances and prospects of metal oxynitrides for supercapacitor. *Prog. Solid State Chem.*, 100381, 68, 2022, https://doi.org/10.1016/j.progsolidstchem.2022.100381.
23. Pal, B., Yang, S., Ramesh, S., Thangadurai, V., Jose, R., Electrolyte selection for supercapacitive devices: A critical review. *Nanoscale Adv.*, 1, 3807–3835, 2019, https://doi.org/10.1039/c9na00374f.
24. Wang, X., Salari, M., en Jiang, D., Chapman Varela, J., Anasori, B., Wesolowski, D.J., Dai, S., Grinstaff, M.W., Gogotsi, Y., Electrode material–ionic liquid coupling for electrochemical energy storage. *Nat. Rev. Mater.*, 5, 787–808, 2020, https://doi.org/10.1038/s41578-020-0218-9.
25. Şahin, M.E., Blaabjerg, F., Sangwongwanich, A., A Comprehensive Review on Supercapacitor Applications and Developments. *Energies*, 15, 1–26, 2022, https://doi.org/10.3390/en15030674.
26. Conte, M., Supercapacitors technical requirements for new applications. *Fuel Cells*, 10, 806–818, 2010, https://doi.org/10.1002/fuce.201000087.

27. Lufrano, F. and Staiti, P., Mesoporous carbon materials as electrodes for electrochemical supercapacitors. *Int. J. Electrochem. Sci.*, 5, 903–916, 2010, https://doi.org/10.1016/s1452-3981(23)15331-4.
28. Yang, H., Kannappan, S., Pandian, A.S., Jang, J.H., Lee, Y.S., Lu, W., Rapidly annealed nanoporous graphene materials for electrochemical energy storage. *J. Mater. Chem. A*, 5, 23720–23726, 2017, https://doi.org/10.1039/c7ta07733e.
29. Sharma, S. and Chand, P., Supercapacitor and electrochemical techniques: A brief review. *Results Chem.*, 5, 100885, 2023, https://doi.org/10.1016/j.rechem.2023.100885.
30. Arunkumar, M. and Paul, A., Importance of electrode preparation methodologies in supercapacitor applications. *ACS Omega*, 2, 8039–8050, 2017, https://doi.org/10.1021/acsomega.7b01275.
31. Yang, H., Kannappan, S., Pandian, A.S., Jang, J.H., Lee, Y.S., Lu, W., Graphene supercapacitor with both high power and energy density. *Nanotechnology*, 28, 445401, 2017, https://doi.org/10.1088/1361-6528/aa8948.
32. Wang, C.M., Wen, C.Y., Chen, Y.C., Chang, J.Y., Ho, C.W., Kao, K.S., Shih, W.C., Chiu, C.M., Shen, Y.A., The influence of specific surface area on the capacitance of the carbon electrodes supercapacitor, in: *Proceedings of the 3rd International Conference on Industrial Application Engineering 2015*, 439–442, 2015, https://doi.org/10.12792/iciae2015.077.
33. Cheng, Q., Tang, J., Ma, J., Zhang, H., Shinya, N., Qin, L.C., Graphene and carbon nanotube composite electrodes for supercapacitors with ultra-high energy density. *Phys. Chem. Chem. Phys.*, 13, 17615–17624, 2011, https://doi.org/10.1039/c1cp21910c.
34. Frackowiak, E., Metenier, K., Bertagna, V., Beguin, F., Supercapacitor electrodes from multiwalled carbon nanotubes. *Appl. Phys. Lett.*, 77, 2421–2423, 2000, https://doi.org/10.1063/1.1290146.
35. Ramachandran, R., Saranya, M., Grace, A.N., Wang, F., MnS nanocomposites based on doped graphene: simple synthesis by a wet chemical route and improved electrochemical properties as an electrode material for supercapacitors. *RSC Adv.*, 7, 2249–2257, 2017, https://doi.org/10.1039/c6ra25457h.
36. Ramachandran, R., Chen, S.M., Gnana Kumar, G.P., Gajendran, P., Devi, N.B., An overview of fabricating nanostructured electrode materials for biosensor applications. *Int. J. Electrochem. Sci.*, 10, 8607–8629, 2015, https://doi.org/10.1016/s1452-3981(23)11122-9.
37. Zhang, G., Huang, X., Ma, J., Wu, F., Zhou, T., Ti/ruo$_2$-iro$_2$-sno$_2$ anode for electrochemical degradation of pollutants in pharmaceutical wastewater: Optimization and degradation performances. *Sustainability*, 13, 1–13, 2021, https://doi.org/10.3390/su13010126.
38. Conway, B.E., Transition from "supercapacitor" to "battery" behavior in electrochemical energy storage. *Proc. Int. Power Sources Symp.*, pp. 319–327, 1991, https://doi.org/10.1149/1.2085829.

39. Chen, Y., Pang, W.K., Bai, H., Zhou, T., Liu, Y., Li, S., Guo, Z., Enhanced Structural Stability of Nickel-Cobalt Hydroxide via Intrinsic Pillar Effect of Metaborate for High-Power and Long-Life Supercapacitor Electrodes. *Nano Lett.*, 17, 429–436, 2017, https://doi.org/10.1021/acs.nanolett.6b04427.
40. Lota, K., Sierczynska, A., Acznik, I., Lota, G., Effect of aqueous electrolytes on electrochemical capacitor capacitance. *Chemik*, 67, 1138–1145, 2013.
41. Liang, C., Wang, S., Sha, S., Lv, S., Wang, G., Wang, B., Li, Q., Yu, J., Xu, X., Zhang, L., Novel semiconductor materials for advanced supercapacitors. *J. Mater. Chem. C*, 11, 4288–4317, 2023, https://doi.org/10.1039/d2tc04816g.
42. Kim, B.C., Hong, J.Y., Wallace, G.G., Park, H.S., Recent Progress in Flexible Electrochemical Capacitors: Electrode Materials, Device Configuration, and Functions. *Adv. Energy Mater.*, 5, 1500959, 2015, https://doi.org/10.1002/aenm.201500959.
43. Brousse, T., Bélanger, D., Chiba, K., Egashira, M., Favier, F., Long, J., Miller, J.R., Morita, M., Naoi, K., Simon, P., Sugimoto, W., *Mater. Electrochem. Capacitors*, Nature materials, 7, 11, pp. 495–561, Springer Handbooks, 2017, https://doi.org/10.1007/978-3-662-46657-5_16.
44. Zhang, L., Yang, S., Chang, J., Zhao, D., Wang, J., Yang, C., Cao, B., A Review of Redox Electrolytes for Supercapacitors. *Front. Chem.*, 8, 1–7, 2020, https://doi.org/10.3389/fchem.2020.00413.
45. Forouzandeh, P., Kumaravel, V., Pillai, S.C., Electrode materials for supercapacitors: A review of recent advances. *J. Catalysts*, 10, 1–73, 2020, https://doi.org/10.3390/catal10090969.
46. Nejati, K., Akbari, A.R., Davari, S., Asadpour-Zeynali, K., Rezvani, Z., Zn-Fe-layered double hydroxide intercalated with vanadate and molybdate anions for electrocatalytic water oxidation. *New J. Chem.*, 42, 2889–2895, 2018, https://doi.org/10.1039/c7nj04469k.
47. Yang, X., Niu, H., Jiang, H., Wang, Q., Qu, F., A high energy density all-solid-state asymmetric supercapacitor based on $MoS_2$/graphene nanosheets and $MnO_2$/graphene hybrid electrodes. *J. Mater. Chem. A*, 4, 11264–11275, 2016, https://doi.org/10.1039/c6ta03474h.
48. Tong, Z., Tian, Y., Zhang, H., Li, X., Ji, J., Qu, H., Li, N., Zhao, J., Li, Y., Recent advances in multifunctional electrochromic energy storage devices and photoelectrochromic devices. *Sci. China Chem.*, 60, 13–37, 2017, https://doi.org/10.1007/s11426-016-0283-0.
49. Forouzandeh, P. and Pillai, S.C., Two-dimensional (2D) electrode materials for supercapacitors. *Mater. Today Proc.*, 41, 498–505, 2020, https://doi.org/10.1016/j.matpr.2020.05.233.
50. Fang, X., Kutkut, N., Shen, J., Batarseh, I., Analysis of generalized parallel-series ultracapacitor shift circuits for energy storage systems. *Renew. Energy*, 36, 2599–2604, 2011, https://doi.org/10.1016/j.renene.2010.05.003.
51. Chen, P., Shen, G., Shi, Y., Chen, H., Zhou, C., Preparation and Characterization of Flexible Asymmetric Supercapacitors. *ACS Nano*, 4, 4403–4411, 2010.

52. Li, J., Guo, J., Zhang, X., Huang, Y., Guo, L., Asymmetric supercapacitors with high energy and power density fabricated using $LiMn_2O_4$ nano-rods and activated carbon electrodes. *Int. J. Electrochem. Sci.*, 12, 1157–1166, 2017, https://doi.org/10.20964/2017.02.41.
53. Raza, W., Ali, F., Raza, N., Luo, Y., Kim, K.H., Yang, J., Kumar, S., Mehmood, A., Kwon, E.E., Recent advancements in supercapacitor technology. *Nano Energy*, 52, 441–473, 2018, https://doi.org/10.1016/j.nanoen.2018.08.013.
54. Zhang, X., Selkirk, A., Zhang, S., Huang, J., Li, Y., Xie, Y., Dong, N., Cui, Y., Zhang, L., Blau, W.J., Wang, J., $MoS_2$/carbon nanotube core–shell nanocomposites for enhanced nonlinear optical performance. *Chem. - A Eur. J.*, 23, 3321–3327, 2017, https://doi.org/10.1002/chem.201604395.
55. Li, M.Y., Chen, C.H., Shi, Y., Li, L.J., Heterostructures based on two-dimensional layered materials and their potential applications. *Mater. Today*, 19, 322–335, 2016, https://doi.org/10.1016/j.mattod.2015.11.003.
56. Jiang, J.W. and Park, H.S., Mechanical properties of $MoS_2$/graphene heterostructures. *Appl. Phys. Lett.*, 105, 3–8, 2014, https://doi.org/10.1063/1.4891342.
57. Toth, P.S., Velický, M., Bissett, M.A., Slater, T.J.A., Savjani, N., Rabiu, A.K., Rakowski, A.M., Brent, J.R., Haigh, S.J., O'Brien, P., Dryfe, R.A.W., Asymmetric $MoS_2$/Graphene/Metal Sandwiches: Preparation, Characterization, and Application. *Adv. Mater.*, 28, 8256–8264, 2016, https://doi.org/10.1002/adma.201600484.
58. Qin, S., Zhang, Q., Yang, X., Liu, M., Sun, Q., Wang, Z.L., Hybrid piezo/triboelectric-driven self-charging electrochromic supercapacitor power package. *Adv. Energy Mater.*, 8, 1–9, 2018, https://doi.org/10.1002/aenm.201800069.
59. Xiao, J., Li, H., Zhang, H., He, S., Zhang, Q., Liu, K., Jiang, S., Duan, G., Zhang, K., Nanocellulose and its derived composite electrodes toward supercapacitors: Fabrication, properties, and challenges. *J. Bioresour. Bioprod.*, 7, 245–269, 2022, https://doi.org/10.1016/j.jobab.2022.05.003.
60. Fleischmann, S., Mitchell, J.B., Wang, R., Zhan, C., Jiang, D.E., Presser, V., Augustyn, V., Pseudocapacitance: from fundamental understanding to high power energy storage materials. *Chem. Rev.*, 120, 6738–6782, 2020, https://doi.org/10.1021/acs.chemrev.0c00170.
61. Li, Z., Huang, T., Gao, W., Xu, Z., Chang, D., Zhang, C., Gao, C., Hydrothermally activated graphene fiber fabrics for textile electrodes of supercapacitors. *ACS Nano*, 11, 11056–11065, 2017, https://doi.org/10.1021/acsnano.7b05092.

# Index

Printed in the USA/Agawam, MA
September 16, 2026

910661.111